Chemistry and Technology of Flavors and Fragrances

Chemistry and Technology of Flavors and Fragrances

Edited by

David J. Rowe
De Monchy Aromatics Ltd
Poole, UK

Blackwell
Publishing

CRC Press

Editorial Offices:
Blackwell Publishing Ltd, 9600 Garsington Road, Oxford OX4 2DQ, UK
Tel: +44 (0)1865 776868
Blackwell Publishing Asia Pty Ltd, 550 Swanston Street, Carlton, Victoria 3053, Australia
Tel: +61 (0)3 8359 1011

ISBN 1–4051–1450–9

Published in the USA and Canada (only) by
CRC Press LLC, 2000 Corporate Blvd., N.W.,
Boca Raton, FL 33431, USA
Orders from the USA and Canada (only) to
CRC Press LLC

USA and Canada only:
ISBN 0–8493–2372–X

First published 2005

Library of Congress Cataloging-in-Publication Data
A catalog record for this title is available from the Library of Congress

British Library Cataloguing-in-Publication Data
A catalogue record for this title is available from the British Library

Set in 10/12 pt Minion
by Integra Software Services Pvt. Ltd, Pondicherry, India
Printed and bound in Great Britain
by MPG Books, Bodmin, Cornwall

The publisher's policy is to use permanent paper from mills that operate a sustainable forestry policy, and which has been manufactured from pulp processed using acid-free and elementary chlorine-free practices. Furthermore, the publisher ensures that the text paper and cover board used have met acceptable environmental accreditation standards.

For further information on Blackwell Publishing, visit our website:
www.blackwellpublishing.com

Contents

Contributors

David Baines	22 Elizabeth Close, Thornbury, Bristol, BS12 2YN, NJ
Neil C. Da Costa	International Flavors and Fragrances Inc., 1515 Highway 36, Union Beach, NJ 07735
Mark L. Dewis	International Flavours & Fragrances Inc., 1515 Highway 36, Union Beach, NJ 07735
Sanja Eri	International Flavors and Fragrances Inc., 1515 Highway 36, Union Beach, NJ 07735
John Grigor	Chemical and Biological Sciences, University of Teesside, Middlesbrough, TS1 3BA
Stephen J. Herman	AFF International, 130 Clinton Road, Fairfield, NJ 07004, USA
Simon B. Jameson	Oxford Chemicals Limited, North Gare, Seaton Carew, Hartlepool, TS25 2DT
Jack Knights	3 Cottage Gardens, Great Billing, Northampton, N3 9YW, UK
Philip Kraft	Givaudan Schweiz AG, Fragrance Research, Überlandstrasse 138, CH-8600 Dübendorf, Switzerland
John Margetts	Ungerer Limited, Sealand Road, Chester, CH1 4LP, UK
Liam O'Hare	Chemical and Biological Sciences, University of Teesside, Middlesbrough, TS1 3BA
David J. Rowe	De Monchy Aromatics Ltd, Blackhill Road, Holton Heath Industrial Park, Holton Heath, Poole, Dorset, BH16 6LS, UK
Luca Turin	Flexitral, Inc., 4313 Walney Rd, Chantilly, VA 20151, USA
Chris Winkel	Quest International, Huizerstraatweg 28, 1411 GP Naarden, The Netherlands
Michael Zviely	Frutarom Ltd, 25 HaShaish Street, Haifa, Israel 26110

Preface

The aim of this book, as I expressed in a letter to the contributors, is to have a work that could be given to a new chemist or technologist joining the industry with the statement 'read this and you'll know what this industry is about'. Several contributors corrected this to '. . . a little of what this industry is about!' Such comments, and those of the anonymous referees whose remarks were passed on to me, helped to crystallise my thinking on the concepts behind the book. One referee referred to Aroma Chemicals as the core of the book; this is true, and as an organic chemist I make no apologies for this. In this context, the structure could be described as Part I: The origins and identification of the important aroma chemicals of nature, Part II: The chemistry and synthesis of aroma chemicals and Part III: The applications of aroma chemicals. This 'chemical' approach is also aimed at demystifying the industry to a neophyte; whilst no-one would wish to deny the essential creativity of perfumers and flavourists, it is also important to realise that, to quote Steve Herman in his Chapter 'Applications II: Fragrance', '. . . the fundamental key to dealing with the technical aspects of fragrance is to treat them as mixtures of chemicals'.

The content of the book, in general terms, was my decision, and hence any complaints should be addressed to me rather than, the authors! A book that would please everyone would run into many volumes, and it was necessary to be focused, even at the risk of omitting topics that others might see as crucial. The omission of essential oils was a conscious decision, as these have been extensively covered elsewhere; the exception is as part of the chapter on natural aroma chemicals where these are isolated from essential oils. With this in mind, each author was asked to ensure that the key essentials, as they saw it, were covered. However, I suggested, where appropriate, a section of the chapter could be devoted to an area, that, whilst firmly within the remit of the chapter, was of especial interest to the author; a 'hot topic', if you will. Beyond that, the choice of the material within a chapter was at the discretion of the authors. Ultimately, however, an editor is responsible for his or her book, and hence the sins of omission and commission are mine, and mine alone.

There are those within the industry who treat the two Fs of the Flavour and Fragrance industry as substantially different topics. Obviously there *are* differences; the specific functions of the perfumer and the flavourist are different, and of course, their palettes, as it were, are different. The markets for a musky flavour or a garlic perfume are somewhat limited! However, both areas ultimately have the same aim; to use aroma chemicals, whether as naturally occurring mixtures or as individual compounds, to create an effect that our chemical senses, taste and smell, find desirable. In this context, the aim of this book is to overcome The Great Divide, or at worst, to ignore it!

Thanks go to the many people with whom I have worked over the past dozen years. Chemistry is a 'social' discipline, and I have learned from many people; I hope that, in a way, this book is giving something back. A particular mention goes to Mark Dewis, Simon Jameson, John Heffernan, Kevin Auty, Peter Setchell, Lee Morgan (the co-designer of 'The devil's flavour wheel'), Gareth Jones, John Johnson, Tracy Brown, Peter Cannon, Rob Gregory, and others who I have now offended by not mentioning, and finally to Duncan Mullis, who first employed me in the Flavour and Fragrance Industry; in a way, Duncan, this book is your fault!

Chapter 1
Introduction

David J. Rowe

> The earliest recognisable chemists were women, the perfume-makers of Babylon, who used the earliest known stills to produce their wares. The first individual chemist known to history was "Tapputi, the perfume-maker", who was mentioned on a cuneiform tablet from the second millennium BCE in Mesopotamia
>
> Paul Strathern, *Mendeleev's Dream*, Penguin Books, London, UK, 2001

Thus, it is not an exaggeration to say that the roots of modern chemistry lie in the flavour and fragrance industry! As an area of the modern chemical industry, it is low profile compared with the 'big boys' of pharmaceuticals and petrochemicals, the areas that the 'general public' recognise as industrial chemistry. Yet it is a multi-billion dollar, global industry that impacts on everyone's life in the developed world.

1.1 History: in the beginning

In a sense the origins of the industry lie in prehistory. At some point our ancestors found there was pleasure in the aroma of a flower, and that mixing certain herbs with food added relish; when this took place will never be known, but it could be said that at that point we became truly human. Certainly our earliest ancestors had uses for aromatic plants; since burial sites have been well preserved, we know that they used aromatic plants as votive offerings to the gods and the supernatural world. By the time that writing developed in the fertile crescent of the Indus, Nile and Tigris valleys, we have many references to the use of herbs and spices. A crucial point is that the uses of flavourings and fragrances, in whatever form, are a feature of a society where at least one social stratum has the 'disposable income' to obtain them; however important they may be to the quality of life, flavours and perfumes are not (despite the efforts of marketing departments to convince us to the contrary) essential to the continuance of life. Thus throughout history we can see the uses of flavours or fragrances as reflecting either the success of a society, or its decadence, depending on the prejudices of the individual. David Pybus [1] and Paul Jose Teisseire [2] have written excellent and amusing accounts of the development of the uses of fragrance, and the interested reader is directed there for more detailed accounts.

1.2 The classical world

The classical world of ancient Greece and Rome was familiar with many oils, spices and perfumes. Some of the key techniques common to the flavour and fragrance industry and the wider chemical industry had been developed, in particular distillation and the concept of extraction. The key concept of separation of materials was introduced to the world. Philosophers such as Aristotle were able to introduce some of the ideas which we might recognise today as science. With the rise of the Roman Empire, the resources of the Mediterranean world became available to the privileged few who could live in perfumed luxury surrounded by slaves. Spices from across the empire and beyond were used to flavour food, both for their inherent flavour and to mask the somewhat dubious taste of meat in a pre-refrigeration age. Pliny the Elder wrote a treatise on 'Natural history' which included observations on aromatic plants and their uses. However, throughout this period there was another feature we frequently see in relation to the uses of flavours and fragrance; for some, such as Pliny, they are an 'honest pleasure', whereas for others they represent decadence and vanity. Those in the latter camp were presumably tutting and saying 'told you so' when, as David Pybus [1] puts it, 'Rome succumbed to the barbarian hordes, the lights went out in all the incense burners in Europe, and the rose petals went out with the bathwater'.

1.3 The mediaeval world

The fall of the Roman Empire in the West ushered in the period commonly known as Western Europe's Dark Ages. The 'one world' of the Roman Empire, stretching from Northern England to the Arabian Peninsula, was broken, and instead we had a period of what might be termed 'localisation', in an analogy to today's globalisation. Petty fiefdoms arose in what were earlier merely provinces of the empire. Nowhere is this clearer than in England, where seven kingdoms, the Heptarchy, arose in what was merely part of a single minor province of the Roman Empire. The end result of this was the loss of raw materials and eventually, the knowledge of what to do with them. The foundations of chemistry were lost into the mysticism of alchemy. Some techniques remained in use, especially distillation, but this also became a mystical affair. The classical world used distillation to refine a crude material into a more valuable one, such as an aromatic plant into an essential oil, a basic concept that we are still familiar with today. In the mediaeval mind this became the search for the Philosopher's Stone, to refine base metal into gold or to confer eternal life. Herbs were still used for flavours, but the breakdown of transport meant that these were restricted to what could be grown in local gardens, the 'parsley, sage, rosemary and thyme' of the English folk song. In parallel with this, the powerful mediaeval church frowned on the use of fragrance as earthly vanity. Cleanliness was not next to godliness, rather 'insanitary' was next to 'holy', from the hovels of the early monks to Thomas Beckett's lice-infested hair shirt. Of course, official disapproval has never prevented those with the finances from doing as they will. Spices and oils still made their way into darkened Europe from the Orient and the Arab world. The latter, in particular, kept some of the flame of the classical world burning, which is reflected in the number of words whose origin is Arabic, including alcohol and chemistry itself. The great trading cities of Venice and Genoa controlled much of this trade, with the Silk Road to the East carrying precious spices and oils via Constantinople, the capital of Byzantium, the Eastern Roman Empire. But disapproval there still was; one of the many things held against England's King John, a man held to be so awful that no other monarch has taken the name in 800 years, was his insufferable vanity – he was said to bathe every month, whether he needed to or not.

1.4 From the Renaissance to the Enlightment

In 1453 Constantinople fell to the Ottoman Turks. The end of the Byzantine Roman Empire, which for a millenium had been the great junction between East and West, had a major impact on trade routes to India and the Orient, the source of the most prized spices. This drove adventurers, especially from the great trading cities of Genoa and Venice, to seek alternative routes to the East. In brief, flavours and fragrances were a major driving force in the European discovery of the Americas (these had been discovered before, of course, by the people who actually lived there, but as they didn't wear trousers, they didn't count). This exposure to new cultures, together with the rediscovery of classical science, often via Arabic texts, led to a new intellectual flowering in Europe. The concepts of science, based on the principles of observation and measurement, were laid down by Galileo. In this context, perhaps the single most important development for the flavour and fragrance industry was the development of the thermometer by Celsius and Fahrenheit. The essentials of modern chemistry were established by the studies of Lavoiser, Davy, Dalton, Priestly, Scheele and others, and the mystical systems the alchemists left behind. It is difficult to overestimate the importance of this change. Though the early chemical laboratory would have looked like an alchemist's workshop with its pots, pans and alembics, chemistry was based on causality; mixing sulfuric acid with sodium chloride always generates hydrogen chloride (or, to use the terminology of the day, mixing oil of vitriol with salt always generates muriatic acid gas) – irrespective of what the operator is wearing, has eaten or has chanted during the process. The difference between alchemy and chemistry is that between astrology and astronomy. Of course, progress is rarely straightforward (witness the number of astrologers in the world); Sir Isaac Newton (1642–1727) believed that his most important work was not on gravity and the calculus but rather his alchemical studies, and in 1794 Antoine Lavoisier found that the Revolution had no need of learned men.

1.5 The industrial age

From the early nineteenth century developments in chemistry, mostly in Great Britain and Germany, began to create the flavour and fragrance industry as we know it. Purification of natural materials, especially essential oils, led to the identification of aroma-active materials such as benzaldehyde, cinnamaldehyde and vanillin. The increased knowledge of organic chemistry, beginning perhaps with the discovery of Wöhler that 'inorganic' and 'organic' materials could be interconverted (i.e. that there was no 'vital force' responsible for creating the complexities of organic materials), meant that these isolated materials could now be synthesised in the laboratory. The great cycle of the chemical industry – identification, laboratory synthesis, large-scale synthesis and commercialisation with falling costs and prices – had now begun. At this time the events and discoveries of importance come so thick and fast that the best way to show them is the timeline (Figure 1.1).

1820 Foundation of Roure by Claude Roure in Grasse under the company name Roure Bertrand
1834 Isolation of cinnamaldehyde
1837 Isolation of benzaldehyde
1852 Cinnamaldehyde synthesised
1852 W.J. Bush established
1858 Microbial resolution of (−)-tartaric acid from a racemic mixture using *Penicillin glaucum*
1859 Synthesis of methyl salicylate
1859 Vanillin purified
1860 Foundation of Fritzsche Brothers
1868 Synthesis of coumarin
1870 Foundation of Haarmann & Reimer
1871 Structure of vanillin determined
1872 Vanillin synthesised
1873 Roure wins its first gold medal at the Vienna International Exposition
1878 Synthesis of cinnamic acid
1879 Saccharin is discovered
1880 Lactic acid, the first known optically active compound to be produced by industrial fermentation
1883 Synthesis of phenylacetaldehdye
1884 Synthesis of cinnamaldehdye
1886 *Bacterium zylinum* identified as cause of vinegar production
1886 Nitromusks discovered
1887 Chirality and taste; enantiomers of asparigine found to have different tastes
1888 First determination of an odour threshold. Ethyl mercaptan by Fischer & Penzoldt
1889 Polak & Schwarz's founded
1889 'Jicky' by Houbigant
1891 Synthesis of the ionones
1892 Allyl disulfide identified in garlic
1894 Fischer's 'Lock & Key' Principle
1896 Foundation of Givaudan by Leon and Xavier Givaudan in Zurich
1898 Foundation of Chuit & Naef by Philippe Chuit & Martin Naef in Geneva
1900 Saccharin first used as a sweetener
1908 MSG identified
1912 Maillard reaction described
1913 Foundation of Fries & Fries in Cincinnati, Ohio by Robert G. Fries and his brother George
1914 Foundation of Polak's Frutal Works (PFW) in the Netherlands by Joseph, Jacob and Henri Polak
1918 A.L. van Ameringen formed
1919 Dragoco founded by Carl-Wilhelm Gerberding
1920 Ruzicka introduces the concept of odourant design
1921 'Chanel No. 5' by Coco Chanel
1922 Takasago founded
1926 Furfuryl mercaptan identified in coffee
1931 Amadori products recognised in the Maillard reaction
1933 Naef, Chuit et Cie. becomes Firmenich & Cie
1937 Non-nutritive sweetener cyclamate discovered
1939 Ruzicka wins Nobel Prize for chemistry
1946 Pauling theorises on the influence on odour of molecular shape and size
1950 Polycyclic musks introduced
1952 GC introduced by James & Martin
1952 Hodge scheme for the Maillard reaction
1954 Roure opens its perfumery school in Grasse
1959 International Flavors & Fragrances formed from van Ameringen-Haebler inc. and Polak & Schwarz NV
1964 GC-Olfactometry introduced
1965 BBA formed from WJ Bush & Co Limited, A. Boake Roberts & Co Limited and Stafford Allen & Sons Limited
1968 Amoore's description of the molecular basis of odour
1969 Cyclamate removed from GRAS list by FDA
1969 Hydroxydimethylfuranone identified in cooked beef
1970 Cyclamate banned in the USA
1970 Damascones identified in rose oil
1972 Sacharin removed from GRAS list by FDA
1981 Aspartame gains FDA approval
1982 Grapefruit mercaptan identified
1989 Solid phase micro-extraction (SPME) introduced
1991 Merger of Givaudan and Roure
1998 Sucralose approved by FDA
2000 Spin off of Givaudan Roure from Roche as Givaudan with a listing on the Swiss Stock Exchange
2001 Bush Boake Allen acquired by IFF

Figure 1.1 Discoveries and foundations from the beginning of the Industrial Age to today.

1.6 The post-war world

> Things are different today, I hear every mother say,
> Cooking fresh food for her husband's just a drag
> So she buys an instant cake and she burns a frozen steak...
>
> The Rolling Stones, *Mother's Little Helper*, 1966

Since the end of Second World War the flavour and fragrance industry has expanded greatly. Several interlocking factors have caused this.

1.6.1 *Technical factors*

1.6.1.1 Analytical

Development of analytical techniques such as Fourier transform nuclear magnetic resonance spectroscopy (NMR), infra-red spectroscopy and, in particular, gas chromatography (GC) and mass spectrometry (MS) has enabled the key odorants of fragrant flowers and foodstuffs to be identified at ever lower levels; the so-called hyphenated techniques of GC-MS and GC-olfactometry, in conjunction with 'trapping' methods such as GC trapping and solid phase microextraction (SPME), means that we can now identify materials present well below the part per billion level and at quantities down to the picogram level. This means that the trace components, which are often the character impact chemicals of a material, can now be identified. Perhaps someone should calculate how many cows would have to be distilled to prepare sufficient 2-methyl-3-furanthiol, the character impact chemical in roast beef, to be identified by the classical techniques of the nineteenth century.

1.6.1.2 Synthetic

The increase in the 'armoury' of synthetic techniques available to organic chemists has made available to the perfumer or flavourist an enormous number of materials. In particular, it has enabled materials of ever-greater complexity to be available to the applications specialists at a price that, even if the cost per kilo is high, still adds value to the end formulation. The cycle of identifying a potentially valuable material, synthesising it in the laboratory and then developing a commercially viable process (at gradually decreasing cost and sales price) has gone on since the nineteenth century, but the cycle time in today's commercial culture is much lesser.

1.6.2 *Social factors*

After the end of Second World War, and especially from the 1960s onwards, the developed world, especially North America, western Europe and Japan, has seen a major increase in the standard of living. This has given the 'disposable income' that we have seen is a key factor in the uses of flavours and fragrances. On top of this these regions have seen an unprecedented increase in the social and geographical mobility of their populations.

This has meant that a wide range of people have been exposed to a wider range of tastes and aroma than previous generations, and this has created a market that the industry has expanded to fill; of course, the industry has been a factor in exposing people to these factors anyway. In short, the public palette, if not more sophisticated, is certainly wider. Much of this change has taken place in the author's (born 1958) lifetime. For example, the simple ersatz flavours of the author's childhood, where soft drinks were defined by their colour rather than their flavour ('raspberry cream soda' was 'red pop' as it did not remotely resemble raspberries; no one could work out what 'green pop' was meant to taste of), and potato crisps (= potato chips for Americans and other non-English speakers) were 'ready salted', or for special gourmet occasions, 'cheese and onion'; 'salt and vinegar' flavour was introduced with great fanfare at some point in the 1960s. We can see this in our potato crisp timeline (Figure 1.2).

The other great change, reflected in the words of Dr Jagger and Professor Richards quoted above, is that 'time is money'; much more prepared and processed food is now eaten than ever before. Again, it is no longer acceptable for this to be the peacetime equivalent of powdered egg and coffee made from acorns. People want, and are willing to pay for, high-quality prepared food, as good as, or better than, they could make themselves if they spent time on it. We can see this in the range of ready-meals and cooking sources now available. Health factors have also played a part; vegetarianism is no longer considered a 'cranky' lifestyle and, as more people have opted for this approach, there has been a rise in the market for vegetarian food that does not taste vegetarian. Hence there is a major demand for savoury flavours for vegetable protein such as soya and mycoprotein.

A similar shift has taken place in the market for fragrances. In the 1960s personal care fragrances were limited to women's perfume and the annual bottle of after-shave that

Prehistory		
1960	Ready-salted cheese and onion	**Then:** Bag of ready-salted and a bottle of red pop
1970	Salt and vinegar	
1980	Smoky bacon Roast chicken Prawn cocktail Roast ox	
1990	Roast beef and mustard	
2000	Sea salt and balsamic vinegar Salsa and mesquite	**Now:** Pack of Monterey Jack kettle chips and a mango and passionfruit smoothie

Figure 1.2 The potato crisp story.

all children bought their father for Christmas. Most household items such as washing powder, detergents and bleach were unfragranced. By the start of the new millennium the situation is reversed; almost everything that *can* be fragranced *is* fragranced, with unfragranced materials carrying a premium price for the specialist market. In personal care, the use of fragrances by both sexes is now the norm; the taboo of 'men wearing perfume' was broken in the 1970s' Britain by an advertising campaign for after-shave where sporting figures such as the boxer Henry Cooper, whose masculinity could not be questioned (at least by anyone not wishing to have their face rearranged), exhorted us to 'splash it all over'. Fragrance is now all around us, from air-fresheners in our cars (though not in mine as it helps to cause my motion sickness) to the bleach in the toilets. A cynic might perhaps suggest we have become afraid of smelling human.

1.7 The future

Does the flavour and fragrance industry have a future? It would be surprising if the editor of a book on flavour and fragrance chemistry did not say yes! Perhaps I should back that up by saying we have used flavours and fragrances for at least five millennia, so we are not likely to stop now. Synthetic chemistry is still developing new methodologies, so materials that are at present interesting, but too expensive to use, may become available at an 'accessible' price due to the discovery of a commercially viable synthetic route. Similarly, analytical work on the examination of new 'exotic' materials (such as IFF's 'Living Flower®' work and Givaudan's TasteTrek™ project) may lead to the identification of exciting new compounds as targets for organic chemists. The other key factor is the development of the global economy, and especially that of the developing world. At present the majority of the world's population has no 'disposable income'; returning to the theme of the market for flavours and fragrance as a mirror of the affluence of a society, we can only hope that these billions can eventually share the living standards of the developed world and, in so doing, will open a market for the industry, even larger than the one it has at present.

1.8 The structure of the flavour and fragrance industry

At the time of writing, the industry is in a state of flux, similar to that in the pharmaceutical industry in the 1990s, and to an extent, any comments I make here may be out-of-date by the time this is published. There have been a number of mergers and acquisitions, which have taken out 'names' that would once have been considered inviolate, most notably the takeover of Bush Boake Allen by International Flavours and Fragrances, and the merger of Haarmann & Reimer with Dragoco to form Symrise. Somehow it is sad to lose the names of Haarmann & Reimer, whose work on Vanillin was such a landmark in the industry. In terms of value, the industry has an estimated value of $15 billion, with an approximately even split between flavours and fragrances. The top five companies, Givaudan, IFF, Firmenich, Symrise and Quest International each have a turnover in excess of $1 billion. Most interesting is the geographical distribution of sales: North America 32%, Asia-Pacific 26% and western Europe 25%, with 'the rest', eastern Europe, South America, the Middle East and Africa, constituting less than 20%. Here we can see again the principle that the flavour and fragrance industry is a consequence and a measure of 'disposable income'; the developed world accounts for 80% of the sales. Since this is necessarily a changing set of figures, readers are directed to the Leffingwell website (www.leffingwell.com) from which these figures were taken in June 2003. Rather than listing further figures and companies here, we can perhaps take a different approach in terms of breaking down the industry. There are four key groupings:

(1) The Big Players, e.g. IFF, Givaudan, Takasago, Firmenich. These are multi-national companies active in both flavours and fragrances. Apart from formulating the final fragrance and flavour they also manufacture aroma chemicals, which they may sell on the open market or keep as 'captive' for internal use only.
(2) The large consumer groups, which may be further subdivided into:
    (a) food groups, e.g. Kerry, Danisco, and Nestle.
    (b) household and consumer goods, e.g. Proctor & Gamble, Unilever, L'Oreal.

These are unlikely to manufacture aroma chemicals, but have active flavourist and perfumer groups, besides utilising flavours and fragrances supplied by the Big Players.

(3) Specialists, e.g. Frutarom, Duckworths, Flavor & Fragrance Specialities. Smaller in total turnover than the big players, these may still exceed them in key areas, e.g. the manufacture of aroma chemicals, domestic fragrances, savoury flavours and others.
(4) 'Mom and Pop Companies' – This common (if faintly derogatory) term covers the large number of companies who, as individuals, are small, but who constitute a large and often neglected area of the industry.
(5) Mainstream chemical companies who manufacture some aroma chemicals as part of their business, e.g. BASF (*cis*-3-hexenol).

Those working in the industry will be an employee of one of these, and will interact with the others as a customer, supplier or both; one peculiarity of the industry is the number of occasions that a customer for one group of materials is a competitor in another area. There is a tendency to 'take in each other's washing'.

1.9 A note on regulations

The flavour and fragrance is subject to a range of regulations, above and beyond those, such as local environmental and safety laws, applied to the chemical industry in general; this is appropriate, given its proximity to the consumer. In this area there is a fundamental divide in the two Fs. In the fragrance area, uses of raw materials and finished products are regulated through two bodies: the Geneva-based International Fragrance Association (IFRA) and the US-founded Research Institute for Fragrance Materials (RIFM). These two organisations, though formally independent of each other, work together closely to ensure the safety of fragrances. The RIFM is responsible for the testing and monitoring of fragrance materials, commissioning and carrying out tests, setting out protocols, and collating data from basic research and consumer studies. If there is concern over a material, for example sensitisation, a particular concern in recent years, the RIFM can review its use. IFRA is the final arbiter, with its Code of Practice representing the 'best practice' in the industry. Steve Meakins has given a good overview of this [3].

In the flavours area the regulations are less international, with more localised legislation. In the European Union there is the so-called white list of approved aroma chemicals and related substances, but the most important world-wide is the US-based Flavor and Essence Manufacturers Association (FEMA). The FEMA GRAS is a list of over 4000 materials that are Generally Recognised As Safe, i.e. they may be used in foodstuffs at set levels, under which circumstances they need not be declared as individual substances; note that flavour materials are not 'additives' or the dreaded E-numbers of EU legislation. The GRAS list is updated on a roughly biannual basis with new materials. In the USA only materials which are on the FEMA GRAS list may be used.

A second concept divides the USA and Europe, and that is the nature identical/natural/artificial definitions. Both regions recognise materials of 'recent biological origin' as natural, but in Europe there is the concept of nature identical, i.e. 'synthetic' ethyl butyrate, made by the reaction of 'synthetic' ethanol (e.g. from ethylene) and 'synthetic' butyric acid (e.g. from the C4 stream from petroleum cracking), is the same as the ethyl butyrate found in nature, and hence the consumer labelling on a product in which this material is used will simply be 'contains flavouring'. In the USA, a pre-Wöhler view seems to hold and such a material is considered to be artificial, i.e. consumer labelling will be 'contains artificial flavouring'. In Europe, the term 'artificial' is reserved for those materials which are not found in nature, and which now are rarely used; in this region, the labelling 'contains artificial flavouring' is a strong negative from the marketing viewpoint. Ironically, since 'contains artificial flavouring' is more commonly seen in the US, it is less 'negative' there, and materials which would be considered artificial in Europe are more widely used in the USA.

A second regulatory area which concerns flavour materials, which reminds us that they are going to be consumed as part of a foodstuff, is that they must conform to dietary regulations. These may be based on faith (Kosher, Halal), health (allergen-free), ethics (vegetarian, vegan) or fashion (GMO-free, 'Organic'). Of these, the longest-established is the Kosher regulations, where for many years Rabbinical bodies have looked at the interaction of Mosaic Law with the modern flavour and chemical industries. Of course, the term Kosher has entered the English language, especially British English, as a general term for the 'real thing', especially as opposed to anything 'hooky', i.e. obtained by dubious and probably illegal means. In this context, it is not surprising that when the author joined the flavour and fragrance industry and saw a file labelled 'Kosher Certificates' he assumed that they were simply certificates of analysis or conformity, i.e. to prove that they were the 'real thing'.

Uwe-Jens Salzer and Kim Jones have given an excellent overview of the complex field of legislation in the flavour industry [4].

1.10 A note on quality

All the materials produced by the flavour and fragrance industry, whether aroma chemicals or finished perfumes and flavours, are ultimately performance chemicals, and the function they perform is to have the correct taste and/or smell. With this in mind, a fundamental quality aspect is their organoleptic properties; above and beyond the chemical specifications of purity, density, refractive index, etc., the material must smell right. Hence the final and defining test which is carried out on a material is an odour assessment; whilst flavour materials may also be tasted (especially final products), the close relationship between taste and smell (see Chapter 9) means that odour is usually used as the test. This assessment is carried out by trained individuals in quality control, usually by smelling the material on a blotter or strip of stiffened paper and by sniffing the material in a bottle or other containers; the latter gives the aroma of the 'head space', the saturated air above the product, which is usually the first impression a customer gets when opening the container of a product. A reference standard of the material is assessed at the same time in the same way in order to get a comparison. The first criterion is that there must be no 'off-notes', i.e. obvious aromas due to different aroma chemicals, either as impurities or contaminants. This is a serious problem, especially for the manufacturers of aroma chemicals, as traces of impurities, well below the level of standard GC detection, may be detectable on odour if the odour threshold of the impurity is lower than that of the desired material; for example a trace of furfuryl mercaptan (odour threshold 0.003 ppb) will dominate the odour of an ester such as ethyl laurate (odour threshold >1000 ppb) even at 0.001%, and will be detected as an off-note well below that. In addition, traces of odorous by-products will have the same effect in creating off-notes. Figure 1.3,

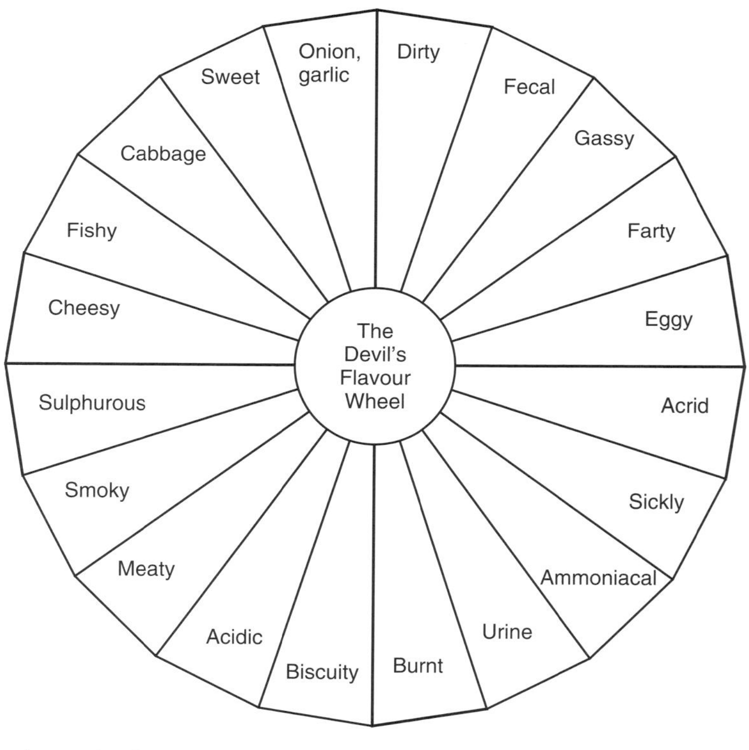

Figure 1.3 The Devil's Flavour Wheel.

'The Devil's Flavour Wheel' [5] summarises just some of the off-notes detected in flavour aroma chemicals.

All this means that cleaning regimes must be rigorous to avoid cross-contamination, and often during a distillation, the odour of individual fractions is assessed and blending carried out accordingly. A second point is that the product must match the reference standard; 'better' does not mean correct in this case. A final point is to refute the claim that odour assessment, as carried out by a trained nose, is 'subjective'. Whilst it can never have the same objectivity as a number coming from a refractometer or from a GC trace, it is as objective as a human assessment can be, and the evidence for the success of this approach is shown in the reliability and consistency that we all expect, and get, from the flavoured and fragranced items that we buy each day.

References

1. Pybus, D. (1999) The history of aroma chemistry and perfume. In D. Pybus and C. Sell (eds), *The Chemistry of Fragrances*. Royal Society of Chemistry, Cambridge, pp. 1–23.
2. Teisseire, P.J. (1994) *Chemistry of Fragrant Substances*, VCH Publishers Inc., NY.
3. Meake, S. (1999) The safety and toxicology of fragrances. In D. Pybus and C. Sell (eds), *The Chemistry of Fragrances*. Royal Society of Chemistry, Cambridge, pp. 174–187.
4. Salzer, U.-J. and Jones, K. (1998) Legislation/toxicology. In E. Ziegler and H. Ziegler (eds), *Flavourings*. Wiley-VCH, NY, pp. 643–698.
5. Rowe, D.J. (2002) High impact aroma chemicals part 2: the good, the bad and the ugly. *Perfumer & Flavorist*, 27(4), 24–33.

Chapter 2
Identification of Aroma Chemicals

Neil C. Da Costa and Sanja Eri

2.1 Introduction

The characterisation of aroma compounds in natural products and foods is still a challenge, despite the sophisticated techniques now available to analysts. Flavour and fragrance compounds are usually present in these matrices at very low concentrations (ppm or even ppb). In addition, they have a wide range of polarities, solubilities, volatilities, and thermal and pH stabilities. The matrices that contain them may be very complex and cause interference with the isolation techniques. Therefore, there is no single, simple method for the identification of aroma compounds. Instead, analysts must first ask themselves what they wish to accomplish with the analysis and then choose an analytical method or combination of methods that they think would work best. Typical analytical goals include obtaining a complete aroma profile of the sample, looking for specific compounds or comparing samples.

In this chapter, a reader will find an overview of the techniques now available to the analyst interested in characterising aroma chemicals. Each technique's practical advantages and disadvantages will also be indicated where relevant. Numerous papers and books have been published on this subject and this overview by no means covers all of them, but it should be a good starting point to a newcomer in the world of flavours and fragrances and a useful reference to those who are already involved in it.

2.2 Isolation of aroma chemicals

To the early organic chemists the only approach available was isolation of pure chemicals and identification by classical means. The development of chromatography, especially gas chromatography (GC), and in particular gas chromatography-mass spectrometry (GC-MS), made identification of compounds in mixtures much easier. However the quality and usefulness of the GC and GC-MS analyses, or any other technique, largely depend upon the method used to prepare the sample. If the sample is concentrated enough it can be injected directly into the GC, such as essential oils or natural extracts. In this case, the only preparation recommended is dilution of the sample with a solvent to obtain better chromatographic results and to reduce the introduction of the contaminants into the GC. Very concentrated aqueous samples can also be injected directly, although the injection volume and split ratio have to be chosen carefully due to the formation of steam in injection port. Problems such as extinguishing the flame ionisation detector (FID), contamination of the mass spectrometer or degradation of non-bonded polar phases may occur.

If, however, the aqueous sample is too dilute or the aroma compounds are contained within a matrix, they need to be isolated. It is beneficial to the analyst to know as much as possible about the matrix (whether it is a fruit, vegetable, flower, thermally processed food, fermentation product, flavour, fragrance, etc.) as well as the aroma chemicals that they expect to find in it, especially the compounds that are the most significant contributors to the aroma. Unfortunately, these are not necessarily the components of highest concentration. Very often, these important components are at trace levels, making isolation and identification more challenging. Once we understand the scope of the problem, we can choose the most appropriate isolation method that will give us the 'truest' extract with minimum loss or distortion of the original aroma and minimal generation of artefacts. Ideally, the isolation method should not discriminate between polar and non-polar compounds, not cause thermal degradation of aroma compounds, oxidation, reduction, pH changes or loss of highly volatile compounds. In addition, if we know that the material we are going to extract has high molecular weight or non-volatile content, we can choose a method that reduces the appearance of these components in the extract, thus reducing the problems of dirty liners, columns and subsequently poor chromatography.

Very often, to achieve the most complete aroma profile, the analyst may have to use more than one extraction technique. For example, headspace techniques obtain a profile of the highly volatile components, which may not have been seen in a liquid extract, as some of them may be too volatile to be captured and others could be obscured by the solvent peak. This kind of approach is illustrated later in Section 2.5.

This latter point is of great importance as solvents may contain impurities, which when concentrated may give misleading results. Aromatic hydrocarbons and phthalates occur almost everywhere; especially if the sample has come into contact with plastic containers, where phthalates are used as plasticisers. Silicones are common impurities as well, arising from oil or grease used in the extraction apparatus, anti-foaming agents and from GC column bleed.

There are many techniques for isolation of aroma chemicals that are currently available and many are still being developed. Once a new method is introduced, it is subsequently modified to increase its efficiency, reproducibility and range of matrices for which it can be used. Many studies that compare the results obtained with different methods using either model solutions or real samples have been published [1–5]. Based on the principle of aroma compound isolation, sample preparation methods for GC analysis can be broadly grouped into solvent extraction methods, steam distillation methods, headspace techniques and sorptive techniques.

2.2.1 Solvent extraction

These techniques are the most widely used and originate from classic organic chemistry extraction work, the concept being the transfer of the aroma chemicals from the sample to the organic solvent. The extract is obtained by mixing and agitating a liquid or solid sample with organic solvent, allowing separation and collection of the solvent phase. The extraction can be done manually or automatically and be in batches or a continuous flow extraction. In any case, if the extraction process is conducted successfully, the analyst should end up with an extract whose sensory properties closely resemble those of the original sample.

The choice of solvent is very important, as it must fulfil certain criteria. It must have a low boiling point so that it can be easily removed from the extract without a significant loss of sample volatiles and so it does not obscure other components during chromatography. It must be able to extract polar and non-polar components. Below are listed some of the most popular solvents used, the top two being most favoured in flavour extraction work.

Solvent	Boiling point (°C)	Miscibility with water
Dichloromethane	45	Immiscible/denser than water
Pentane/Diethyl ether	35	Immiscible/less dense than water
Freons (halogenated mixture)	<45	Immiscible/denser than water
Hexane – concretes extraction	69	Immiscible/less dense than water
Ethanol – absolutes extraction	78	Miscible with water
Acetone – some fragrance bases	56	Miscible with water

In addition, a new breed of solvents has been used for the extraction of aroma compounds; namely supercritical fluids such as carbon dioxide and nitrogen. Supercritical fluid extraction (SFE) will be covered later in this chapter.

Sometimes, the material from which we are planning to extract aroma compounds may require additional care before we mix it with a solvent (homogenising, grinding, milling, degassing, enzyme deactivation, etc.). For example, in the case of fresh fruits, enzyme deactivation can be easily effected using fresh saturated calcium chloride solution or methanol.

Advantages

- Good recovery of aroma compounds; the most quantitative of all isolation methods.
- Relatively easy to perform.

Disadvantages

- Solvent removal may cause loss of some of the more volatile compounds.
- Use of solvents which may be toxic and/or flammable and which may introduce contaminants.
- A need for large amounts of sample in order to produce a strong extract.
- Obtained extracts may contain high-boiling and non-volatile materials and colour components, which may lead to further problems in analysis and chromatography.
- Emulsion formation during extraction may occur.
- Solvent peak may cover early eluting volatiles in the chromatogram.

If a sample is carefully concentrated, frozen and the top liquid layer decanted, it is possible to reduce or even eliminate very high-boiling, non-volatile or unwanted co-eluting components in an extract (such as fat, preservatives like benzoic and sorbic acids and caffeine).

Other possible ways to reduce contamination of a GC when constantly dealing with samples containing high-boilers and non-volatile materials, are the use of a 'guard' column of deactivated fused silica tubing, or a retention gap (a small length of the normal column) between the injector and the column. The 'guard' column or retention gap can be replaced once they get dirty.

2.2.1.1 Extraction of liquid samples

If the sample is an aqueous liquid, extraction with an organic solvent can be easily carried out in a separatory funnel or commercially available extractor. However, the separation of an aqueous and organic solvent layer may be complicated by emulsion formation. In this case, centrifugation is often required to break down emulsions and aid in layers separation. Using a saturated salt solution (sodium chloride), instead of pure distilled water to dilute the sample, may also prevent emulsion formation. The use of solid phase extraction (SPE) columns [6–8] and solid phase disc extraction (SPDE) methods can also prevent emulsions formation. These last two approaches can also aid in pre-concentration and cleaning the samples of non-volatiles.

2.2.1.2 Extraction of solid samples

If the sample is solid, it can either be extracted as it is or dissolved in water and extracted as an aqueous sample. If the sample is to be extracted as it is (such is usually the case with solid foods, non-encapsulated powders or in the manufacture of concretes), the sample can be simply soaked in the solvent for a fixed period of time, after which the solvent is decanted, filtered and the extract concentrated. In the case of encapsulated samples, it is preferable to use water to release flavour before the solvent is added.

Other techniques available for solid sample extraction are Soxhlet extraction, accelerated solvent extraction (ASE) and SFE.

2.2.1.3 Soxhlet extraction

Soxhlet extraction is a continuous solvent extraction method that can be used for solid, semi-solid or viscous liquid samples. In this method, a sample is placed in a porous thimble and continuously extracted over several hours by allowing a low-boiling organic solvent to flow through it. New automated versions of Soxhlet, which can extract and concentrate the sample in one step, are now available on the market. Semi-solid and viscous liquid samples need to be mixed with a solid support such as diatomaceous earth (e.g. Celite®) and/or sodium sulfate or magnesium sulfate to absorb water and make the sample solid. A disadvantage of this method is the presence of non-volatile components (especially fats) in the resulting extract.

2.2.1.4 Accelerated solvent extraction

A method that has been used more recently for extracting solids and semi-solids is ASE [9]. In this method, the sample is packed into a pressure-tight cell, topped up with solvent and subjected to an elevated temperature and pressure. The cell is then purged using nitrogen gas to send the extract into a collection vial. As in Soxhlet extraction, semi-solid samples need to be mixed with a support. Typical solvents used are methanol, isopropyl alcohol, hexane and water. The advantages of this method are that it uses relatively small amounts of solvent and that extraction times can be as short as 15 min. The disadvantage is that very volatile compounds can be lost during the elevated temperature extraction. Also, on transfer from the cell to the collection vial, it is possible that the sample may precipitate, thus clogging the line.

2.2.1.5 Supercritical fluid extraction

One of the most innovative methods of solvent extraction is SFE, where the solvents used are supercritical fluids [10–14] such as carbon dioxide. By changing the pressure during the extraction, the extracting power of the supercritical fluid can be modified, hence the extract profile tailored. It works best on dry solids or viscous liquids. In practice, SFE has been shown as particularly useful for extracting fragrances from products with high-boiling materials and bases, such as shampoos and conditioners. Liquids again need to be mixed with a solid support and packed inside cells for extraction. The cells are pressurised and extracted with supercritical fluid, and the extract collected in small amounts of solvent in vials.

Advantages of SFE include use of non-toxic solvents, selectivity of extraction, short extraction times, and less reaction of thermally labile and oxygen-sensitive compounds. The main problem with the apparatus is clogging of the transfer line with extraction material. Supercritical solvents tend to discriminate against polar components but this can be overcome by using modifiers like methanol or ethanol. Additional disadvantage is the difficulty in balancing temperature, pressure and flow rate for each type of sample. This makes method development very time consuming. Also, the sample capacity is relatively limited.

2.2.1.6 Fractionation of solvent extracts

The extracts obtained by solvent extraction, especially ones from natural products, can be very complex and the GC chromatogram may contain many co-eluting compounds, which make identification of individual aroma compounds difficult. One way to overcome this problem is to fractionate the extract, and hopefully move co-eluting peaks into the different fractions for better identification. This is particularly important for gas chromatography-olfactometry (GC-O) analysis. In this case, the analyst's goal is to separate co-eluting peaks as best as possible to reveal organoleptically important compounds that may be hidden under bigger, less-odorous peaks. GC-O techniques will be covered later in the chapter.

Fractionation of the extracts can be accomplished in several ways. Manipulation of the extract by washing it with dilute acid, dilute base, and either sodium metabisulfite or 2,4-dinitrophenylhydrazine can lead to the elimination of acids, bases or carbonyl compounds from the extract, respectively [15]. If each of the wash solutions is then re-extracted with solvent, the fractions containing only acids, bases or carbonyls can be recovered.

Column chromatography or SPE can also be used. Fractionation by column chromatography involves aqueous or solvent extracts on a silica gel column using a polar/non-polar solvent system. It can remove colours and high boilers. The concentrated fractions may give more detail than the whole extract. A typical solvent system used is pentane/ether. If the extract volume is small, commercially prepacked SPE tubes (containing packings with different functionalities) are very convenient.

Fractionation of the solvent extract, however, can be very time consuming, and due to the multiple manipulations of the extract may cause the loss of some very volatile compounds.

2.2.1.7 Concentration of solvent extracts

The extract or combined extracts are typically dried over anhydrous sodium sulfate or magnesium sulfate, to remove any water present, and filtered. At this stage the concentration of volatiles in the extract is usually too low for immediate analysis, therefore most of the solvent needs to be removed. The initial concentration of the extract (down to 10–15 ml) may be accomplished by techniques such as a Vigreaux column distillation, rotary evaporation or by use of a Kuderna-Danish concentrator. This can be followed by passing a fine stream of pure nitrogen gas over the surface of the extract to concentrate to a final volume (1–2 ml). Automated solvent evaporation systems that can be programmed to stop evaporation after a fixed time or after reaching a certain extract volume (i.e. 1 ml) are also available; for example, the Zymark™ system. Note that the extract must always be concentrated with great care, as many hours of extraction work can be ruined in seconds by too rapid a solvent reduction or by reducing too far, thus losing important volatiles.

2.2.1.8 Solvent assisted flavour evaporation (SAFE)

To overcome some of the disadvantages of solvent extraction techniques, the extraction may be followed by the SAFE technique. The SAFE technique can also be used as a sole extraction technique for aqueous foods, such as milk or fruit pulps or matrices with high oil content [16]. This technique takes a solvent extract or food matrix and strips off the volatiles with low heat and high vacuum. The extract is collected in flasks, which are cryogenically cooled with liquid nitrogen. The extracts should be representative of the original sample, not ‘cooked’, but without high boilers and colour. Figure 2.1 shows the crude liquid extract funnel (upper left), the heated sample flask (lower left), condensing system (centre) and cryogenic traps (right). Note that the upper trap leads to the high vacuum source and acts as a guard for the pump.

Disadvantages of a SAFE technique are its complicated set-up and time-consuming cleaning. Considering the alternative (dirty liners and columns), it is well worth putting some extra time into setting it up and cleaning afterwards. Also, high fat content can cause skewing of the flavour profile.

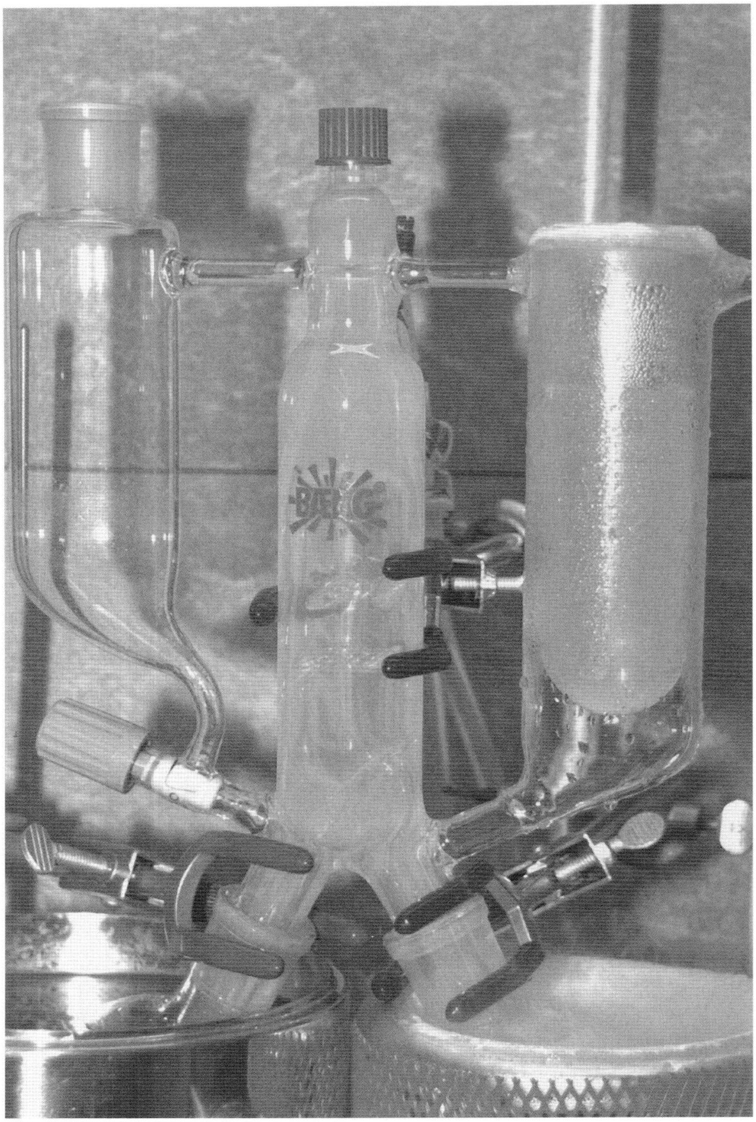

Figure 2.1 SAFE apparatus.

2.2.2 *Steam distillation methods*

This group of techniques includes simple steam distillation methods, which generate an aqueous extract that needs to be extracted with solvent, and simultaneous distillation-extraction (SDE) methods in which the final product is a solvent extract [17, 18]. The main advantage of steam distillation techniques is that resulting extracts do not contain any non-volatiles.

In a simple steam distillation, the sample is dispersed in water and placed in a round-bottom flask, which is heated either directly or indirectly by steam. Vapours are condensed, collected and extracted with organic solvent.

SDE is also known as the Likens–Nickerson steam distillation [19]. In a typical set-up (Figure 2.2), sample is added to distilled water in a larger round-bottom flask and stirred constantly throughout the distillation to avoid bumping. The smaller flask is filled with solvent (typically dichloromethane or freon) and both flasks are heated simultaneously. The sample is heated to 100°C and solvent to its boiling point, 45°C for dichloromethane. The vapours rise in each wing of the apparatus, combine centrally, and the volatiles are transferred between the condensing liquids. The water and the solvent then return

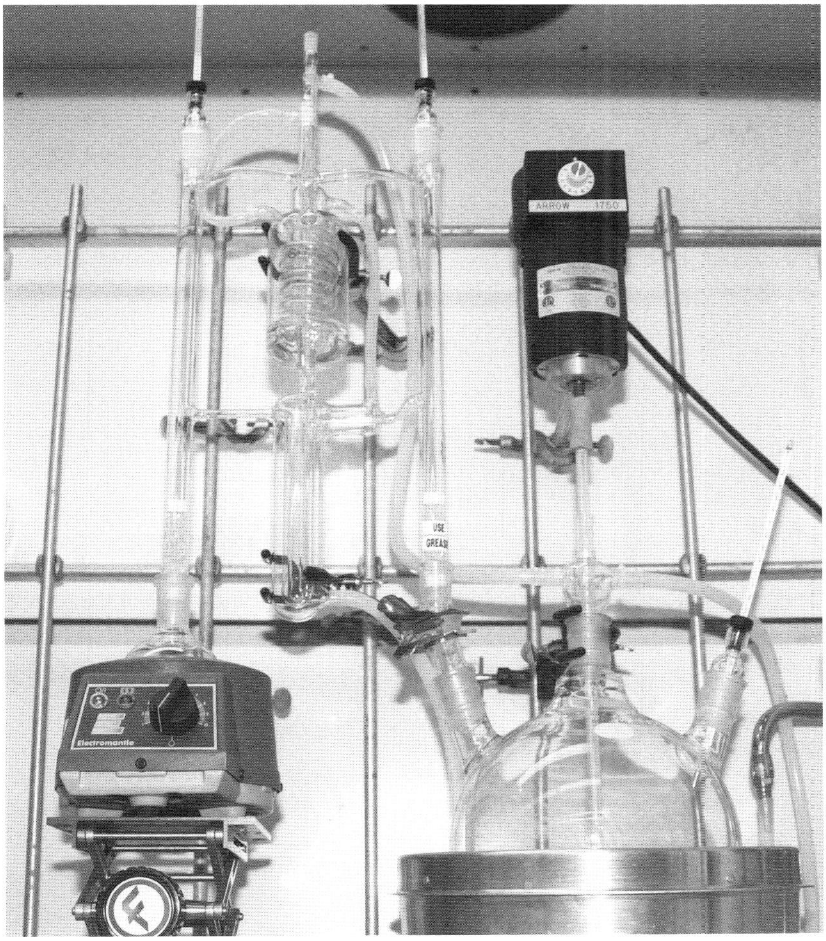

Figure 2.2 Simultaneous distillation-extraction apparatus.

to their respective flasks, in this continuous extraction process. The extraction time is usually about 2–4 h, dependent upon the matrix.

Advantages

- The extracts produced do not contain any high-boiling or non-volatile materials, and therefore, there is no danger of contaminating GC liners and columns.
- Can concentrate the sample, further enabling trace component detection.

Disadvantages

- Highly polar or hydrophilic compounds such as acids and alcohols extract poorly and may not even appear in the final extract.
- Not good for fresh materials that were not previously thermally processed such as fruits and vegetables, as the extract is more akin to a cooked aroma, rather than fresh fruit or vegetable aroma.
- Artefact formation is possible due to the thermal degradation of the sample components.
- Less quantitative extracts than solvent extraction methods; recoveries vary.
- Problems with foaming require use of anti-foaming agents, which can introduce silicone contaminants.

To reduce the thermal degradation of the sample components during the extraction, steam distillation methods can be carried out under reduced pressure and thus closer to room temperature [20]. This set-up has the advantage of generating fewer artefacts and breakdown products, but in the case of SDE it is difficult to balance the boiling of two flasks, avoid evaporation of the lower boiling solvent and keep pressure constant.

2.2.3 Headspace techniques

Headspace collection is probably the easiest way to capture and detect aroma compounds, since they exist in the space above the sample. It is simple and convenient and it has been used for all kinds of materials [21–23]. It is especially useful for matrices that give off a lot of odour, such as flowers and fruits. For matrices that are not as odorous, gentle heating can be applied to aid in release of volatiles. Due to the fact that headspace techniques detect highly volatile compounds, they can be used to help identify compounds that may be hidden by solvent peaks in liquid extracts. These two techniques compliment each other very well.

Advantages

- Simple and quick.
- Solventless technique.
- Requires small amounts of sample.
- No artefacts are formed, and if properly used, no contaminants introduced.

Disadvantages

- The relative concentration of components in headspace does not reflect the concentration in the sample due to the differences in volatility of aroma compounds (to overcome this disadvantage, the results can be vapour pressure adjusted using calculated or experimentally determined values).
- Aroma profile is dependent upon the sampling temperature.

Headspace-sampling techniques include static headspace and dynamic headspace ('purge and trap').

2.2.3.1 Static headspace

In classic static headspace, the sample is put into a sealed headspace vial, left to equilibrate and the atmosphere above the sample is sampled with a gas tight syringe (usually 0.1–2.0 ml) and subsequently injected into a GC. This technique is easily automated, and very convenient for screening.

The biggest disadvantage of static headspace is that this technique discriminates on volatility, since more volatile compounds will fill the space above the matrix more readily. The aroma profile will depend on the temperature of the sample, sampling time, volume of the vial being used and solubility of the compound in the matrix. Due to the limited space above the sample to which volatiles can escape, this technique is not a very sensitive one. Samples can be heated and/or stirred during collection for better release of volatiles.

2.2.3.2 Dynamic headspace

In dynamic headspace, the volatiles above the sample are swept away by a carrier gas, usually helium or nitrogen, onto a trap (glass or glass-lined stainless steel tube containing porous polymers such as Tenax™ (Figures 2.3 and 2.4)). The other option is the use of a vacuum pump that draws the air above the sample onto the trap (the air from the atmosphere coming into the sampling jar goes through a carbon filter). The carrier gas or air is vented through the trap while volatiles remain on it. The traps containing volatiles can either be washed with the solvent or desorbed into the GC by using a thermal desorber.

In dynamic headspace, the volume of the vial used is not a factor since headspace volumes much larger than the vial can be pulled through the trap. As a result, volatiles collected on the trap are concentrated which results in improved sensitivity as compared to static headspace method. To increase the release of volatiles, samples can be heated and/or stirred during collection. It is possible to overload the trap capacity, and if this 'breakthrough' occurs, very volatile components can be lost.

When this technique is used for liquid samples, the carrier gas is introduced into the sample, and 'purges' the volatiles out into the headspace and onto the trap. Therefore, for liquid samples, the term 'purge and trap' instead of 'dynamic headspace' is usually used.

Dynamic headspace, although a little bit more complicated than static headspace, is still quick and easy to use.

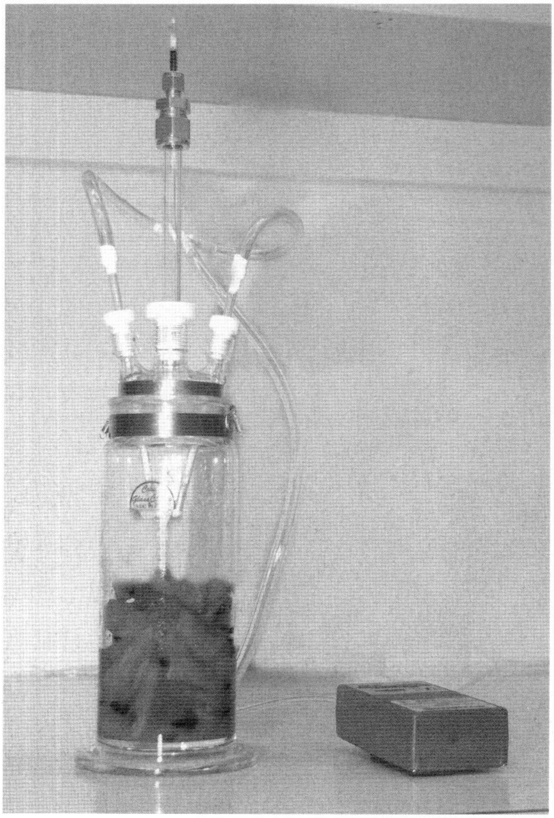

Figure 2.3 Headspace sampling of watermelon using Tenax™ traps and a pump.

2.2.4 Direct thermal desorption (DTD)

Probably the most appreciated advantages of the DTD (also called short path thermal desorption) method are its simplicity and rapidity. DTD permits the analysis of solid samples without complicated or time-consuming sample preparation and without the use of a solvent. In this technique, solid samples (such as spices) are placed directly into a thermal desorption tube (the same glass or glass-lined stainless steel tube used for dynamic headspace analysis) between glass wool plugs. The tube is put onto a thermal desorption unit connected to a GC. The desorber subjects the sample to controlled heating in a flow of inert carrier gas, thus driving the volatiles onto the column. The desorbed volatiles are transferred directly into the GC for analysis in a one-step process. Sample preparation needed includes grinding/milling of the sample and mixing it with a porous polymer such as Chromosorb™ (to prevent clogging) before filling the desorption tube. For the purpose of quantitation, a small bed of porous polymer, such as Tenax™, can also be put in the tube so that an internal standard can be added. Descriptions of the different applications of DTD have been published in the last few years [24–26].

Figure 2.4 Headspace sampling of wisteria using Tenax^{TM} traps.

2.2.5 Sorptive techniques

Sorptive techniques allow rapid and solventless extraction and pre-concentration of aroma compounds. They are based on the partitioning of organic components between aqueous or vapour phase and thin polymeric films. This group includes solid phase microextraction (SPME), headspace sorptive extraction (HSSE) and stir bar sorptive extraction (SBSE).

2.2.5.1 Solid phase microextraction (SPME)

Solid phase micro extraction has been widely used for the analysis of aromatic and fragrant materials [27]. This technique uses a fused silica fibre (1 or 2 cm in length) coated with a polymeric film to collect the volatiles from the sample. A range of polar, non-polar and mixed fibres are available in the market. The fibre is contained within a needle, which is placed into a SPME holder for sampling and desorbing purposes (Figure 2.5). The sample is placed in a SPME vial and sealed with a septum cap. The fibre is introduced (through the needle) into the headspace above the sample or can be

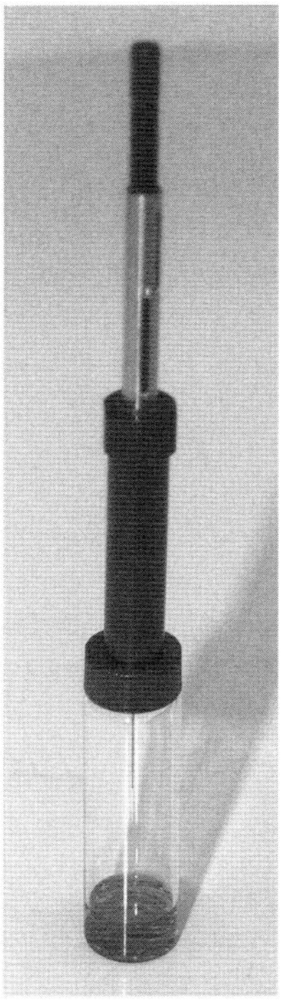

Figure 2.5 Solid phase microextraction sampling.

immersed in liquid samples. After a fixed sampling time, in which volatiles are absorbed on the fibre coating, the fibre is withdrawn and desorbed directly into a GC or GC-MS. The SPME fibre is reconditioned by heating in a GC injection port for 5–15 min.

It takes time to determine the most suitable fibre type, as well as sampling time and temperature, to analyse a particular matrix. Once the optimal conditions are determined, as long as they are kept constant, the analyst should be able to obtain reproducible results.

Advantages

- Quick and easy to use.
- Solventless.
- Good technique for a quick sample comparison or identification of an off-odour in a sample.

Disadvantages

- Aroma profile of collected volatiles is dependent upon the type, thickness and length of the fibre used as well as on the sampling time and temperature.
- Early fibres discriminated against polar components, but mixed fibres have largely overcome this problem.
- For comparison the best results are obtained if the exact same fibre is used on all the samples.

2.2.5.2 *Headspace sorptive extraction (HSSE) and stir bar sorptive extraction (SBSE) (commercially available as Twister™)*

Headspace sorptive extraction (HSSE) and stir bar sorptive extraction (SBSE) are two recently introduced techniques for sample preparation and analysis [28–30]. Both methods make use of a glass-jacketed magnetic stirrer bar, coated with a film of adsorbent polymer (polydimethylsiloxane, PDMS, as used for SPME technology). The stir bar has been commercialised under the name Twister™ and is available in 1 or 2 cm lengths.

HSSE is done by pouring sample into a vial and suspending the Twister™ above it. For SBSE, Twister™ is immersed in the liquid sample and allowed to spin during volatile collection. After a fixed sampling time, the Twister™ is carefully removed from the sample (if it was immersed in the sample, it needs to be washed with distilled water and patted dry) and volatiles collected are either extracted by solvent or, more commonly, recovered by desorbing into a GC. The Twister™ can be regenerated by heating it at high temperatures (325°C) for up to an hour.

Advantages

- Quick and easy to use.
- Solventless.
- Little sample required for analysis.
- Good for alcoholic beverages, as the alcohols which mask the early part of the chromatogram on a non-polar phase appear at much lower concentrations revealing many trace components, if present.

Disadvantages

- Polar chemical discrimination due to the use of non-polar (PDMS) coating.
- Need to desorb soon after sampling due to potential loss of adsorbed volatiles.

2.3 Gas chromatography-olfactometry

The aromas of natural products are usually due to complex mixtures of hundreds of individual compounds. The contribution of each of these compounds to the overall aroma profile varies greatly. As mentioned before, compounds present in trace amounts can be more important contributors than those present in higher concentrations. How does one determine which compounds are the most important contributors to the aroma of a particular sample? The best current answer is GC-O. In this technique, in addition to a FID detector, a gas chromatograph is fitted with an odour port. After injection of the sample and passing through the column, the sample is split into two, going to the FID and column vent respectively. The column vent is often fitted with a funnel-shaped glass piece for sniffing, and a humidifier system keeps the air moist around it. A trained flavourist or perfumer sits by the GC and sniffs the carrier gas as it flows from the chromatograph column. The odour of each compound as it elutes from the GC is recorded and an 'aromagram' is created. If the 'aromagram' is matched with the GC-FID and GC-MS chromatogram of the sample, it is possible to identify the compounds that are the most potent and/or have the most interesting odours. GC-O on a sample enables work to be focused on parts of the chromatogram where compounds of greatest interest are suspected. Subsequently it is possible to identify these key components by mass spectrometry or isolation by preparative GC, followed by nuclear magnetic resonance spectroscopy (NMR) elucidation for unknowns.

Different techniques for designing GC-O experiments, recording the data and estimating the sensory contribution of individual aroma components have been developed; for example, Aroma Extraction Dilution Analysis, CharmAnalysis and Osme. Several papers and books that review these different techniques have been published [31–33].

2.4 Techniques for identification of aroma compounds

As mentioned before, the instrumentation technique most widely used for characterisation of flavour and fragrance compounds is gas chromatography-mass spectrometry (GC-MS) [34–38]. While GC provides separation and quantitation of complex aroma mixtures, mass spectrometry is used for identification of individual compounds.

The data obtained by GC alone can also be used towards the identification of aroma compounds (by the compound's relative retention times), although it needs to be combined with an identification technique for an unambiguous identification. Relative retention times (retention indices) based on a mixture of *n*-paraffins (Kovats indices) [39, 40] are available for both polar and non-polar columns for a great number of aroma compounds. The relative retention times can also be established with respect to a standard mixture of ethyl esters [35].

The data obtained by GC is a valuable pointer to the possible identification of aroma compounds. The retention index of a compound can greatly narrow down the number of possible candidates. In addition to the FID, which is the most widely used for GC analysis, specific detectors such as the flame photometric detector (FPD), atomic emission detector (AED) and chemiluminiscence detector can be used to highlight the areas in a chromatogram where compounds containing sulfur and nitrogen elute [35]. This is very important data since sulfur and nitrogen compounds are among the most potent odour compounds known and are usually present in very low concentrations in a sample.

To confirm the identity of an aroma compound, its mass spectrum and retention index on polar (such as Carbowax®) as well as non-polar (such as OV-1) capillary GC columns have to match those of a known reference compound. However, if there is a need to distinguish between isomers, or if a sample contains an unknown compound for which there is no reference data (neither mass spectrum nor retention indices), other analytical techniques, namely infrared spectroscopy (IR) and NMR, have to be employed. While the mass spectrum provides information about the molecular weight and fragmentation pattern of the unknown molecule, IR spectrum can reveal functional groups and isomerism information and NMR spectrum will tell us the number and positions of the carbon and hydrogen atoms in the molecule [41–44].

Collecting enough of the unknown compound for these analyses can be time consuming. An extract with a high concentration of the unknown is a good start, and preparative GC has to be employed to separate the compound from the rest of the extract. For that purpose, large injections of the extract are made into a GC usually fitted with a packed or megabore column. This manual technique requires sitting next to the GC and waiting for the component of interest to elute before collecting into a glass capillary tube, with cooling. Automated fraction collectors that can be programmed to collect fractions coming from the GC at certain time intervals into glass tubing or polymer traps are available in the market. Considering that usually 100–1000 injections are required to obtain enough of the unknown, the use of an automated fraction collector is the better option.

Another approach towards the identification of an unknown is the use of multidimensional GC technique (MDGC) [35, 44]. In this technique, the GC contains two columns, the first of which is usually a packed or wide-bore capillary column and second is a high-resolution capillary column. The other option is that both columns are high-resolution columns but with different polarities. In MDGC, selected fractions from the first column are directed onto the second column. The second column can provide better resolution of the compounds within that fraction. Also, if multiple injections and collections are performed, trace components can be significantly enriched on a second column and the possibility for their identification by MS increased.

Among aroma compounds it is often the case that different enantiomers of the same compound have different odours. For example L-carvone is known to smell like caraway and D-carvone, like spearmint [35]. Separation of these isomers by GC could be done either by using chiral columns or by forming diastereoisomeric derivatives and using an achiral phase [35, 45–48].

When considering the flavour of the food, some non-volatile materials such as sugars may play an important role in the overall flavour perception. Also, some polar aroma compounds such as vanillin, furaneol®, maltol tend to stay in the polar phase during extractions with non-polar solvents which makes it difficult to determine their true concentrations in the sample by GC analysis of the extract. For the identification and quantitation of non-volatile and polar compounds, high performance liquid chromatography (HPLC) and liquid chromatography-mass spectrometry (LC-MS) have been employed [35, 49].

2.5 A case study: Generessence®

The concept of Generessence® is to re-create as exactly as possible the flavour as nature intended. Only component compounds found in the target food are allowed in the flavour formulation. This enforced criterion forces the flavourist to be truly creative and not rely on traditional flavour ingredients. Once the qualitative part is correct, the quantitative part follows and the result is a truly natural flavour character which is not distorted by dosage, water and/or oil solubility and can be utilised in many applications.

2.5.1 *Sample preparation*

It is usually necessary to acquire large quantities of the natural product or food for extraction. The more the sample extracted, the better the chance of finding novel flavour and fragrance chemicals. Sometimes it is possible to obtain unique extracts from the processing plant, where tonnes of natural product have been processed. Below are three examples of projects and the different techniques used in each case.

2.5.2 Valencia orange (Figure 2.6)

First press oil
This oil was obtained at an orange processing plant. It was produced from the peel of large quantities of oranges.

Oil distillate fraction
This fraction was produced from the condensate obtained after heat treatment process in the orange processing plant.

Water distillate fraction
This water sample was also obtained from the orange processing plant. It was produced from the condensate obtained after heat treatment process. Three litres of this fraction were extracted three times with dichloromethane (200 ml). The combined extracts were dried over sodium sulfate, filtered, reduced by rotary evaporator and further concentrated to 1 ml under a nitrogen gas stream.

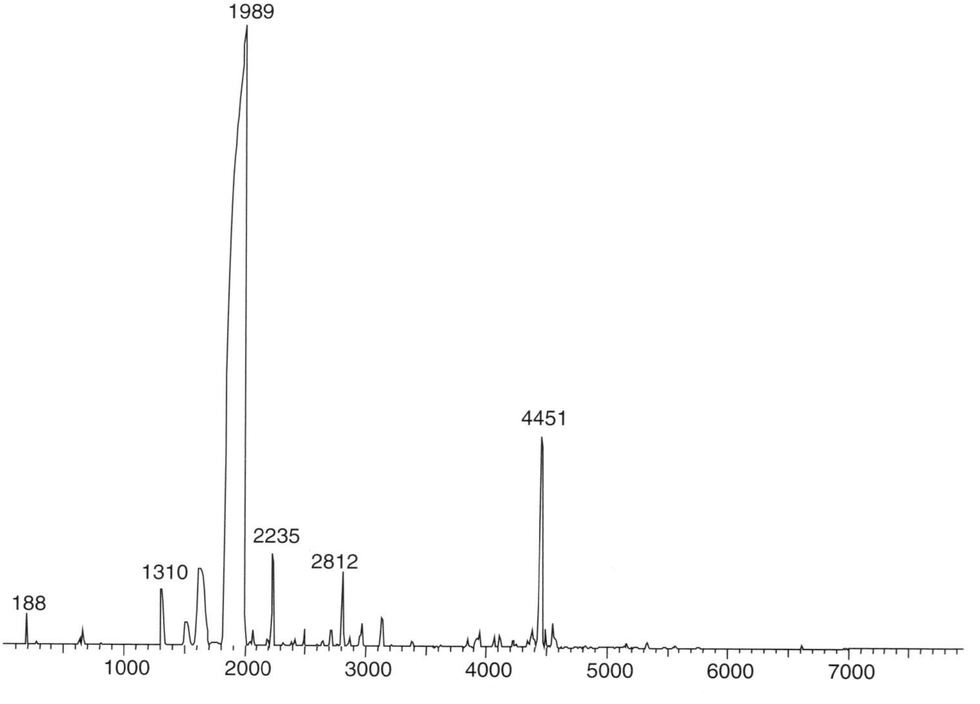

Figure 2.6 Mass spectral total ion current of Valencia orange oil.

Liquid/liquid of juice
Approximately 20 kg of oranges were peeled and the juice was extracted using a kitchen juicer to yield 10.5 l of juice. Litre volumes of juice were transferred to a separator and extracted twice with dicholoromethane (200 ml). Centrifugation was required to aid separation. The combined extracts were dried over anhydrous sodium sulfate, filtered, reduced by rotary evaporator and further concentrated to 1 ml under nitrogen.

Headspace
Headspace-sampling kits were set up in two of the most aromatic parts of the orange processing plant and sampled for 1.5 h. The pairs of traps were brought back to the laboratory and desorbed onto a GC and GC-MS.

All the extracts were analysed by GC, GC-MS and by a GC fitted with a sulfur detector.

Highlights of the volatile components identified
A total of 471 components were identified and quantified in the five analyses, including seven components which were sulfur and/or nitrogen-containing. The main components were limonene (94%), myrcene (2%), α-pinene (0.5%), linalol (0.5%), decanal (0.4), sabinene (0.3%) and valencene (0.04%). Trace quantities of the following sulfur compounds were found: methionol, 2-(methylthio) ethanol, mint sulfide and 1-(methylthio) propan-2-one.

The water distillate provided the most unusual and revealing extract. The main components were linalol (15%), acetaldehyde diethyl acetal (7%), hexanal (7%), α-terpineol (6%), ethyl 3-hydroxyhexanoate (4%), ethyl butyrate (4%).

2.5.3 Roast chicken (Figure 2.7)

Steam distillate of meat
One medium-sized chicken was oven-roasted until golden brown. The chicken skin was removed, the meat was chopped up into small pieces and one litre of water was added for steam distillation. Distillation time using a Likens–Nickerson apparatus was 3 h, into 80 ml of dichloromethane solvent, which was subsequently dried over anhydrous sodium sulfate and concentrated by distillation.

Steam distillate of skin
Two medium-sized chickens were oven-roasted until golden brown. The chicken skins were removed, chopped into small pieces and added to one litre of water for steam distillation. The steam distillation conditions and times were the same as previously described.

Headspace
Two medium-sized chickens were each divided into three pieces and oven-roasted until golden brown. The pieces, including skin, were roasted such that one was ready for

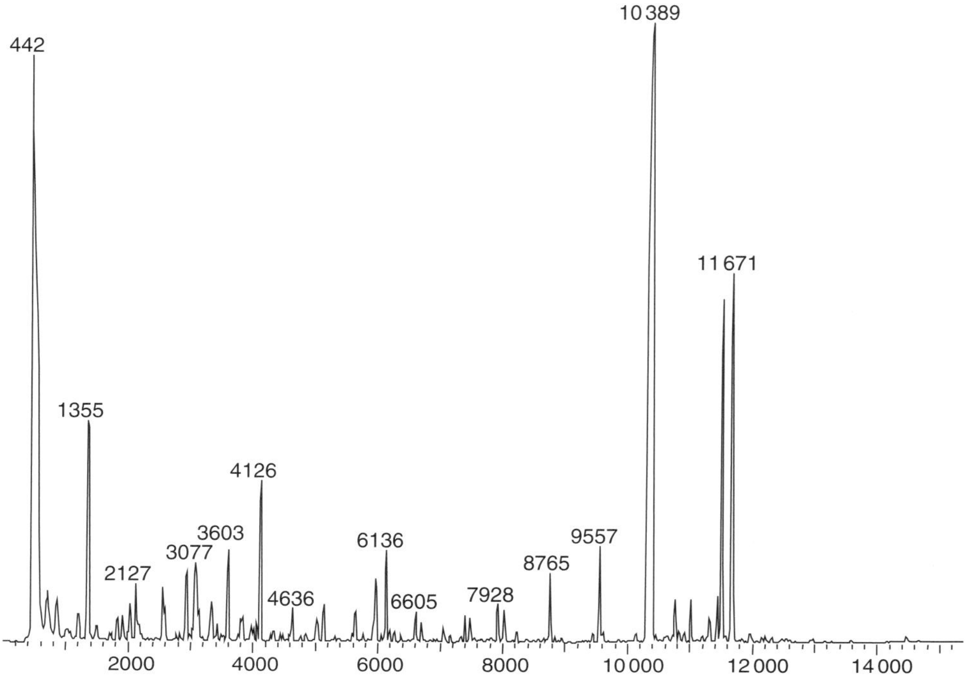

Figure 2.7 Mass spectral total ion current of roast chicken steam distillate.

headspace sampling every 45 min. Each chicken piece was placed inside a headspace-sampling jar fitted with a condenser and Tenax sampling traps. Total sampling time was 4 h 40 min.

All the extracts were analysed by GC, GC-MS and by a GC fitted with a sulfur detector.

Highlights of the volatile components identified

A total of 460 components were identified and quantified in the three analyses, including 97 components which were sulfur- and/or nitrogen-containing. The main components found in the meat analysis were valeraldehyde (2%), hexanal (12%), octadecanal (3%), hexadecanal (50%). Trace amounts of the following sulfur chemicals were found methanedithiol, 3-mercaptobutan-2-one, 2-methylfuran-3-thiol, 3-mercaptopentan-2-one, thialdine. The skin was found to contain largely unsaturated aldehydes such as 2,4-decadienal (21%).

2.5.4 Narcissus

Concrete

The picked flowers (500 g) were placed in a large beaker and soaked in hexane (1 l) for 30 min. This was repeated two times. The combined extracts were dried over anhydrous sodium sulfate, filtered and stripped of hexane at 35°C on a rotary evaporator under reduced pressure. Further concentration was under nitrogen.

Headspace
The flowers were sampled in the field using pairs of Tenax^{TM} traps for 1.5 h and 3 h respectively. The traps were brought back to the laboratory and desorbed into a GC-MS and GC.

Highlights of the volatile components identified
A total of 239 components were identified and quantified in two analyses, including four components which were sulfur- or nitrogen-containing. The main components were benzyl benzoate (3%), *cis*-3-hexenyl hexanoate (3%), isoamyl docosanoate (4%) and tetracosanal (3%). Other components of interest were dillapiole, coumarin, elemicin, methyl jasmonate and methyl vanillate.

2.5.5 Post-analysis work

The liquid analyses results are usually complimented by the headspace results in producing a full flavour or fragrance profile for each sample. If the headspace results are vapour pressure adjusted they can be translated more meaningfully into formulations.

Each project was assigned a flavourist or perfumer from the start. They confirmed that the extract smelled the same as the original sample and performed GC-O work on the liquid extract. The GC-O work helped to focus on parts of the chromatograms where important components existed and where potentially novel aroma components could be identified.

Synthesised new molecules can be confirmed in the extracts by mass spectra and standard retention data. These new molecules can be registered with the flavour and fragrance legal bodies and cleared for use in new aromas. Thus a truly unique flavour or fragrance can be created.

References

1. Leahy, M.M. and Reineccius, G.A. (1984) Comparison of methods for the isolation of volatile compounds from aqueous model systems. In P. Schreirer (ed.), *Analysis of Volatiles: Methods-Applications*. Walter de Gruyter, New York, pp. 19–49.
2. Cai, J., Liu, B. and Su, Q. (2001) Comparison of simultaneous distillation extraction and solid-phase microextraction for the determination of volatile flavor components. *J. Chromatogr. A*, **930**, 1–7.
3. Dron, A., Guyer, D.E., Gage, D.A. and Lira, C.T. (1997) Yield and quality of onion flavor obtained by supercritical fluid extraction and other methods. *J. Food Proc. Eng*, **20**, 107–124.
4. Tanchotikul, U. and Hsieh, T.C.W. (1991) Analysis of volatile flavor components in steamed rangia clam by dynamic headspace sampling and simultaneous distillation and extraction. *J. Food Sci.*, **57**(2), 327–331.
5. Eichner, K., Lange-Aperdannier, M. and Vossman, U. (1985) Analysis of volatile Maillard reaction products by different methods. In D.D. Bills and C.J. Mussinan (eds), *Characterization and Measurement of Flavor Compounds*. American Chemical Society, Washington DC, pp. 124–141.

6. Mangas, J.J., Gonzalez, M.P., Rodriguez, R. and Blanco, D. (1996) Solid-phase extraction and determination of trace aroma and flavour components in cider by GC-MS. *Chromatographia*, **42**(1), 101–105.
7. Wang, Z.H., Dou, J., Macura, D., Durance, T.D. and Nakai, S. (1998) Solid phase extraction for GC analysis of beany flavours in soymilk. *Food Res. Inter.*, **30**(7), 503–511.
8. Adahchour, M, Vreuls, R.J., Van der Hayden, A. and Brinkman, U.A.Th. (1999) Trace-level determination of polar flavour compounds in butter by solid phase extraction and gas chromatography – mass spectrometry. *J. Chromatogr. A*, **844**, 295–305.
9. Richter, B.E., Jones, B.A., Ezzel, J.L., Porter, N.L., Avdalovic, N. and Pohl, C. (1996) Accelerated solvent extraction: a technique for sample preparation. *Anal. Chem.*, **68**, 1033–1039.
10. Anklam, E., Berg, H., Mathiasson, L., Sharman M. and Ulberth, F. (1998) Supercritical fluid extraction (SFE) in food analysis: a review. *Food Additives and Contaminants*, **15**(6), 729–750.
11. King, J.W. (2002) Supercritical fluid extraction: present status and prospects. *Grasas y Aceites*, **53**(1), 8–21.
12. McHugh, M. and Krukonis, V. (1986) Supercritical Fluid Extraction, Principles and Practice. Butterworth, USA.
13. Hawthorne, S.B., Krieger, M.S. and Miller, D.J. (1988) Analysis of flavour and fragrance compounds using supercritical fluid extraction coupled with gas chromatography. *Anal. Chem.*, **50**, 472–477.
14. Sinha, N.K., Guyer, D.E., Gage, D.A. and Lira, C.T. (1992) Supercritical carbon dioxide extraction of onion flavors and their analysis by gas chromatography–mass spectrometry. *J. Agric. Food Chem.*, **40**, 842–845.
15. Parliment, T.H. (1998) Applications of a microextraction class separation technique to the analysis of complex flavor mixtures. In C.J. Mussinan and M.J. Morello (eds), *Flavor Analysis: Developments in Isolation and Characterization.* American Chemical Society, Washington DC, pp. 8–22.
16. Engel, W., Bahr, W. and Schieberle, P. (1999) Solvent assisted flavour evaporation – a new versatile technique for the careful and direct isolation of aroma compounds from complex food matrices. *Eur. Food Res. Tech.* **209**, 237–241.
17. Parliment, T.H. (1997) Solvent extraction and distillation techniques. In R. Marsili (ed.), *Techniques for Analyzing Food Aroma.* Marcel Dekker, Inc., New York, pp. 1–27.
18. Blanch, G.P., Tabera, J., Herraiz, M. and Reglero, G. (1993) Preconcentration of volatile components of foods: optimization of the steam distillation–solvent extraction at normal pressure. *J. Chrom.*, **628**, 261–268.
19. Siegmund, B., Leitner, E., Mayer, I., Farkas, P., Sadecka, J., Pfannhauser, W. and Kovac, M. (1996) Investigation of the extraction of aroma compounds using the simultaneous distillation extraction according to Likens-Nickerson method. *Dtsch. Lebensm. Rundsch.* (German), **92**(9), 286–290.
20. Maignial, L., Pibarot, P., Bonetti, G., Caintreau, A. and Marion, J.P. (1992) Simultaneous distillation-extraction under static vacuum: isolation of volatile compounds at room temperature. *J. Chrom.*, **606**, 87–94.
21. Wampler, T.P. (1997) Analysis of food volatiles using headspace-gas chromatographic techniques. In R. Marsili (ed.), *Techniques for Analyzing Food Aroma.* Marcel Dekker, Inc., New York, pp. 27–59.
22. Badings, H.T. and de Jong, C. (1984) Headspace analysis for the study of aroma compounds in milk and dairy products. In P. Schreirer (ed.), *Analysis of Volatiles: Methods-Applications.* Walter de Gruyter, New York, pp. 401–419.
23. Dirinck, P., de Pooter, H., Willaert, G. and Schamp, N. (1984) Application of a dynamic headspace procedure in fruit flavour analysis. In P. Schreirer (ed.), *Analysis of Volatiles: Methods-Applications,* Walter de Gruyter, New York, pp. 381–401.

24. Eri, S., Khoo, B.K., Lech, J. and Hartman, T.G. (2000) Direct thermal desorption-gas chromatography and gas chromatography-mass spectrometry profiling of hop (*Humulus lupulus* L.) essential oils in support of varietal characterization. *J. Agric. Food Chem.*, **48**, 1140–1149.
25. Hartman, T.G., Overton, S.V., Manura, J.J., Baker, C.W. and Manos, J.N. (1991) Short path thermal desorption: food science applications. *Food Tech.*, **45**, 104–105.
26. Grimm, C.C., Lloyd, S.W., Miller, J.A. and Spanier, A.M. (1997) The analysis of food volatiles using direct thermal desorption. In R. Marsili (ed.), *Techniques for Analyzing Food Aroma.* Marcel Dekker, Inc., New York, pp. 59–81.
27. Adahchour, M., Beens, J., Vreuls, R.J.J., Batenburg, A.M., Rosing, E.A.E. and Brinkman, U.A.T. (2002) Application of solid-phase micro-extraction and comprehensive two-dimensional gas chromatography (GC × GC) for flavour analysis. *Chromatographia*, **55**(5/6), 361–367.
28. Sandra, P., David, F. and Vercammen, J. (2002) New developments in sorptive extraction for the analysis of flavours and fragrances. In A.D. Karl (ed.), *Advances in Flavours and Fragrances: From the Sensation to the Synthesis.* The Royal Society of Chemistry, UK, pp. 27–39.
29. Bicchi, C., Iori, C., Rubiolo, P. and Sandra, P. (2002) Headspace sorptive extraction (HSSE), stir bar sorptive extraction (SBSE), and solid phase microextraction (SPME) applied to the analysis of roasted arabica coffee and coffee brew. *J. Agric. Food Chem.*, **50**, 449–459.
30. Bicchi, C., Cordero, C., Iori, C. and Rubiolo, P. (2000) Headspace sorptive extraction (HSSE) in the headspace analysis of aromatic and medicinal plants. *J. High Res. Chrom.*, **23**(9), 539–546.
31. Leland, J.V., Schieberle, P., Buettner, A. and Acree, T.E. (2001) *Gas Chromatography-Olfactometry: The State of Art.* American Chemical Society, Washington DC.
32. Drawert, F. and Christoph, N. (1984) Significance of the sniffing technique for the determination of odour thresholds and detection of aroma impacts of trace volatiles. In P. Schreirer (ed.), *Analysis of Volatiles: Methods-Applications.* Walter de Gruyter, New York, pp. 269–293.
33. Acree, T.E., Barnard, J. and Cunningham, D.G. (1984) The analysis of odor-active volatiles in gas chromatographic effluents. In P. Schreirer (ed.), *Analysis of Volatiles: Methods-Applications.* Walter de Gruyter, New York, pp. 251–269.
34. Shibamoto, T. (1984) Applications of high-resolution capillary columns on flavor and fragrance analysis. In P. Schreirer (ed.), *Analysis of Volatiles: Methods-Applications.* Walter de Gruyter, New York, pp. 233–251.
35. Mussinan, C.J. (1993) Instrumental analysis in the flavor industry. In T.E. Acree and R. Teranishi (eds), *Flavor Science: Sensible Principles and Techniques.* American Chemical Society; Washington DC, pp. 169–225.
36. Takeoka, G., Ebeler, S. and Jennings, W. (1985) Capillary gas chromatographic analysis of volatile flavor compounds. In D.D. Bills and C.J. Mussinan (eds), *Characterization and Measurement of Flavor Compounds.* American Chemical Society, Washington DC, pp. 95–109.
37. Busch, K.L. and Kroha, K.J. (1985) Tandem mass spectrometry applied to the characterization of flavor compounds. In D.D. Bills and C.J. Mussinan (eds), *Characterization and Measurement of Flavor Compounds.* American Chemical Society, Washington DC, pp. 121–138.
38. Gilbert, J. (1987) *Applications of Mass Spectrometry in Food Science.* Elsevier Applied Science, New York.
39. Kovats, E. (1958) Gas-chromatographic characterisation of organic compounds. I. Retention indexes of aliphatic halides, alcohols, aldehydes and ketones. *Helv. Chim. Acta*, **41**, 1915–1932.
40. McReynolds, W.D. (1966) *Gas Chromatographic Retention Data.* Preston Technical Abstract Company, Evanston, Illinois.
41. Kubeczka, K.H. and Formacek, V. (1984) Application of direct carbon-13 NMR spectroscopy in the analysis of volatiles. In P. Schreirer (ed.), *Analysis of Volatiles: Methods-Applications.* Walter de Gruyter, New York, pp. 219–233.
42. Herres, W. (1984) Capillary GC-FTIR analysis of volatiles: HRGC-FTIR. In P. Schreirer (ed.), *Analysis of Volatiles: Methods-Applications.* Walter de Gruyter, New York, pp. 183–219.

43. Idstein, H. and Schreirer, P. (1985) High-resolution gas chromatography-Fourier transform IR spectroscopy in flavor analysis; limits and perspectives. In D.D. Bills and C.J. Mussinan (eds), *Characterization and Measurement of Flavor Compounds*. American Chemical Society; Washington DC, pp. 109–201.
44. Schomburg, G., Husmann, H., Podmaniczky, L., Weeke, F. and Rapp, A. (1984) Coupled gas chromatographic methods for separation, identification, and quantitative analysis of complex mixtures: MDGC, GC-MS, GC-IR, LC-GC. In P. Schreirer (ed.), *Analysis of Volatiles: Methods-Applications*. Walter de Gruyter, New York, pp. 121–151.
45. Konig, W.A. (1984) New development in enantiomer separation by capillary gas chromatography. In P. Schreirer (ed.), *Analysis of Volatiles: Methods-Applications*. Walter de Gruyter, New York, pp. 77–93.
46. Tressl, R. and Engel, K.H. (1984) Formation and analysis of optically active aroma compounds. In P. Schreirer (ed.), *Analysis of Volatiles: Methods-Applications*. Walter de Gruyter, New York, pp. 323–343.
47. Mosandl, A. and Heusinger, G. (1984) Stereoisomers of fruit flavour substances – some aspects of synthesis and analysis. In P. Schreirer (ed.), *Analysis of Volatiles: Methods-Applications*. Walter de Gruyter, New York, pp. 343–357.
48. Tressl, R., Engel, K.-H., Albrecht, W. and Bille-Abdullah, H. (1985) Analysis of chiral aroma components in trace amounts. In D.D. Bills and C.J. Mussinan (eds), *Characterization and Measurement of Flavor Compounds*. American Chemical Society, Washington DC, pp. 43–61.
49. Rouseff, R.L. (1985) The use of high-performance liquid chromatography in flavor studies. In D.D. Bills and C.J. Mussinan (eds), Characterization and measurement of flavor compounds. American Chemical Society, Washington DC, pp. 79–95.

Chapter 3
Flavor Generation in Food

Liam O'Hare and John Grigor

3.1 Introduction

This chapter will concentrate on two key topics within the subject that are characteristic and illustrative, essentially as case studies. These two topics are:

(1) The development of flavour in cooked meat. This topic has been chosen as it is well studied and illustrates most of the key aspects of flavour development in cooked foods. In particular, the fundamental reactions occurring in cooked and stored foods are all represented.
(2) The development of flavour in cheese. This topic has been chosen as it is one of the best-studied examples of ripened fermented foods. Again, all of the important fermentation processes are represented.

3.2 Taste and aroma

This chapter concentrates on the production of aroma compounds during cooking or fermentation, but it should be remembered that both meat and fermented products are rich in compounds which contribute to taste, and in that slightly mystical group of compounds the 'flavour enhancers'. Cooked meat and cheeses both contain significant levels of salt which undoubtedly contributes to the flavour. Sugars may not be significant in either case, but acids, including free fatty acids, may contribute a note of sourness. Both meats and cheeses also contain high levels of free amino acids and peptides which may contribute salty, sweet, brothy or bitter tastes. These tastes may be especially important in boiled meats. The flavour enhancers (for example glutamate, inosine-5′-monophosphate and guanosine-5′-monophosphate) are also significant contributors to flavour [1, 2]. For a discussion of this, see Chapter 9.

3.3 Cooked meat

While raw meat is relatively bland, with few volatile flavour compounds, over 1000 aroma compounds have been reported in cooked meat [1, 3]. The flavour of cooked meat is derived predominantly from the Maillard reaction and the degradation and oxidation of lipids. These processes encompass a complex and only partly charted array of reactions. The exact path through these reactions depends firstly on the particular reactants which are present, as determined by

- the species, sex and age of the animal;
- its diet;
- its condition at, and just prior to, slaughter;
- aging and storage of the meat;
- preparation of the meat and the addition of other ingredients and, possibly most significantly;
- the conditions of cooking such as boiling, roasting or frying.

The complexity of the reactions is increased as many of the products can subsequently take part as reactants – in particular, aldehydes produced during lipid degradation are significant participants in the Maillard reaction [1].

3.3.1 Flavour precursors

The huge variety of volatile compounds found in cooked meats are formed in a complex web of interdependent reactions. Many of the reactive compounds which instigate this web are derived from four well-known and well-studied processes. These are:

- Lipid oxidation and degradation;
- The Maillard reaction;
- Strecker degradation of amino acids;
- Caramelisation of sugars (ribose).

The reactive compounds formed in these initial reactions then further react to form the important classes of flavour compounds.

3.3.1.1 Lipid oxidation and degradation

Several hundred of the volatile compounds which have been reported in cooked meat are thought to be derived from lipid degradation [3]. These include aliphatic (and some aromatic) hydrocarbons, aldehydes, ketones, alcohols, fatty and other carboxylic acids, esters, lactones and alkylfurans.

3.3.1.1.1 Hydrolysis

Lipid degradation can involve hydrolysis to release free fatty acids. While short-chain fatty acids have distinctive odours (Chapter 2), the long-chain fatty acids which are more common in meat fat are not significant aroma chemicals. Hydrolysis is favoured by high water activity and acid/base catalysis.

3.3.1.1.2 Oxidation

Lipid oxidation is a more significant contributor to cooked meat aroma. The mechanisms of lipid oxidation have been frequently reviewed (see for example Mistry and Min [4] and Baron and Anderson [5]). The lipids which contribute flavour precursors include triglycerides found in subcutaneous fat, but possibly more important are the membrane phospholipids, which contain a much higher proportion of unsaturated fatty acids than the triglycerides. This makes them more susceptible to oxidation during cooking [4]; thus lipid-derived volatiles are quantitatively important even in cooked lean meat [6]. Some lipid oxidation products have strong and characteristic aromas, and even very low levels of compounds such as *trans,cis*-2-4-decadienal, *cis*-hex-3-enal and penta-2,4-dienal contribute the prominent and unpleasant rancid aroma of oxidised fats. However the formation of these compounds is favoured by low temperatures; at high temperatures thermolytic and oxidative degradation of lipids produces a different balance of compounds, which is normally considered more beneficial, especially in the context of cooked meat [1, 4].

3.3.1.1.3 Fatty aldehydes as flavour compounds

Fatty aldehydes themselves, and most of the alkyl derivatives of heterocyclic compounds discussed below, may not be significant contributors to the flavour of cooked meats. Aldehydes with 6–10 carbons are perceived as having green, fatty or tallow aromas (Chapter 4). Unsaturated aldehydes may have more interesting aromas, for example 2,4-decadienal has an aroma of fat-fried food. Alkyl derivatives of heterocyclics also have fried or fatty flavours.

3.3.1.1.4 Fatty aldehydes (pentanal and hexanal) as flavour precursors

The products of reactions between the aldehydes produced during lipid oxidation and compounds produced in other reactions such as the Maillard reaction may be of note [3]. Fatty aldehydes may, for example, contribute the alkyl groups to alkylthiazoles and alkylpyridines [6]. Butyl and pentyl pyrazines could result from the reaction of lipid-derived pentanal or hexanal with a dihydropyrazine, formed by the condensation of two aminoketone molecules [7]. Pentanal may also be involved in the formation of 5-butyl-3-methyl-1,2,4-trithiolane [3]. The reaction of pentanal or hexanal with hydrogen sulfide and acetaldehyde has been suggested as the route to butyl and pentyl trithiolanes [7, 8].

3.3.1.1.5 Other lipid-derived flavour precursors

One significant alkylpyridine is 2-pentyl pyridine which has been found in meat from each of the main species. It may be formed by a ring-closing reaction of deca-2,4-dienal with ammonia [3]. The heterocyclic compound 2-methyl-3-(methylthio)furan is an important aroma character-impact compound in meat and has a very low odour threshold value [9]. It has been identified in model systems containing phospholipid but could not be detected in systems without phospholipid. It has been suggested that the choline moiety of phosphatidyl choline is important in formation of this product [10].

Plasmalogens, which contain long-chain alkenyl ether substituents, are another potential source of fatty aldehydes; long-chain alkylthiazoles with alkyl moieties up to C15 have been found in cooked meats and it has been suggested that the plasmalogens are the most likely source of these alkyl groups. Heart muscle contains high concentrations of plasmalogens and this may be linked to high levels of alkylthiazoles found in cooked beef heart [3].

3.3.1.1.6 Fatty aldehydes as 'flavour moderators'

Although few immediate products of lipid degradation, such as fatty aldehydes or the compounds which are directly derived from them, appear to be major contributors to meat flavour, Mottram and Edwards [11] found that when meat was cooked following removal of all lipids by chloroform/methanol extraction, a biscuit-like aroma and no meaty aroma was obtained. Removal of trigylcerides only (using hexane extraction) had little effect on the aroma produced. The pattern of aroma volatiles from the control meat and the meat extracted with hexane was dominated as expected by fatty aldehydes and alcohols, but the pattern obtained for the meat extracted with chloroform/methanol was very different. There were no lipid oxidation products and a large increase in the amounts of alkyl pyrazines. This has been taken to imply that fatty aldehydes derived from degradation of phospholipids limit formation of heterocyclic aroma compounds from the Maillard reaction [3]. Subsequent work with model systems has largely supported this theory. The aromas of Maillard reaction mixtures which gave rise to high concentrations of meaty aroma compounds, such as 2-methyl-3-furanthiol, were sulphurous and rubbery, with only an underlying meaty aroma, rather than clearly meaty. Inclusion of beef triglycerides did not significantly affect the pattern of volatiles produced or the aroma. Inclusion of beef phospholipids led to lower concentrations of the meaty aroma compounds but produced an intense meaty aroma and less-pronounced sulphurous notes. This was because at high concentrations furanthiol derivatives have strong, sulphurous odours, and meaty aromas are only observed at low concentrations. It was suggested that meat phospholipids are important in the development of typical cooked meat flavour, not principally because the products derived from them have typical aromas, but because they limit the formation of high-impact heterocyclic compounds in the Maillard reaction [3].

3.3.1.2 Strecker degradation

The Strecker degradation of amino acids often occurs in conjunction with the Maillard reaction, and is particularly important in the context of cooked meat. In the classic Strecker degradation an amino acid is decarboxylated and deaminated in the presence of a dicarbonyl compound forming a 'Strecker' aldehyde and an aminoketone (Figure 3.1). The aldehyde may often be an important reactive carbonyl which can take part in the Maillard reaction, while the aminoketone may take part in cyclisation reactions to produce heterocyclics such as pyrazines (Chapter 5). Dicarbonyl compounds which participate in the Streker degradation reactions may be derived from lipid oxidation or from the Maillard reaction [12].

Of particular note is the degradation of cysteine. The Strecker aldehyde produced from cysteine is 2-thioethanal (mercaptoacetaldehyde). Further degradation releases carbon dioxide and hydrogen sulfide which reacts with compounds such as aldehydes, furfurals and furanones to produce a wide range of sulfur compounds, many of which have been described as 'meaty' [13]. Although they are found at low concentrations many of them have very low odour thresholds, suggesting that they make a significant contribution to the aroma.

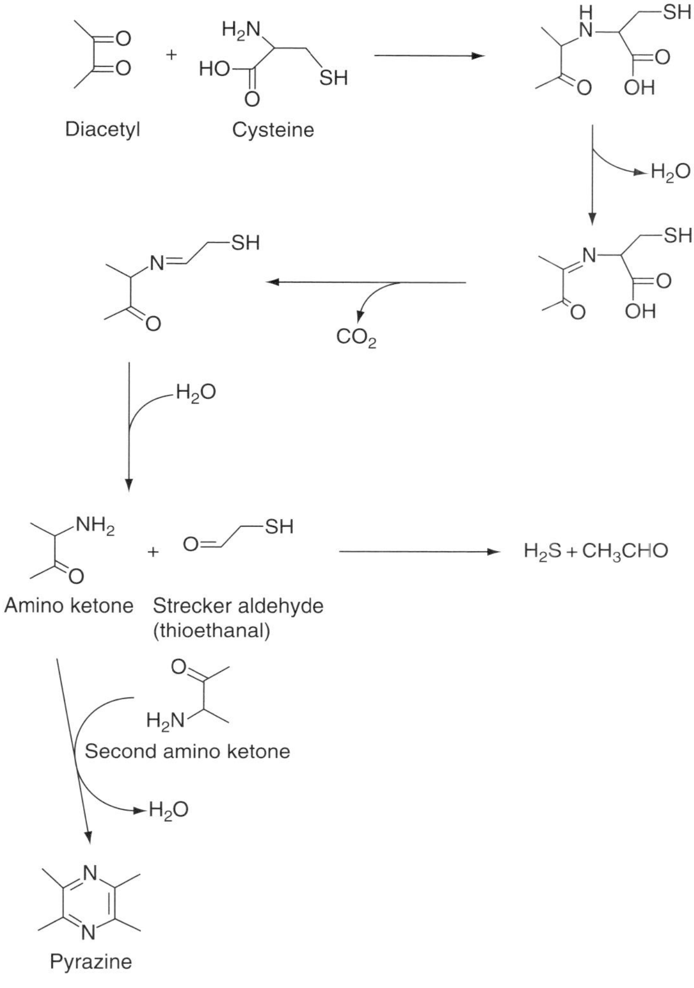

Figure 3.1 The Strecker degradation (illustrated by the reaction of diacetyl with cysteine).

Like caramelisation, the Strecker degradation is favoured by the conditions found on the surface of meat during grilling, rather than the lower temperature, moister conditions found in the middle of large pieces and boiled meats.

3.3.1.3 The Maillard reaction

The Maillard reaction, beginning with reactions between carbonyls and amino groups, is the best-known and the most-studied aspect of flavour chemistry. While it is important in the formation of most cooked flavours, it is perhaps most significant in the context of flavour generation in cooked meat. The reaction is named after Louis Maillard, who reported on the formation of melanoidins following the reaction of glycine with glucose [14], but it is most famously represented by the classic scheme (Figure 3.2)

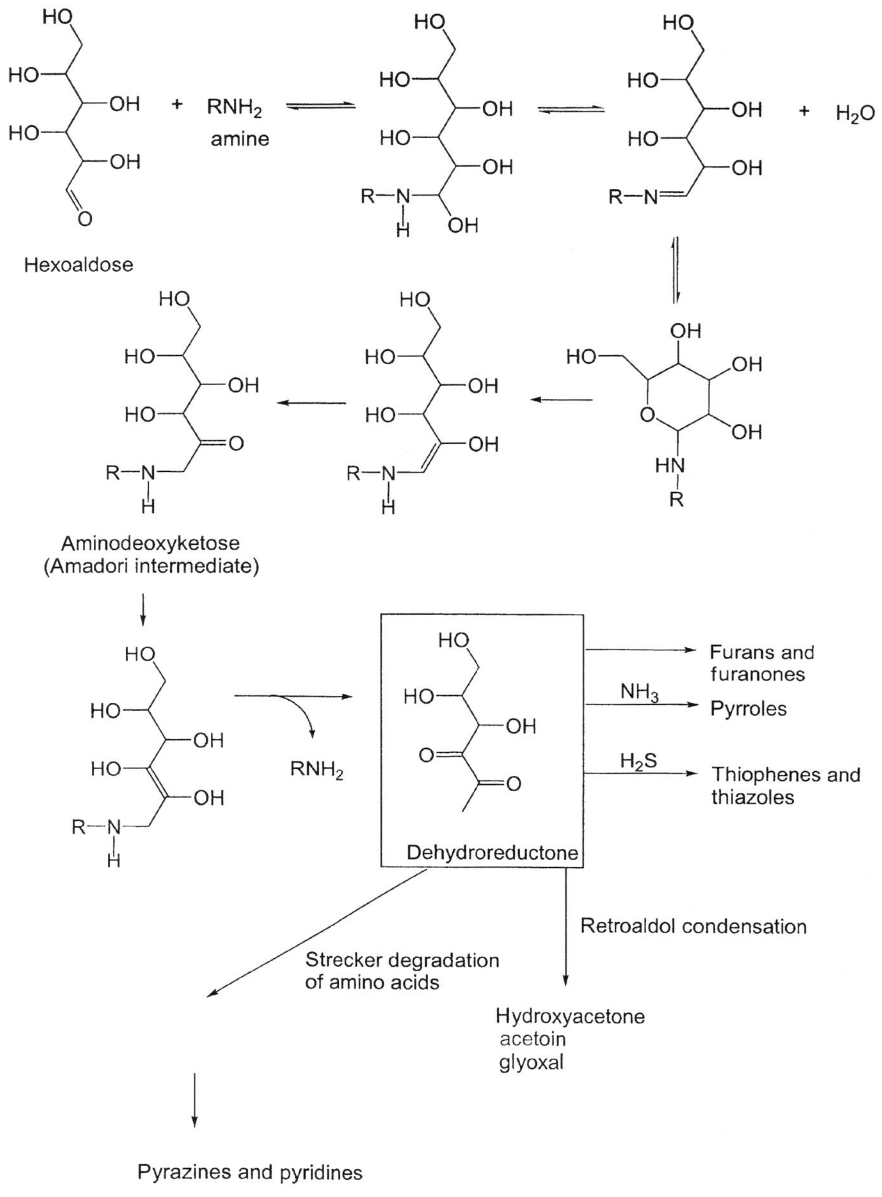

Figure 3.2 The Maillard reaction (adapted from Hodge, 1953).

proposed by Hodge [15]. This scheme is still useful and much used, though it is necessarily incomplete and a great deal simplified. It has been reproduced in many reviews and textbooks and has been adapted and expanded [12, 16]. The contribution of the Maillard reaction to flavour has been well reviewed (see for example the recent review by Yaylayan [17]). The initial stages of the reaction involve the condensation of a carbonyl group (classically found on a reducing sugar such as glucose) with an amino compound (classically an ε-amino group of a protein-bound lysine residue, or the α-amino group of a free amino acid such as glycine) to give a Schiff's base and then a glycosylamine. Subsequent stages include rearrangements (the Amadori rearrangement if the starting sugar was an aldose, or the Heyns rearrangement if the starting sugar was a ketose) which give rise to thermally unstable intermediates which then undergo dehydrations and deaminations to give various products such as hydroxyketones, dicarbonyl compounds and derivatives of furfural and furanone. Thus, even relatively simple starting materials can rapidly give rise to a wide range of reactive compounds. In more complex systems the carbonyl and the amine groups may be found on a much wider range of reactants; for example, the aldehydes produced during lipid degradation. The Strecker degradation, and the Maillard reaction itself, are also prime sources of reactive carbonyls. A whole series of analogous reactions can occur with thiols (for example, derived from cysteine or methionine) replacing amino groups as the nucleophile. Water activity, pH and temperature are key parameters in determining the progress of the Maillard reaction. Many of the steps are dehydrations or condensations and are favoured by low water activity but other steps are promoted by acid or base catalysis and go only very slowly at extremely low water activity. Water also acts as a solvent, promoting the reaction by allowing mobility of the reactants [16, 17].

3.3.1.4 Caramelisation

The formation of anhydrides when sugars are heated to temperatures above 150°C has been well studied. This process is favoured by low water activity. The anhydrides further dehydrate (Figure 3.3) to form furfural (from pentoses) or 5-hydroxymethylfurfural (from hexoses). Continued reaction at high temperature gives rise to furans, aldehydes and ketones (including dicarbonyls which may take part in the Strecker degradation) and aliphatic and aromatic hydrocarbons [3, 16]. Under certain conditions complex alcohols and lactones may be formed – the best-studied example is in the caramelisation of maple syrup, but similar reactions may occur in cooked meats. Sugars (e.g. honey) are often added to meats before cooking, but one source of sugars found in all meats is ribose. van den Ouweland and Peer [18] suggested that dephosphorylation and dehydration of ribose phosphate (derived from ribonucleotides particularly inosine monophosphate) forms 4-hydroxy-5-methyl-3(2*H*)-furanone which then readily reacts with hydrogen sulfide (possibly formed by the Strecker degradation), and that this may be the principal precursor of the furan and thiophenethiols found in roasted and other cooked meats. The high temperatures and low water activity which favour classic caramelisation reactions mean that they are more likely on the surface of grilled, fried and roasted meats. Boiled meats and the centre of large pieces of meat cooked in other ways are not likely to get hot enough or dry enough to promote significant caramelisation.

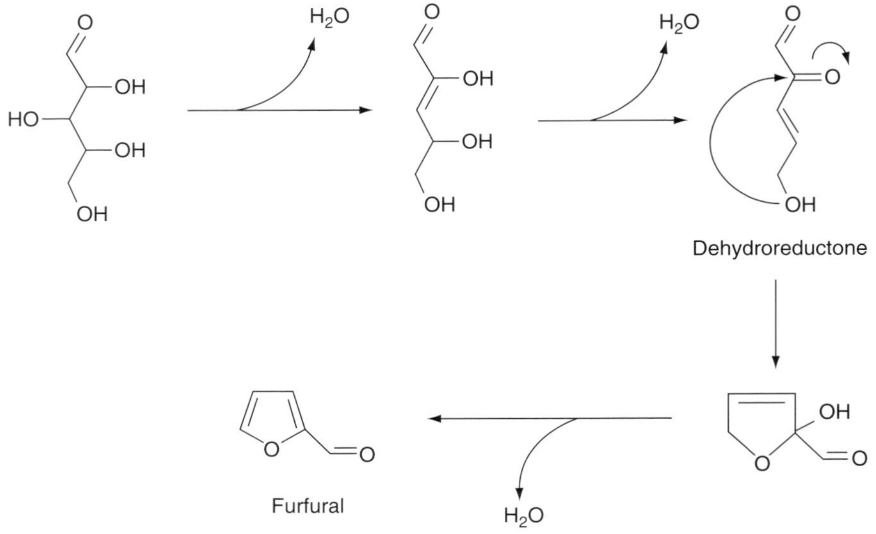

Figure 3.3 Caramelisation.

3.3.2 *Influence of method of cooking*

3.3.2.1 *Roasting, frying and grilling*

Roast flavours are usually attributed to heterocyclic compounds such as pyrazines, thiazoles and oxazoles [3].

3.3.2.1.1 *Pyrazines*

Many different alkyl pyrazines have been found in meat volatiles [1, 3]. They may be formed from the condensation of two α-aminoketone molecules produced in the Strecker degradation of amino acids [3]. Very high temperatures and low water activity are attained on the surface of meat during grilling. This can be expected to promote Strecker degradation and caramelisation reactions. In well-done grilled meat, pyrazines were reported to be the major class of volatiles [6].

3.3.2.1.2 *Thiazoles*

Alkyl-substituted thiazoles tend to have lower odour thresholds than pyrazines. Vernin and Parkanyi [19] proposed a mechanism for the formation of thiazoles which also involves products formed by the Strecker degradation of cysteine.

3.3.2.1.3 *Thiophenes and furans*

Thiophenes and furans with a thiol group in the 3-position, which possess strong meat-like aromas and exceptionally low odour threshold values, have been found in cooked meat. A well-known example is 2-methylfuran-3-thiol which has a meaty aroma at low concentration, while the methylthioether derivative of this compound has a 'roasted' aroma [20]. Oxidation of these thiols results in mixtures of symmetrical and unsymmetrical disulfides which also contribute to aroma; the disulfide of one thiol in this group (bis (2-methyl-3-furanyl) disulfide) has an intense beefy aroma, and has been found to have one of the lowest-known odour threshold values – 0.00002 ppb [21]. At low concentrations these compounds, and their derivatives, have pleasant meaty and roasted aromas, but at higher concentrations these compounds may have less-desirable sulphurous or burnt odours.

3.3.2.2 Boiling

Boiling does not allow very high temperatures or low water activities to be attained. This can be expected to rule out significant caramelisation and limit the Strecker degradation. This restricts the range of potential reactants which can participate in the Maillard reaction, which can occur, though it, too, is not favoured by high water activity. Lipid degradation, in particular hydrolysis, also proceeds under these conditions. Volatile sulfur compounds appear to be very important in boiled meat flavours. Both aliphatic thiols and heterocyclic compounds with up to three sulfur atoms are found more commonly and at higher levels in boiled than in roasted meat [3]. The flavour of boiled meat is thought to be characterised by three compounds: 2-methyl-3-furanthiol, bis(2-methyl-3-furyl) disulfide and 2-furfurylthiol [22].

3.3.2.3 Reheating

An undesirable 'warmed-over' flavour develops when cooked meat is stored and then reheated; this has been associated with oxidation of highly unsaturated fatty acids, such as arachidonic acid (20:4), which are derived from structural phospholipids [23]. When the meat is cooked initially the membranes are disrupted and dehydrated. This makes the phospholipids more susceptible to oxidation which proceeds rapidly while cooked meat is stored. Oxidation may be further enhanced due to catalysis by traces of iron which have been released from haem during cooking.

3.3.2.4 Species differences

The distinctive characters of cooked meats from different species are to a large extent determined by the precursors derived from the lipid fraction. Branched-chain aldehydes with between 11 and 17 carbon atoms have been reported at detectable levels in the volatiles of cooked beef, but much lower levels were found in veal, lamb and deer and only trace amounts were found in non-ruminant meats such as pork, chicken and turkey [24]. Some of these compounds are reported to have a tallowy, beef-like aroma which may play a role in determining characteristic beef flavour. The oxidation products of triply unsaturated fatty acids have been reported to be mainly responsible for the characteristic flavour of grass-fed lamb [25]. Microbes in the rumen produce significant quantities of branched-chain fatty acids (including 4-methyloctanoic acid and 4-methylnonanoic acid) which find their way into sheep meat and have been linked with the characteristic flavour of lamb [20], though these and their degradation products have also been linked with undesirable aspects of mutton flavour [26]. Phenols and thiophenols may also contribute to lamb and mutton flavour [27]. Pork has a relatively high proportion of unsaturated fatty acids in the subcutaneous fat, and degradation of these may give rise to high levels of unsaturated aldehydes which may contribute to the specific cooked pork aroma [3]. Chicken also has a relatively high proportion of linoleic acid in the fat, and its degradation products, such as *trans,trans*-2-4-decadienal and *trans*-undecenal, may contribute more to cooked chicken flavour than the heterocyclic compounds associated with beef flavour [1].

3.3.2.5 Fermented foods

The strict definition of fermentation refers to the metabolic pathways which allow organisms to obtain energy from simple substrates such as sugars, citrate or glycerol, without net oxidation of the fuel (no net production of NADH). These pathways lead to acids such as lactate, formate, acetate, propionate and butyrate, alcohols such as ethanol, 1,3-propandiol and butanol, carbonyls such as acetaldehyde and butandione (diacetyl) and gases such as carbon dioxide and hydrogen. Since fermentation by non-pathogens mops up the substrates on which pathogens might grow, and usually causes a decrease in the pH, it acts to preserve perishable foodstuffs, and consequently, fermentations are widespread and well established. They also convert starting materials which are often bland or even inedible into interesting, highly flavoured and relatively stable foodstuffs. The potential to produce alcohol by fermentation may also be valued. Campbell-Platt [28] has described the wide range of fermented foods which are produced throughout the world. Important examples include production of cheese, yoghurt, soy sauce, wine, beer and sauerkraut. Other foods, such as risen bread, tea and chocolate also involve a fermentation step in their production. The flavours produced by fermentation are just as complex as those produced by cooking and, due to involvement of micro-organisms, they may be subject to additional subtle variation. This makes them still more difficult to define and elucidate. The main pathways of fermentation by the principal organisms are fairly well worked out but in the most interesting cases these pathways are merely the 'A roads' which produce a fairly simple blend of flavour precursors which are then converted by a more complex and less well-understood pattern of B roads and uncharted tracks to give the particular balance of compounds that gives rise to the typical flavour of a traditional fermented food.

3.4 Cheese

Flavour development in cheese has been well studied and reviewed (see, for example, McSweeney & Sousa [29]). The fresh milk of most mammals has a bland, slightly sweet flavour. The only additional ingredients which are added during the making of a typical cheese are salt, an enzyme coagulant and a culture of the starter organisms; but these transform the bland milk flavour into some of the most stimulating, distinctive and desirable of food flavours. The processes by which these flavours are formed are typical of the many examples of the generation of flavour compounds by fermentation. Many potent flavour volatiles have been found in cheese. Most of these may be considered to be derived either from lactose and citrate fermentation, protein degradation and amino acid catabolism or from lipid degradation. While heterocylic aroma compounds, such as 2,6-dimethyl pyrazine, attributed to heat treatments (see, for example, Bellesia *et al.* [30]), and compounds such as α-pinene and D-carvone, derived from pasture (see for example Papademas and Robinson [31]), have been found in cheese and may contribute to the flavour, the discussion below will concentrate on aroma compounds formed by enzymic and microbial action.

All cheeses are made by the same basic mechanism:

- A culture of a starter organism is added to milk and this ferments the milk sugar (lactose) primarily producing lactic acid. The most noticeable effect of this is to reduce the pH, so that the main group of milk proteins (the caseins) coagulate forming a loose curd, or if a coagulant protease is also included, forming a firm gel. A high proportion of any milk fat present is usually trapped in the curd.
- The curd is then separated from the fraction which remains soluble (the whey).
- The separated curd is usually salted and may be pressed to remove additional whey.
- Normally the cheese then requires an extended period of storage to mature (ripen).
- Sometimes additional organisms or enzymes are included to promote ripening.

3.4.1 Lactose and citrate fermentation

Flavour generation begins immediately on addition of the starter organisms. A variety of organisms are used but they all share the ability to ferment lactose primarily to lactic acid. The clean simple taste of the lactic acid is the main flavour of many fresh cheeses, but other transformations of lactose are possible. Some starter organisms (heterofermentative starters) can use other substrates (for example citrate) for fermentation, and produce a wider range of products, including acetaldehyde, ethanol, butan-1,2-diol and butandione (Figure 3.4), giving a rounder, more buttery flavour. One of the most noticeable effects of the use of heterofermentative starters is the production of carbon dioxide which gives an open texture. A particular example is the use of *Propionbacterium freudenreichii* subsp. *shermanii* which ferments lactose, producing, amongst other products, propionic acid and carbon dioxide. The propionic acid contributes to the typical taste of Swiss cheese; carbon dioxide creates the holes [32]. Even when only homofermentative starters are used, other fermentation products are possible, especially in matured cheeses.

(1) Other 'adventitious' or non-starter organisms may ferment lactose and other substrates to produce a wide variety of products. Of particular importance is the 'coliform' or 'mixed acid' fermentation which produces formic acid, carbon dioxide and hydrogen, the 'butyric' fermentation which produces butan-1-ol, butyric acid, acetone, propan-2-ol and carbon dioxide and the alcohol (yeast) fermentation which produces ethanol and carbon dioxide. A consortium of organisms is often established; for example, while some organisms can ferment lactose, others can only metabolise the glucose moiety, leaving the galactose as a substrate for less-fastidious organisms.
(2) When the supply of lactose is depleted, and as the pH, water activity and reducing potential move away from the optimum for growth, surviving homofermentative

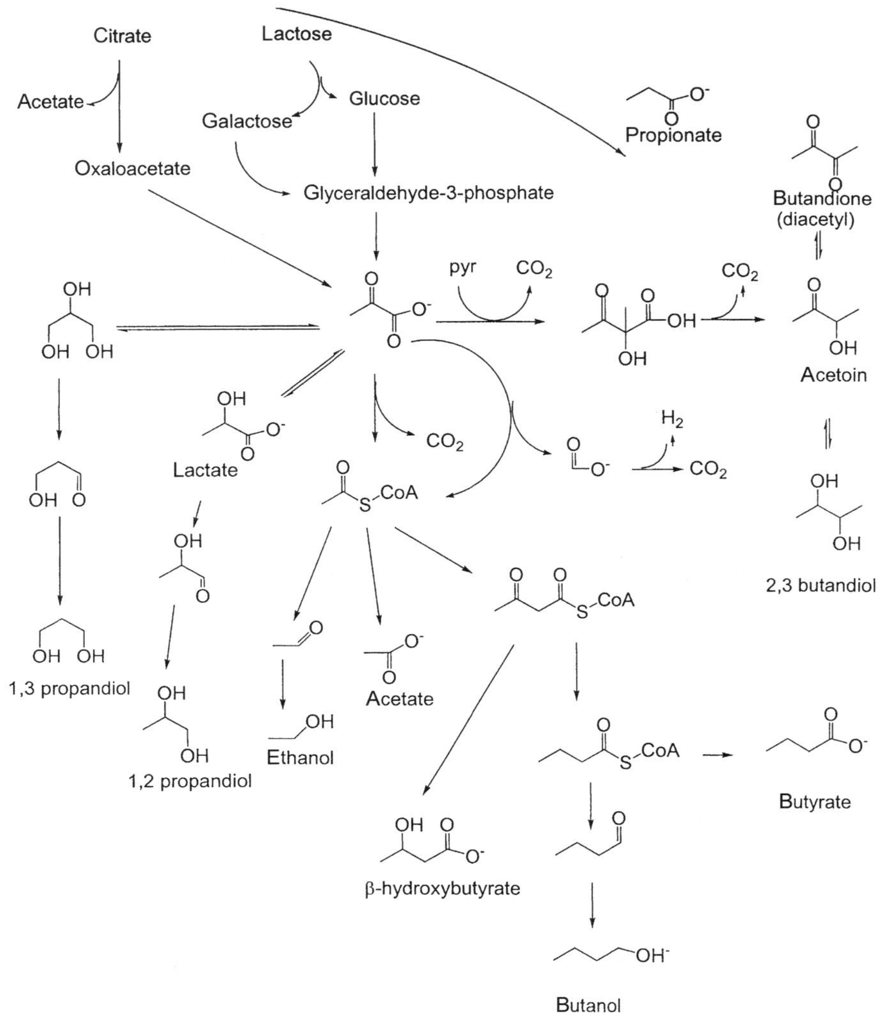

Figure 3.4 Some fermentation pathways of lactose and citrate. L = Homolactic; HL = Heterolactic; CL = Butyric; CF = Coliform; Y = Ethanolic; WW = Propionic.

organisms may become less fastidious and widen their range of fermentative activity. The rate at which lactose is depleted from young cheese varies considerably between different cheeses; from 24 h in Swiss-type cheeses to up to 20 days in Cheddar [33].

(3) Lactic acid can itself be a substrate for fermentation, particularly by yeasts which are common adventitious organisms, and may in some cases be deliberately added [34]. Many bacteria are also able to degrade lactic acid if an electron acceptor such as oxygen is available, or under anoxic conditions in the presence of alternative electron acceptors such as citrate, for example, producing succinic acid, acetate, formate and carbon dioxide [35]. Others may degrade lactic acid using glycerol as

the electron acceptor, producing acetate, 1,3-propanediol and carbon dioxide [36]. Some organisms can ferment lactic acid without requiring an external electron acceptor, forming acetic acid, ethanol, carbon dioxide and hydrogen or acetic acid, 1,2-propanediol and traces of ethanol [37].

(4) Citrate may act a fermentation substrate. An important example is the formation of diacetyl by some strains of lactic acid bacteria (also known as butandione) which is associated with a significant pleasant buttery aroma in dairy products [38].

3.4.2 Protein degradation

Some of the raw materials for flavour generation are low molecular weight compounds such as citrate and lactose, but many of the most important aroma compounds are derived from high molecular weight proteins and lipids.

3.4.2.1 Proteolysis

As cheese matures the protein is degraded by proteases to give firstly peptides and then free amino acids. The enzymes responsible for proteolysis may include

(1) any protease initially used to promote gel formation;
(2) extracellular proteases produced by adventitious and some starter organisms;
(3) intracellular proteases released from lysed cells [39];
(4) indigenous milk proteases;
(5) proteases from organisms, such as psychrotrophs, which often grow in raw milk, which may not be completely inactivated by pasteurisation, even though the organism itself is killed.

The traditional coagulant enzyme used in production of most cheeses is rennin, derived from calf stomach. Cheaper plant and microbial proteases have been used but these are thought to lead to production of an undesirable flavour, possibly due to excessive production of bitter hydrophobic peptides [40]. Rennin has now been widely replaced by a recombinant enzyme which has activity and specificity identical to the natural enzyme and produces cheese indistinguishable from the traditional product. The use of a crude rennet paste (which may contain other enzymes including lipolytic enzymes) is common in southern Europe [41].

3.4.2.2 Metabolism of free amino acids

Proteolysis proceeds slowly, especially in cheeses with a low moisture or high salt content, and this was considered to control the rate of maturation. The concentration of free amino acids is found to be closely linked to the maturity of cheeses such as Cheddar, but it is unlikely that these contribute more than a characteristic background to cheese flavour; rather it is likely that they serve as precursors for aroma compounds. The limiting step in flavour formation is now thought to be the conversion of the liberated amino acids to flavour compounds [42]. Many organisms, both starter and adventitious, can use free amino acids as substrates, forming important aroma compounds and precursors, such as amines and Strecker aldehydes.

3.4.2.2.1 Cysteine and methionine

Any route to formation of sulfur-containing aroma compounds is likely to involve cysteine or methionine as precursors. It has been suggested that the major aroma compounds produced from methionine are methional and methanethiol [42]. The oxidation products of methanethiol, dimethyldisulfide (DMDS) and dimethyltrisulfide (DMTS) may also be significant. Methionine γ-lyase, cystathionine lyases and other elimination enzymes expressed by some starter and non-starter organisms may cleave the side chain from methionine to produce methanethiol directly. An alternative route to methanethiol follows transamination of methionine with α-ketoglutarate as the acceptor, forming glutamate and α-ketomethylthiobutyrate which can be degraded to methanethiol or methional [42, 43]. It has been reported that methionine is converted by starter organisms to furanones, pyrrolodines and pyrazines. The action of cystathione γ-lyase on cysteine produces pyruvate with release of NH_3 and H_2S which are well-known potent aroma compounds in their own right, as well as being highly reactive precursors which may lead to formation of others [42].

3.4.2.2.2 Threonine

Threonine aldolase has been found in a number of organisms commonly found in cheese, and can lead to degradation of threonine to acetaldehyde [44].

3.4.2.2.3 Arginine

It has been suggested that fermentation of arginine may promote the development of a secondary microflora in cheeses. It may be converted to citrulline (releasing NH_3). Citrulline may then be converted to ornithine with the production of carbamyl phosphate, which is fermented giving ammonia and carbon dioxide as end products [33]. This may act as an energy source in many organisms, allowing a variety of other reactions, including some of those detailed below, as well as releasing ammonia which is reactive and a powerful odourant in its own right.

3.4.2.2.4 Valine, leucine and isoleucine

Transamination of leucine forms α-ketoisocaproic acid which may then be decarboxylated to form 3-methylbutanal (Figure 3.5). This compound has an unpleasant acrid or pungent aroma at high concentrations, but a more pleasant fruity aroma at low concentrations. It has been associated with unclean and harsh flavours when found at elevated concentrations in Cheddar [45]. In the same way isoleucine is degraded to 2-methylbutanal and valine forms 2-methylpropanal [42]. 3-Methyl butanal may be oxidised (catalysed by aldehyde dehyrogenases) to isovaleric acid which has been found in Camembert, Cheddar and Swiss cheese. This has a rancid, cheesy, sweaty and putrid odour that probably contributes to ripe cheese aroma, though it may have a detrimental effect at high concentrations [46]. Analogous pathways lead to formation of 2-methylbutyric acid and isobutyric acid from isoleucine and valine.

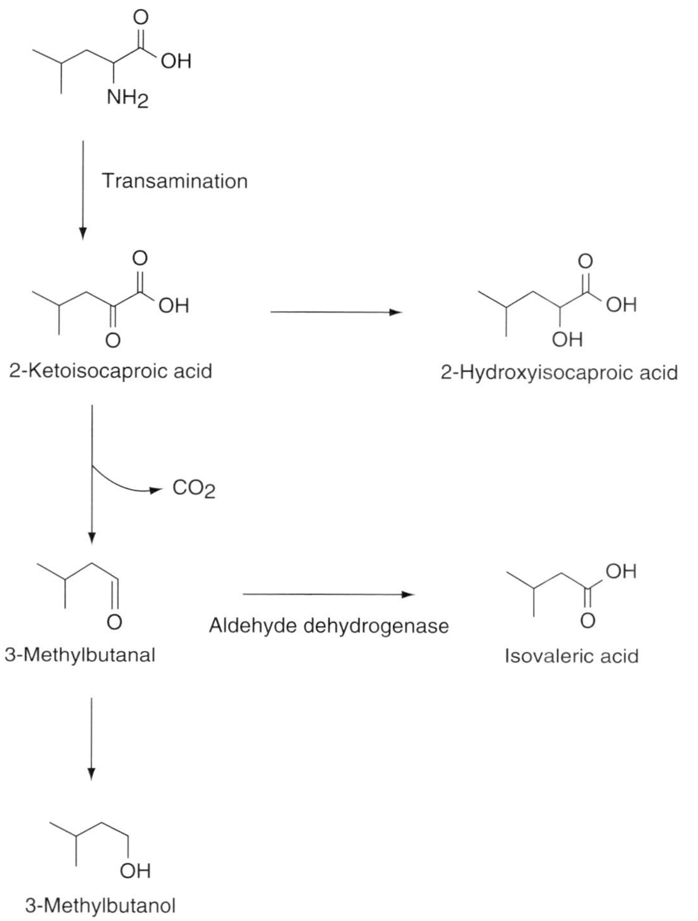

Figure 3.5 Metabolism of branched-chain amino acids (illustrated by leucine).

3.4.2.2.5 Phenylalanine and tyrosine

The action of transaminase on phenylalanine produces phenylpyruvate which may then be decarboxylated to form phenylacetaldehyde and then further converted to 2-phenylethanol and phenethyl acetate, which all have floral aromas and contribute positively to the aroma of cheeses such as Camembert [47] but are associated with unclean and 'utensil' off-flavours in Cheddar [45]. Phenylpyruvate may also be chemically converted to benzaldehyde [48] which has a well-known aroma. Transamination of tyrosine forms 4-hydroxyphenylpyruvate which then in the same way leads to formation of 4-hydroxyphenylacetaldehyde, 4-hydroxyphenylethanol and hydroxyphenethyl acetate. Tyrosine lyase, expressed by some adventitious organisms, produces phenol directly by elimination of the side chain from tyrosine. Phenol has been linked with typical Cheddar flavour.

3.4.2.2.6 Tryptophan

Transamination of tryptophan produces indole-3-pyruvate which may then be decarboxylated to indoleacetaldehyde or oxidatively decarboxylated to indoleacetic acid (Figure 3.6). A further decarboxylation produces skatole, while indole may be produced directly by the action of tryptophan-indole lyase on tryptophan [42]. Indole and skatole have also been were associated with unclean and 'utensil' taint in Cheddar [45].

3.4.3 Lipid degradation

Experiments with cheeses made from skim milk, or milk in which milk fat has been replaced by other lipids, have demonstrated that milk fat is essential for the development of correct flavour in ripened cheeses [49]. For example, the correct balance of odd-chain ketones, including heptan-2-one and nonan-2-one, has been considered the key contributor to blue cheese flavour. Although the catabolic reactions which lead to formation of flavour compounds from lipids have been less well studied than the

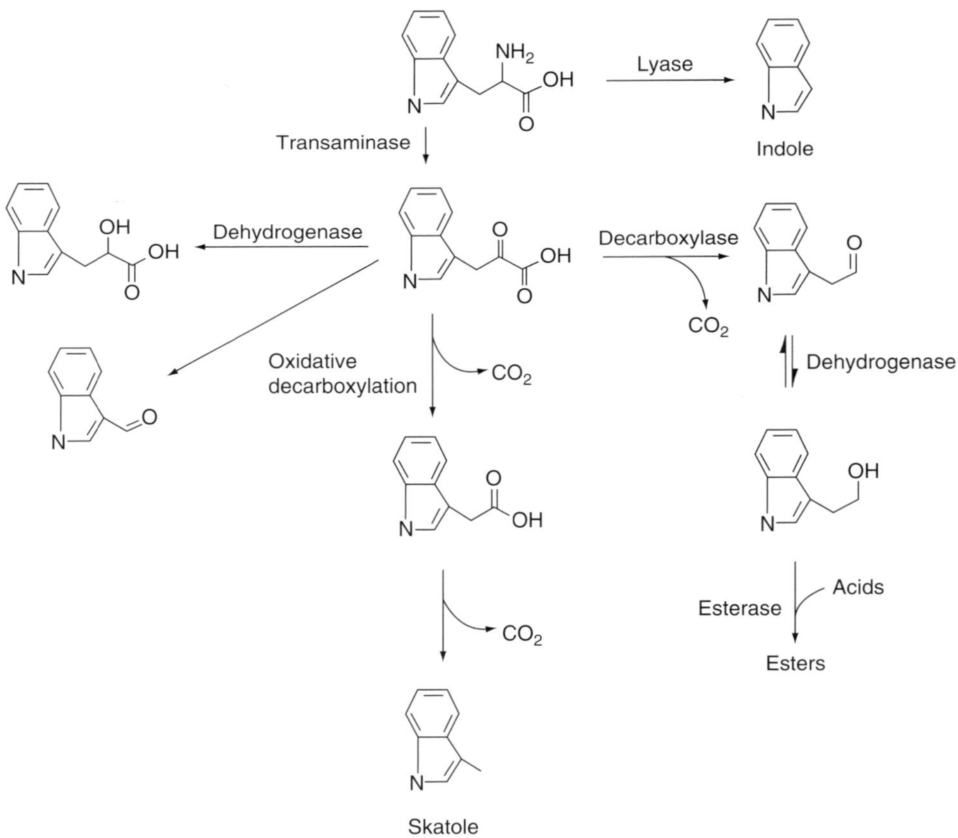

Figure 3.6 Metabolism of aromatic amino acids (illustrated by tryptophan).

formation of protein-derived aroma compounds, and are less well understood [29, 50] some of the main pathways have been elucidated. These are mainly microbial pathways accomplished by living organisms. A few steps may involve extracellular enzymes and fewer are non-enzymic.

3.4.3.1 Oxidative degradation

Whereas oxidative degradation of lipids is key to the development of flavour in cooked meats, it has been suggested that it is not important in the development of cheese flavour [50]. This may be because the predominant lipids in milk fat are relatively resistant to oxidation, because of the protection afforded by the milk fat globule membrane and high levels of vitamin E, or because of the reducing potential which is found in ripening cheese.

3.4.3.2 Hydrolytic degradation

On the other hand, hydrolytic degradation of lipids is thought to be the key to flavour development in cheeses. The sources of lipases in maturing cheese are similar to the sources of proteases outlined above. The main difference is that lipases are not normally added as part of the coagulation process, although animal lipases may be added to milk in production of cheeses such as Parmesan, and rennet paste (which is widely used in southern Europe) contains lipolytic enzymes such as pregastric esterase (PGE) [41, 51]. These various lipolytic enzymes produce free fatty acids as the cheese ripens. The extent of lipolysis varies considerably between different types of cheese, being the highest in mould ripened cheeses. Free fatty acids released by lipolysis are flavour compounds in their own right and act as precursors for reactions leading to the production of other flavour compounds including esters, thioesters, lactones, methyl ketones and alcohols [50] .

3.4.3.2.1 Acids

Milk fat contains relatively high levels of short-chain fatty acids [52] which are potent aroma chemicals, and as they are found in milk triglycerides mainly at the sn-3 position, they are preferentially released by lipases [50]. In most circumstances these short-chain fatty acids would produce off-flavours or taints, but in the context of a ripened cheese they contribute desirable notes to the overall aroma [41, 53]. It is thought that the very high levels of short-chain fatty acids found in mould ripened cheeses do not contribute an off-flavour as they are largely found in the unprotonated form due to the relatively high pH of these cheeses [50]. Butanoic acid is also produced by the butyric fermentations known to occur in cheese, and some acids may also be derived from oxidation of aldehydes and alcohols [46]. Branched-chain fatty acids, derived from the milk fat, appear to contribute directly to characteristic flavours of sheep's and goats' milk cheeses [54].

3.4.3.2.2 Esters

Esterifications, catalysed by esterases and lipases, occur between alcohols produced predominantly by fermentations and acids predominantly released by lipolysis. Ethyl esters of many short-chain fatty acids are found in most cheeses, and have been associated predominantly with fruity, but also sweet and floral aromas in cheese [55]. In some cheeses high levels of these esters is considered a defect associated with excessive growth of yeasts during fermentation, although they may also contribute desirable characteristics [34]. Other alcohols may also be involved in esterification reactions, including for example methanol, propanol and butanol, the complex alcohols derived from amino acid degradation or secondary alcohols derived from β-oxidation of fatty acids.

3.4.3.2.3 Thioesters

Lipases can also catalyse thioester formation between free fatty acids and compounds with free sulfhydral groups [56] such as those formed by degradation of sulfur-containing amino acids, especially methanethiol. The thioesters produced may have intense flavours and may contribute to aroma of various cheeses.

3.4.3.2.4 Methyl ketones and secondary alcohols

Methyl ketones are formed from free fatty acids by the β-oxidation pathway; they are first oxidised to α-ketoacids. These are then decarboxylated to form the ketones with one carbon less than the parent acid; hence odd-chain ketones predominate [29]. A final, reversible, reduction to the secondary alcohol may also occur. Methyl ketones may also be formed directly by decarboxylation of endogenous ketoacids present at low concentrations in milk fat or, alternatively, by oxidation of long-chain monounsaturated fatty acids [50]. They have been associated with the 'mouldy' or 'blue' flavour of blue cheeses [55].

3.4.3.2.5 Lactones

Lactones are formed by internal esterification of hydroxyacids to produce cyclic structures. The five-membered rings of δ-lactones and the six-membered rings of γ-lactones are relatively stable, and these compounds have been found in cheese [50, 57]. The hydroxyacids which are precursors for lactone formation may be found naturally in milk triglycerides, as oxidation products of unsaturated fatty acids, or following reduction of ketoacids produced, for example, by transamination of acidic amino acids. Lactones may have low flavour thresholds, and have been implicated as major contributors to cheese flavour [50].

Acknowledgements

Dr K.B. O'Hare for reading and improving the manuscript.

References

1. Farmer, L.J. (1999) Poultry flavour: a review. In R.I. Richardson and G.C. Mead (eds), *Poultry Meat Science*. CAB International, UK, pp. 127–158.
2. Cambero, M.I., Perineira-Lima, C.I., Ordonez, J.A. *et al.* (2000) Beef broth flavour: relation of components with the flavour developed at different cooking temperatures. *J. Sci. Food Agric.*, **80**, 1519–1528.
3. Mottram, D.S. (1998) Flavour formation in meat products: a review. *Food Chem.*, **62**, 415–424.

4. Mistry, B.S. and Min, D.B. (1992) Oxidised flavor compounds in edible oils. In G. Charalambous (ed.), *Off Flavors in Foods and Beverages*. Elsevier, Amsterdam, pp. 171–210.
5. Baron, C.P. and Anderson, H.J. (2002) Myoglobin induced lipid oxidation: a review. *J. Agric. Food Chem.*, **50**, 3887–3897.
6. Mottram, D.S. (1985) The effect of cooking conditions on formation of volatile heterocyclic compounds in pork. *J. Sci. Food Agric.*, **36**, 377–382.
7. Ho, C.T., Carlin, J.T. and Huang, T.C. (1987) Flavour development in fat fried food. In M. Martens, G.A. Dalen and H. Russwurm (eds), *Flavour Science and Technology*. Wiley, Chichester, UK, pp. 35–42.
8. Boelens, M., van der Linde, L.M., de Valois, P.J. *et al.* (1974) Organic sulfur compounds from fatty aldehydes, hydrogen sulphide, thiols and ammonia as flavour constituents. *J. Agric. Food Chem.*, **22**, 1071–1076.
9. MacLeod, G. and Ames, J.M. (1986) 2-Methyl-3-(methylthio)furan: a meaty character impact compound aroma compound identified from cooked beef. *Chem. Ind.*, 175–176.
10. Mottram, D.S. and Whitfield, F.B. (1995) Maillard-lipid interactions in non-aqueous systems: volatiles from the reaction of cysteine, ribose with phosphatidylcholine. *J. Agric. Food Chem.*, **43**, 1302–1306.
11. Mottram, D.S. and Edwards, R.A. (1983) The role of triglycerides and phospholipids in the aroma of cooked beef. *J. Sci. Food. Chem.*, **34**, 517–522.
12. Martins, S.I.F.S., Jongen, W.M.F. and Van Boekel, M.A.J.S. (2001) A review of Maillard reaction in food and implications to kinetic modeling. *Trends Food Sci. Tech.*, **11**, 364–373.
13. Rizzi, G.P. (1999) The Strecker degradation and its contribution to food flavor. In R. Teranishi, E.L. Wick and I. Hornstein (eds), *Flavor Chemistry: 30 Years of Progress*. Kluwer Academic/Plenum Pubs, NY, pp. 335–344.
14. Maillard, L.C. (1912) Action des Acides Amines sur les Sucres; formation des melanoidins par voie methodique. *CR Hebd. Seances Acad. Sci.*, **154**, 66–68.
15. Hodge, J.E. (1953) Chemistry of browning reactions in model systems. *J. Agric. Food Chem.*, **1**, 928–943.
16. Tressl, R. and Rewicki, D. (1999) Heat generated flavors and precursors. In R. Teranishi, E.L. Wick and I. Hornstein (eds), *Flavor Chemistry: 30 Years of Progress*, Kluwer Academic/Plenum Pubs, NY, pp. 305–326.
17. Yaylayan, V.A. (2003) Recent advances in the chemistry of Strecker degradation and Amadori rearrangement: implications to aroma and color formation. *Food Sci. Tech. Res.*, **9**(1), 1–6.
18. van den Ouweland, G.A.M. and Peer, H.G. (1975) Components contributing to beef flavour: Volatile compounds produced by the reaction of 4-hydroxy-5-methyl-3(2H)-furanone and its thio analogue with hydrogen sulphide. *J. Agric. Food Chem.*, **23**, 501–505.
19. Vernin, G. and Parkanyi, C. (1982) Mechanisms of formation of heterocyclic compounds in Maillard and pyrolysis reactions. In G. Vernin (ed.), *Chemistry of Flavours and Aromas*. Ellis Horwood, Chichester, UK, pp. 151–207.
20. Rowe, D.J. (2002) High Impact Aroma Chemicals. In K.A. Swift (ed.), *Advances in Flavours and Fragrances: From the Sensation to the Synthesis*. RSC, Cambridge, UK.
21. Buttery R.G. (1999) Flavour chemistry and odour thresholds. In R. Teranishi, E.L. Wick and I. Hornstein (eds), *Flavor Chemistry: 30 Years of Progress*, Kluwer Academic/Plenum Pubs, NY, pp. 353–365.
22. Grosch, W., Sen, A., Guth, H. *et al.* (1990) Quantification of aroma compounds using a stable isotope dilution assay. In Y. Bessiere and A.F. Thomas (eds), *Flavour Science and Technology*, Pub. John Wiley & Sons, NY, pp. 191–194.
23. Asghar, A., Gray, J.L., Buckley, D.J. *et al.* (1988) Perspectives on warmed-over flavour. *Food Tech.*, **42**(6), 102–108.
24. Guth, H. and Grosch, W. (1993) 12-Methyltridecanal, a species specific odorant of stewed beef. *Lebensm.-Wiss. U.-Technol.*, **26**, 171–177.

25. Priolo, A., Mikol, D. and Agabaiel, J. (2001) Effects of grass feeding systems on ruminant meat colour and flavour: A review. *Animal Res.*, **50**, 185–200.
26. Wong, E., Nixon, L.N. and Johnson, C.B. (1975) Volatile medium chain fatty acids and mutton flavour. *J. Agric. Food Chem.*, **23**, 495–498.
27. Sutherland, M.M. and Ames, J.M. (1996) Effect of castration and slaughter age on fatty acids and phenols in sheep adipose tissue. In A.J. Taylor and D.S. Mottram (eds), *Flavour Science: Recent Developments*, RSC, Cambridge, UK, pp. 19–24.
28. Campbell-Platt, G. (1987) *Fermented Foods of the World: A Dictionary and Guide.*, Butterworths, London, UK.
29. McSweeney, P.L.H. and Sousa, M.J. (2000) Biochemical pathways for the production of flavour compounds in cheeses during ripening: A review. *Le Lait*, **80**, 293–324.
30. Bellesia, F., Pinetti, A., Pagnoni, U.M. *et al.* (2003) Volatile components of Grana Parmigiano-Reggiano type hard cheese. *Food Chem.*, **83**, 55–61.
31. Papademas, P. and Robinson, R.K. (2002) Some volatile plant compounds in Halloumi cheeses made from ovine or bovine milk. *Lebensm.-Wiss. U.-Technol.*, **35**, 512–516.
32. Vangtal, A. and Hammond, G. (1986) Correlation of the flavour characteristics of Swiss-type cheeses with chemical parameters. *J. Dairy Sci.*, **69**, 2982–2993.
33. Laht, T., Kask, S., Elias, P. *et al.* (2002) Role of arginine in the development of secondary microflora in Swiss type cheese. *Int. Dairy J.*, **12**, 831–840.
34. Ferreira, A.D. and Viljoen, B.C. (2003) Yeasts as adjunct starters in matured Cheddar cheese. *Int. J. Food Microbiol.*, **86**, 131–140.
35. Cselovszky, J., Wolf, G. and Hammes, W.P. (1992) Production of formate, acetate, and succinate by anaerobic fermentation of *Lactobacillus pentosus* in the presence of citrate. *Appl. Microbiol. Biotechnol.*, **37**, 94–97.
36. Veiga da Cunha, M. and Foster, M.A. (1992) Sugar–glycerol cofermentations in lactobacilli: the fate of lactate. *J. Bacteriol.*, **174**, 1013–1019.
37. Oude Elferink, S.J.W.H., Krooneman, J., Gottschal, J.C. *et al.* (2001) Anaerobic conversion of lactic acid to acetic acid and 1,2-propanediol by *Lactobacillus buchneri*. *Appl. Environ. Microbiol.*, **67**, 125–132.
38. Hugenholtz, J. (1993) Citrate metabolism in lactic acid bacteria. *FEMS Microbiol. Rev.*, **12**, 165–178.
39. Hannon, J.A., Wilkinson, M.G., Delahunty, C.M. *et al.* (2003) Use of autolytic starter systems to accelerate the ripening of Cheddar cheese. *Int. Dairy J.*, **13**, 313–323.
40. Lemieux, L. and Simard, R.E. (1992) Bitter flavour in dairy products. II. A review of bitter peptides from caseins: their formation, isolation and identification, structure masking and inhibition. *Le Lait*, **72**, 335–382.
41. Virto, M., Chavarri, F., Bustamante, M.A. *et al.* (2003) Lamb rennet paste in ovine cheese manufacture. Lipolysis and flavour. *Int. Dairy J.*, **13**, 391–399.
42. Yvon, M. and Rijnen, L. (2001) Cheese flavour formation by amino acid catabolism. *Int. Dairy J.*, **11**, 185–201.
43. Engels, W.J.M., Alting, A.C., Arntz, M.M.T.G.J. *et al.* (2000) Partial purification and characterization of two aminotransferases from *Lactococcus lactis* subsp. *cremoris* B78 involved in catabolism of methionine and branched-chain amino acids. *Int. Dairy J.*, **10**, 443–452.
44. Ott, A., Germond, J.E. and Chaintreau, A. (2000) Origin of acetaldehyde during milk fermentation using ^{13}C-labelled precursors. *J. Agric. Food Chem.*, **48**, 1512–1517.
45. Dunn, H.C. and Lindsay, R.C. (1985) Evaluation of the role of Strecker derived aroma compounds in unclean type flavours of Cheddar cheese. *J. Dairy Sci.*, **68**, 2859–2874.
46. Marilley, L. and Casey, M.G. (2004) Flavours of cheese products: metabolic pathways, analytical tools and identification of producing strains. *Int. J. Food Microbiol.*, **90**(2), 139–159.
47. Kubickova, J. and Grosch, W. (1997) Evaluation of potent odorants of Camembert cheese by dilution and concentration techniques. *Int. Dairy J.*, **7**, 65–70.

48. Nierop Groot, M.N. and de Bont, J.A. (1999) Involvement of manganese in conversion of phenylalanine to benzaldehyde by lactic acid bacteria. *Appl. Environ. Microbiol.*, **65**, 5590–5593.
49. Wijesundera, C. and Watkins, P. (2000) Milkfat is essential for the development of Cheddar cheese flavour. *Aust. J. Dairy Technol.*, **55**, 86–91.
50. Collins, Y.F., McSweeney, P.L.H. and Wilkinson, M.G. (2003) Lipolysis and free fatty acid catabolism in cheese: a review of current knowledge. *Int. Dairy J.*, **13**(11), 841–866.
51. Bustamante, M., Chavarri, F., Santisteban, A. *et al.* (2000) Coagulating and lipolytic activies of artisanal lamb rennet pastes. *J. Dairy Res.*, **67**, 393–402.
52. Jensen, R.G., Gander, G.W. and Sampugna, J. (1962) Fatty acid composition of the lipids from pooled, raw milk. *J. Dairy Sci.*, **45**, 329–331.
53. Bills, D.D. and Day, E.A. (1964) Determination of the major free fatty acids of Cheddar cheese. *J. Dairy Sci.*, **47**, 733–738.
54. Ha, J.K. and Lindsay, R.C. (1993) Release of volatile branched-chain and other fatty acids from ruminant milk fats by various lipases. *J. Dairy Sci.*, **76**, 677–690.
55. Lawlor, J.B., Delahunty, C.M., Sheehan, J. *et al.* (2003) Relationships between sensory attributes and the volatile compounds, non-volatile and gross compositional constituents of six blue-type cheese. *Int. Dairy J.*, **13**, 481–494.
56. Cavaille-Lefebvre, D., Combes, D., Rehbock, B. *et al.* (1998) A chromatographic and mass-spectrometric approach for the analysis of lipase produced thioester derivatives. *Appl. Microbiol. Biotechnol.*, **49**, 136–140.
57. Eriksen, S. (1976) Flavours of milk and milk products. I. The role of lactones. *Milchwissenschaft* **31**, 549–550.

Chapter 4
Aroma Chemicals I: C, H, O Compounds

David J. Rowe

4.1 Introduction

This group of compounds includes both the simplest materials used in the flavour and fragrance industry and some of the most complex ones. It is a very large group, and so in places will constitute more of a survey than a detailed discussion. There are a number of texts [1–3] and reviews [4, 5] that have covered some of the areas, and hence these will be referred to as appropriate. In addition, cyclic oxygen compounds, especially furans and pyrans, are covered in Chapter 5.

4.2 Alcohols

4.2.1 Saturated alkyl alcohols

The simplest alcohol used is ethanol (**1**) itself. Whilst it is rarely considered as a flavour chemical, it constitutes much of the taste of simple brewed beverages (as those forced to drink alcohol-free beer find to their cost) and is the dominant flavour in distilled spirits. It is also a common solvent for fragrances, especially those for fine fragrances. Both synthetic (hydration of ethylene) and natural (fermentation) materials are available.

A second material is 'isoamyl alcohol'. This material is usually obtained by distillation of fusel oils, i.e. the residue from distillation of spirits after the main ethanolic fraction has been removed, and is actually a mixture of two alcohols, 2-methylbutan-1-ol (**2**) and 3-methylbutan-1-ol (**3**), which are not separable by distillation. It is also a major contributor to the flavour of the Italian brandy 'grappa'.

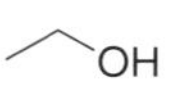

(1)

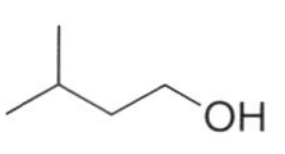

(2)

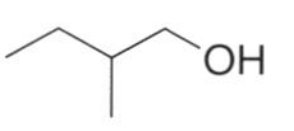

(3)

Other simple alcohols are used at a much lower level in fruit and 'alcoholic' flavours. An interesting branched alcohol used in fragrances is the 'woody-amber' 3,4,5,6,6-trimethyl-2-heptanol (Kohinool®) (**4**). There are two alcohols which are very widely used as solvents in the industry: propylene glycol, PG (**5**) (in flavours) and dipropylene glycol, DPG (**6**) (in fragrances).

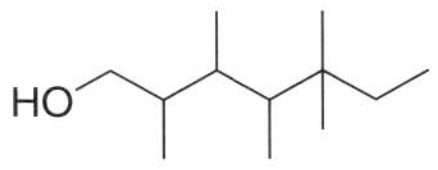

(**4**)

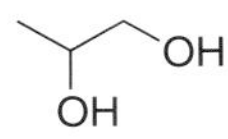

(**5**)

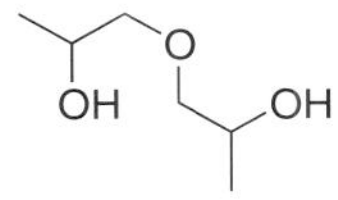

(**6**)

4.2.2 *Unsaturated alkyl alcohols*

The most important here is *cis*-3-hexenol (leaf alcohol) (**7**). This is one of the so-called C6 wound compounds produced when green vegetation is cut; oxygen is 'mopped up' by reaction with linoleic acid to avoid peroxide damage, and in doing so is cleaved to give the unstable *cis*-3-hexenal, which is reduced to *cis*-3-hexenol. As expected, it has the fresh green smell of cut grass, making it essential for adding freshness to both fragrances and flavours. It was first used, together with galbanum oil, in the famous 'Vent Vert' (Balmain, 1945) by Germaine Cellier. Its synthesis makes use of acetylene chemistry. The isomeric *trans*-2-hexenol (**8**) has a similar, if rather sweeter, note and is obtained by the reduction of *trans*-2-hexenal.

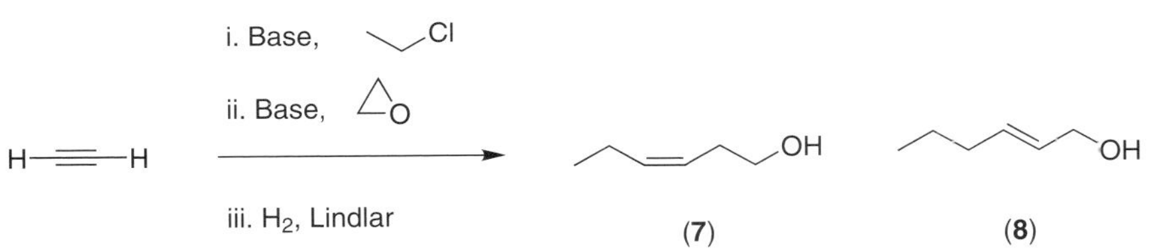

(**7**) (**8**)

1-Octen-3-ol (mushroom alcohol, amyl vinyl carbinol, AVC) (**9**) contributes the familiar earthy note to mushrooms. It is used in both flavours and fragrances for its earthy note. Again, its industrial synthesis makes use of acetylene chemistry.

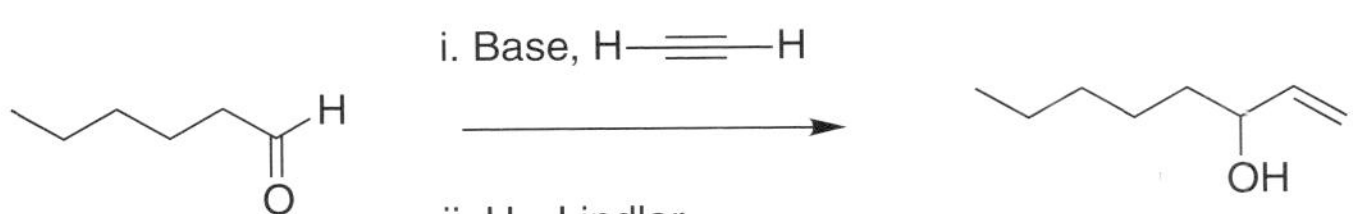

(**9**)

The terminally unsaturated 10-undecan-1-ol (**10**) is prepared from the acid, which is in turn obtained by the 'cracking' of castor oil; the polyunsaturated molecule *trans*-2-*cis*-6-nonadienol (violet-leaf alcohol) (**11**) is found in both violets and cucumber and is prepared from the corresponding aldehyde, *vide infra*. 4-Methyl-3-decen-5-ol (Undecavertol®) (**12**) is a powerful synthetic used to add green-floral notes to fragrances:

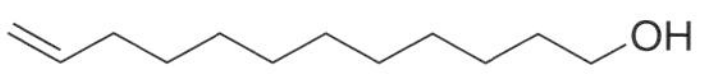

(**10**)

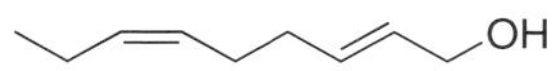

(**11**)

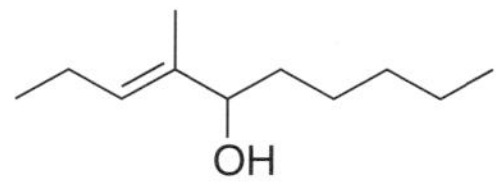

(**12**)

Terpenic alcohols are the largest group of unsaturated alcohols. The most important are doubly unsaturated C10 alcohols (monoterpenes) or reduced derivatives: geraniol (**13**), its stereoisomer nerol (**14**), linalool (**15**), myrcenol (**16**), citronellol (**17**), dihydromyrcenol (**18**), tetrahydrogeraniol (**19**) and tetrahydrolinalool (**20**). These are often high-volume aroma chemicals. Bauer *et al.* [1] and others [2, 6] have done excellent reviews of the industrial synthesis of these materials.

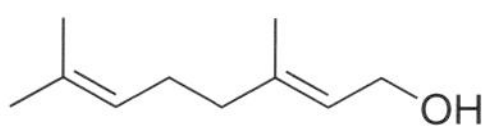

(**13**)

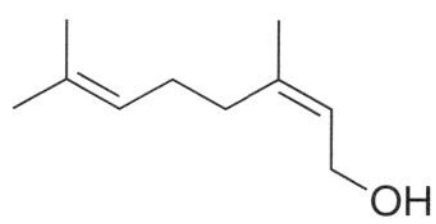

(**14**)

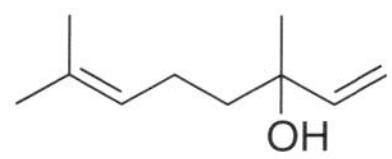

(**15**)

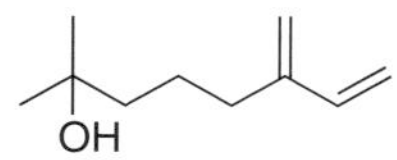

(**16**)

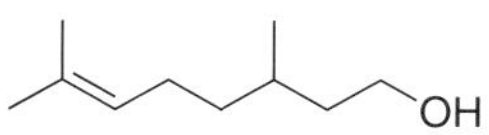

(**17**)

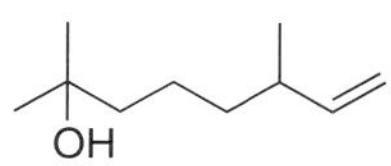

(**18**)

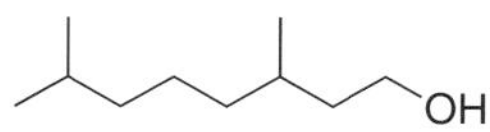

(**19**)

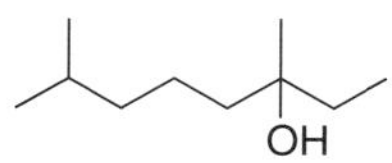

(**20**)

The most important cyclic terpene alcohols are menthol (**21**) (see Chapter 9) and the terpineols, especially α-terpineol (**22**).

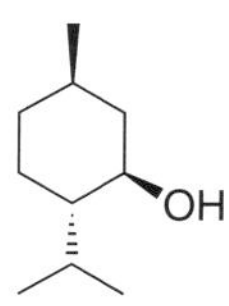

(**21**)

(**22**)

4.2.3 Complex fragrance alcohols

There are a large number of complex alcohols which are found as key odorant in naturally occurring fragrant substances. In some cases they are used in their own right; more commonly they have served as 'concept' molecules with commercial materials based on them. This can be illustrated by an example from sandalwood chemistry. An important odorant in sandalwood is β-santanol (23). Total synthesis of this has been reported, but this is not commercially viable. Instead a range of artificial materials have been developed based on syntheses from α-campholenic aldehyde (24), which is in turn obtained from the readily available β-pinene (25), and is available in both stereoisomers. The scheme below illustrates the range of structures that have been developed based on this.

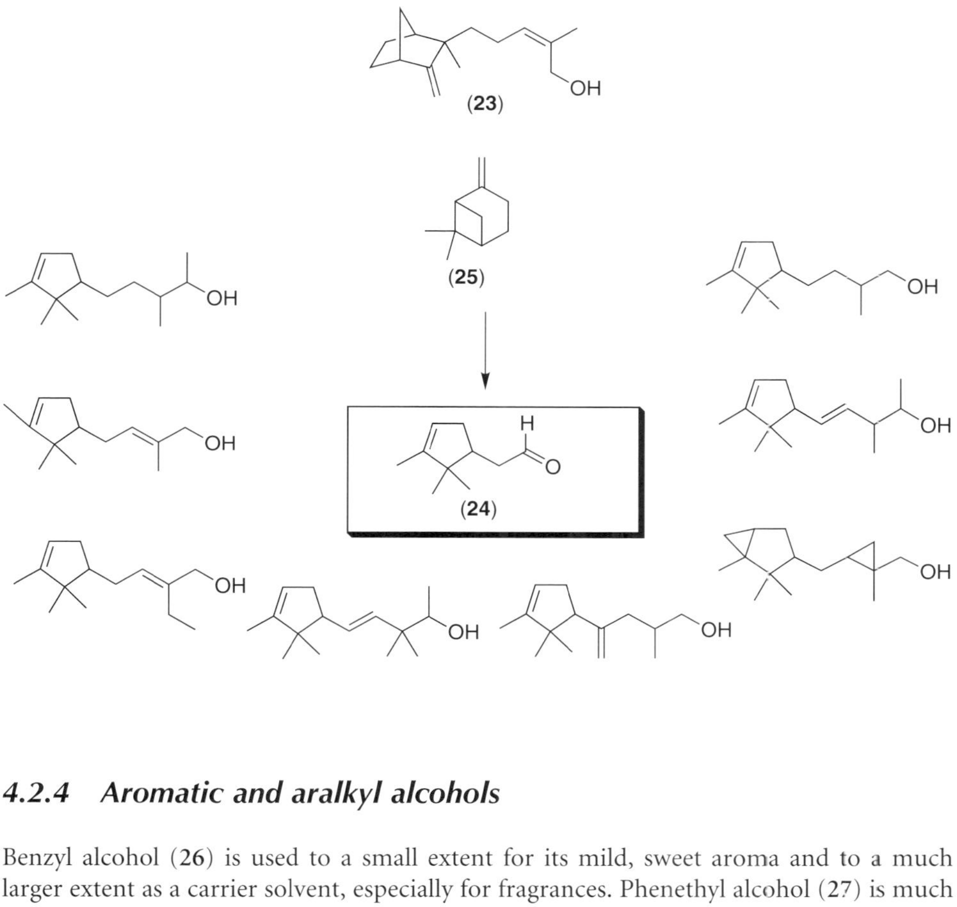

4.2.4 *Aromatic and aralkyl alcohols*

Benzyl alcohol (26) is used to a small extent for its mild, sweet aroma and to a much larger extent as a carrier solvent, especially for fragrances. Phenethyl alcohol (27) is much more important as an aroma chemical. Its synthesis was a milestone in the flavour and fragrance industry, and it is extensively used for its sweet, rose-like aroma. It can be prepared by the classical Grignard route, and an important alternative industrial synthesis is from styrene (28) via oxidation to its epoxide followed by hydrogenation [7]:

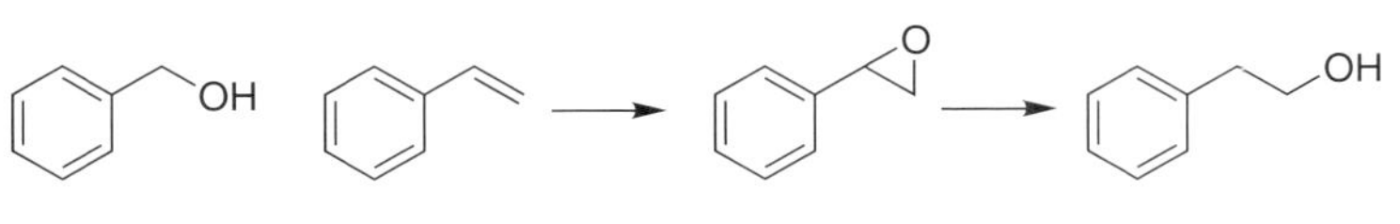

(26) (28) (27)

Both cinnamyl (**29**) and dihydrocinnamyl (**30**) alcohols are obtained by the reduction of cinnamaldehyde (**31**). Both are naturally occurring, and are used in fragrances for floral compositions, as well as the obvious balsamic and cinnamon notes.

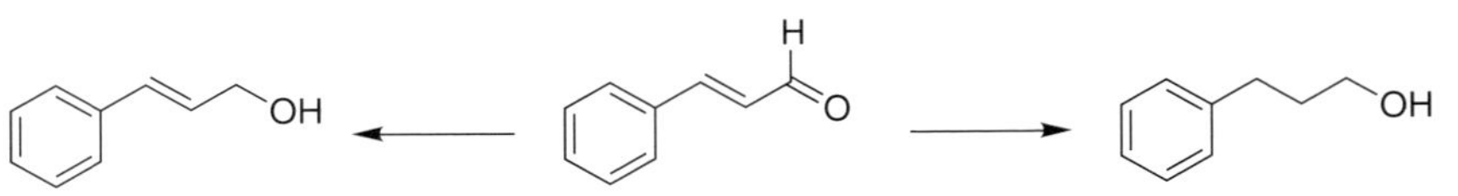

(**29**) (**31**) (**30**)

Aromatic alcohols with a saturated side-chain are very stable and are valuable for adding floral notes in soaps and domestic fragrances; these include the tertiary 'carbinols' α,α-dimethyl benzyl carbinol (**32**), α,α-dimethyl phenethyl carbinol (**33**), and phenethyl methyl ethyl carbinol (**34**), the secondary alcohol 3-methyl-4-phenyl-2-butanol (**35**) (Muguesia®), and the primary alcohols 2-methyl-5-phenylpentanol (**36**) (Rosaphen®) and 3-methyl-5-phenylpentanol (**37**) (Phenoxanol®, Phenylhexanol®).

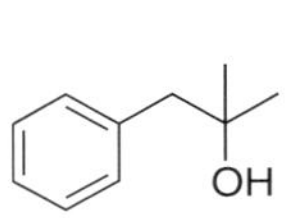

(**32**)

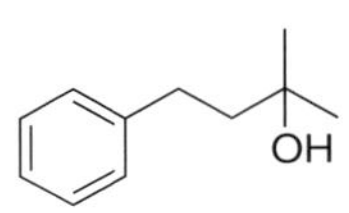

(**33**)

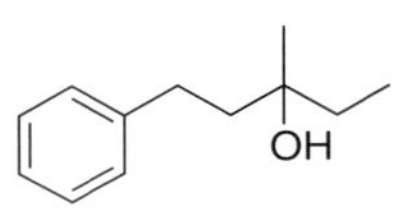

(**34**)

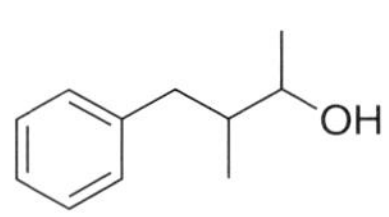

(**35**)

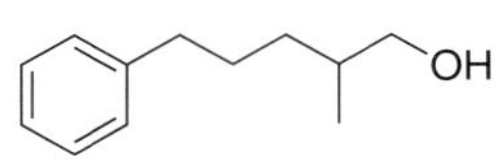

(**36**)

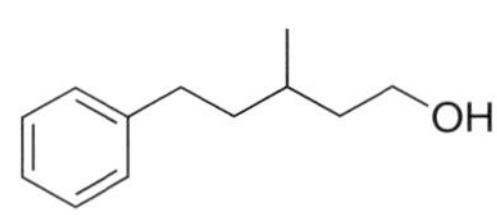

(**37**)

4.2.5 *Phenolics*

These must be used with care as 'phenolic' is often encountered as an off-note. The simplest phenol of importance is thymol (**38**), which is used for both its odour and antiseptic properties in personal care products.

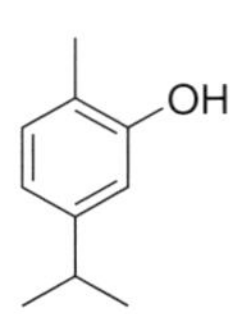

(**38**)

More commonly the phenol derivatives of importance are derivatives of guaiacol, i.e. 2-methoxyphenol. The alkylguaiacols 4-methylguaiacol (**39**) (Creosol) and 4-ethylguiacol (**40**) retain a medicinal odour and are important in smoke flavours, whereas 4-vinylguaiacol (**41**) (2-methoxy-4-vinylphenol, MVP, Varamol®) is an aroma chemical for all seasons; to savoury flavourists it has a smoky note, to sweet flavourists a vanilla note and to perfumers a spicy, clove note. It has also been identified as the second most important aroma chemical in the aroma of roasted coffee [8]. Its homologue eugenol (**42**) is a component of clove and cinnamon oil and has a spicy note; its use in perfumery has been restricted due to concerns about sensitisation.

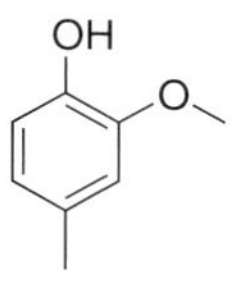

(39)

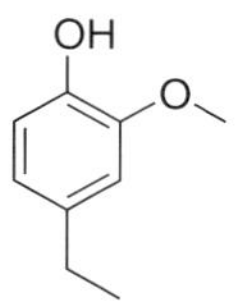

(40)

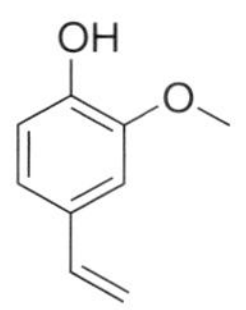

(41)

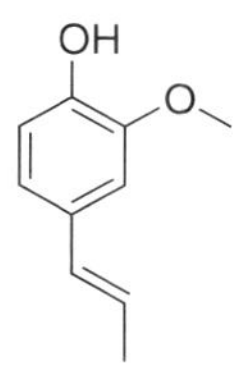

(42)

The most important phenolic in flavours and fragrance is guaiacol-4-carboxaldehyde, better known as Vanillin, which will be discussed under aldehydes.

4.3 Acids

4.3.1 Saturated aliphatic acids

Simple acids, especially acetic acid are important components of flavours, especially savoury and fruit formulations. Simple acids have sharp flavours, with the aroma decreasing with chain length. Acetic acid (**43**) has the characteristic aroma of vinegar, butyric acid (**44**) an aroma best described, accurately and graphically, as 'baby sick' and valeric acid (**45**), a sweaty, cheesy smell. Despite being obviously 'pungent', the acids actually have rather high odour thresholds: propionic acid 20 000 ppm, valeric acid 3000 ppm and butyric acid 240 ppm. In dilution acids such as 2- and 3-methylbutryic acids (**46**), (**47**) have a fruity aroma. Long-chain acids such as lauric acid (C12) (**48**) are virtually odourless. Citric acid (**49**) is also essentially odourless, and this makes it valuable in adding an acid taste, or as an acidity modifier, without adding an aroma.

Two interesting acids used in flavours are 2-methyloctanoic and 2-methylnonanoic acids (**50**), (**51**). These are found in lamb, and have the slightly sharp, fatty aroma of that meat.

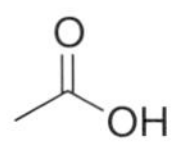

(43)

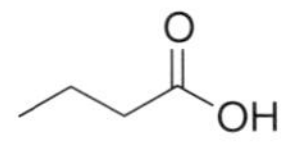

(44)

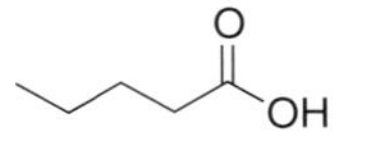

(45)

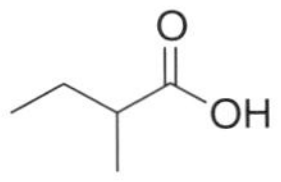

(46)

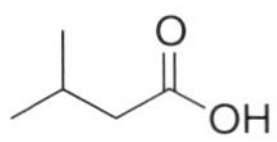

(47)

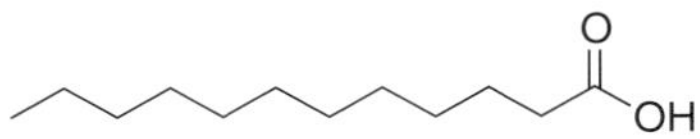

(48)

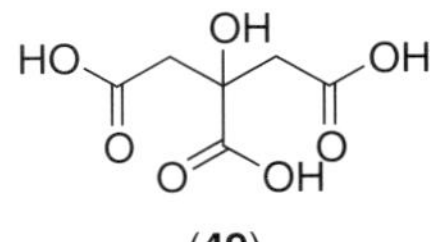

(49)

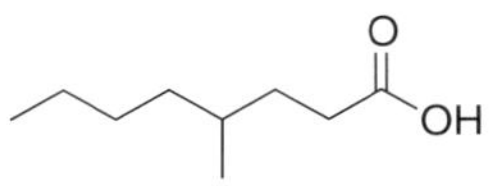

(50)

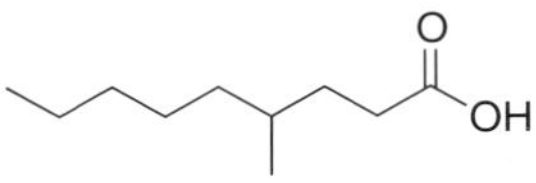

(51)

4.3.2 Unsaturated acids

These tend to have stronger, sharper odours than their saturated counterparts. *trans*-2-Butenoic acid (52) (crotonic acid) is very 'cheesy', and 2-methyl-2-pentenoic acid (53) (Fragarone, Strawberiff®) has a fruity note, as the name implies.

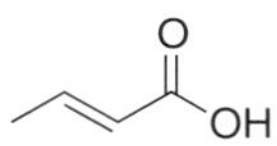

(52)

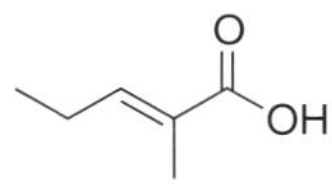

(53)

4.3.3 Aromatic acids

Phenylacetic acid (54) is used for its fruity, honey note, but in general aromatic acids are not used in flavours or fragrances, though sodium and potassium benzoates (55) are extensively used as preservatives.

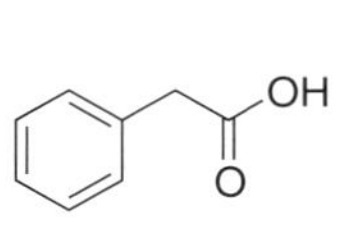

(54)

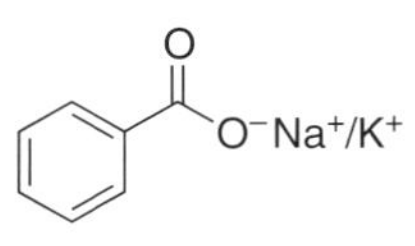

(55)

4.4 Esters

These are the workhorses of the flavour and fragrance industry; after all, the one property of esters that all chemistry students learn is their fruity odour – though this character is actually limited to the simpler esters. I use this term because, although other materials are of more obvious importance in individual areas (e.g. Hedione® and the musks in fragrances), esters are used in both flavours and fragrances, often at high levels, and are large-volume products; to borrow a term from the mainstream chemical industry, many of them are 'commodity chemicals'. To an extent, here one may perm any acid referred to above with any alcohol from above! A list of all the esters used would be astonishingly dull; however, there are some points that can be made.

4.4.1 Saturated esters

Ethyl esters (**56**) are the single most important group of esters used, from ethyl acetate to longer-chain materials such as ethyl laurate; the characteristic fruity aroma falls in intensity as the volatility drops. The second most important are probably the acetates (**57**); since many of these are also used as solvents, this is an area where the flavour and fragrance industry overlaps with the wider chemical industry. Isoamyl esters, prepared from 'isoamyl alcohol', i.e. the mixture of isomeric alcohols, are also very important; isoamyl acetate (**58**), (**59**), has the familiar 'pear drops' aroma.

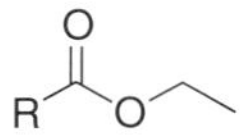

(**56**)

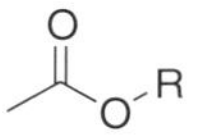

(**57**)

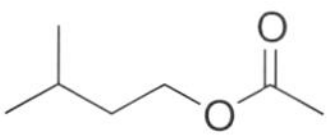

(**58**)

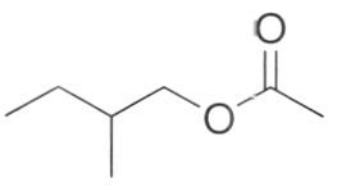

(**59**)

One should also mention a few aliphatic esters which are used as carriers and solvents precisely due to their lack of odour: triacetin (**60**) (glyceryl triacetate), triethyl citrate (**61**) (TEC) and 'medium chain triglycerides' (**62**) (MCT) which is essentially a mixture of glyceryl esters of longer-chain fatty acids, is sold under a variety of trade names (e.g. Mygliol®, Neobee®, Olembicol®).

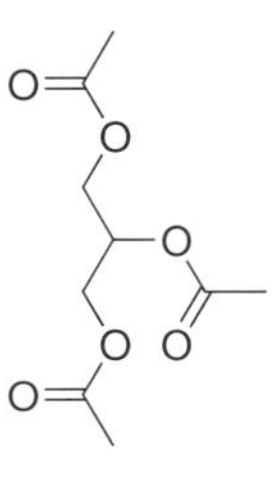

(**60**)

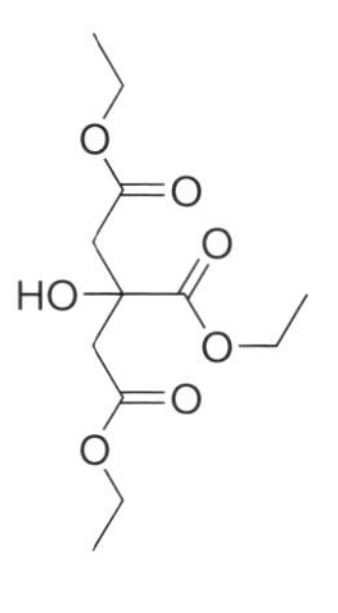

(**61**)

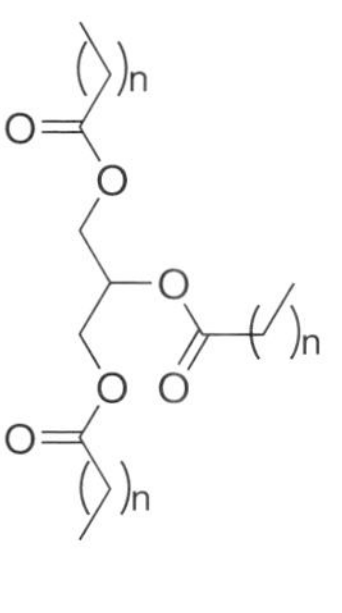

(**62**)

4.4.2 Unsaturated esters

Esters of unsaturated acids tend to have more powerful odours than their saturated counterparts. Ethyl *trans*-2-butenoate (**63**) (ethyl crotonate) has a powerful rum-like aroma. Methyl *trans*-2-nonenoate (**64**) (Neofolione) has a powerful green and floral note. The polyunsaturated ethyl *E*, *Z*-deca-2,4-dienoate (**65**) is a component of the flavour of Williams pears and hence is known as pear ester.

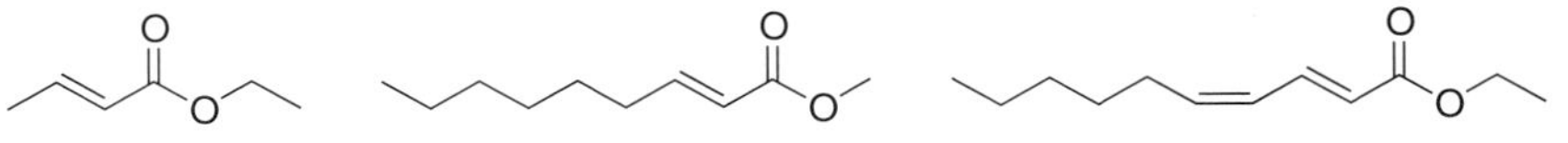

(**63**) (**64**) (**65**)

The synthetic analogue methyl *E*-octa-4,7-dienoate (**66**) (Anapear®) has been prepared by a Claisen rearrangement [9].

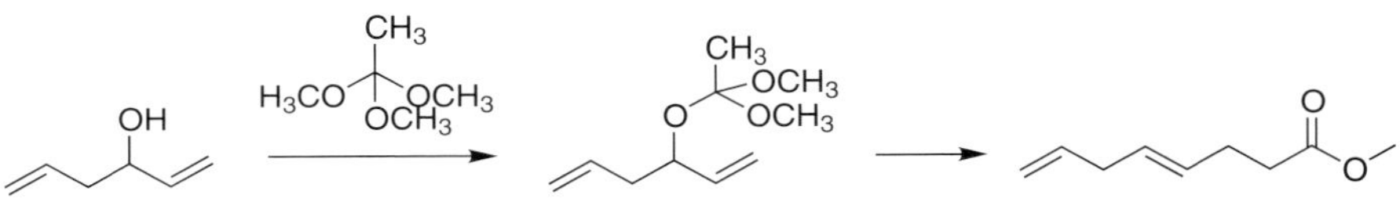

(**66**)

Acetylenic compounds are not generally used in the flavour and fragrance industry, though a number are important precursors. An exception to this is the two acetylenic esters, methyl-2-octynoate (**67**) and methyl-2-nonynoate (**68**), commonly known as Folione and methyl octine carbonate, which are used for green, violet-leaf, floral notes in fragrances.

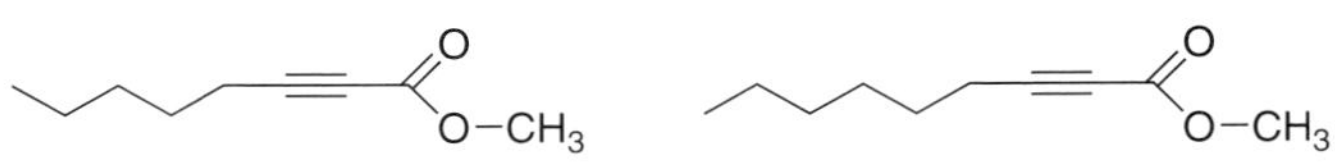

(**67**) (**68**)

Whilst, in general, the alcohol precursor of an ester is also an important aroma chemical in its own right, an important exception occurs with allyl esters. Allyl alcohol is a toxic, lachrymatory material which must be handled with great care. However, several esters are important; allyl hexanoate (**69**) has a powerful pineapple aroma, as to a lesser extent does its homologue allyl heptanoate (**70**), making them useful for fruit flavours and fruity notes in perfumery. Allyl 3,5,5-trimethylhexanoate (71), allyl cyclohexylpropionate (72) and allyl cyclohexyloxyacetate (73) also have pineapple notes; they are not found in nature and so are not used in the flavour sector but can be used in fragrances.

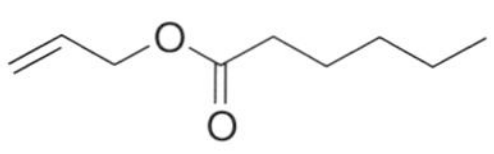

(69)

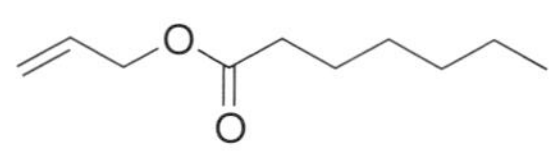

(70)

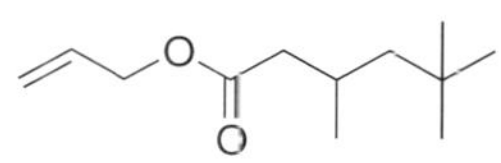

(71)

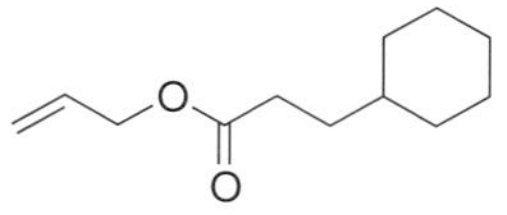

(72)

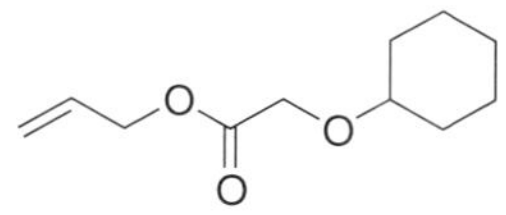

(73)

Ester of the C6 wound compounds, i.e. esters of *cis*-3-hexenol and *trans*-2-hexenol, and terpene esters tend to have similar, if milder, aromas to their corresponding alcohol; e.g. geranyl isovalerate (74) has a floral aroma and *cis*-3-hexenyl hexanoate (75) a sweet, green aroma, less harsh than *cis*-3-hexenol.

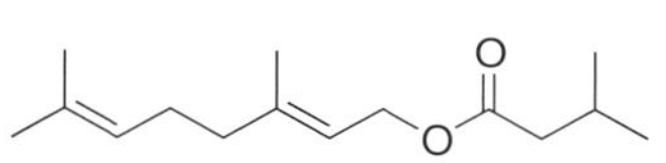

(74)

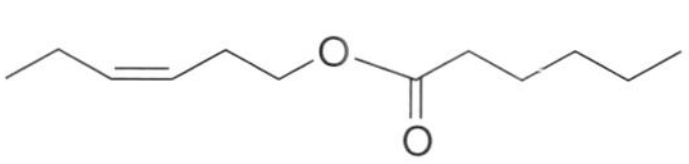

(75)

4.4.3 Aromatic esters

Simple esters such as benzyl acetate (76), phenethyl acetate (77) and methyl cinammate (78) have fruity-floral aromas and are used in fragrances and fruit compositions for flavours. The higher molecular weight esters formed from aromatic alcohols and aromatic acids are of low odour, often described as 'balsamic', and are used as 'fixatives' or as carrier solvents, especially benzyl benzoate (79).

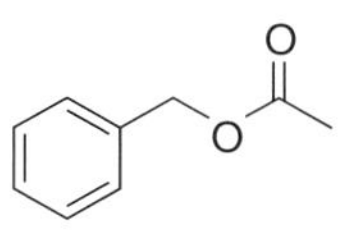

(76)

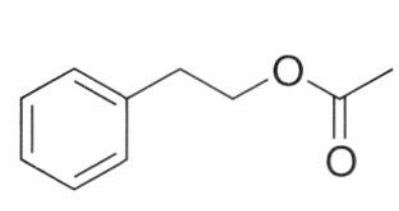

(77)

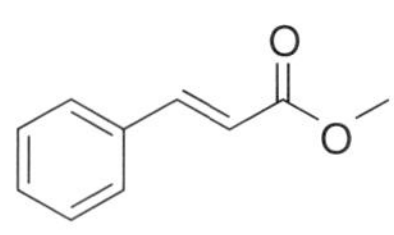

(78)

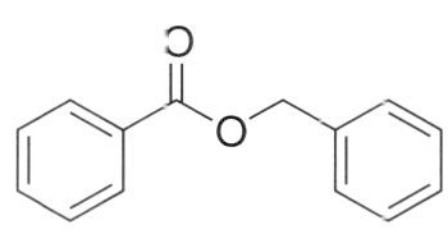

(79)

Salicylates are found in a number of essential oils. Methyl salicylate (80) is found in oil of wintergreen and is responsible for the character of that oil; it is used in fragrances, both for its character and antiseptic properties. Its use in flavours is largely restricted to the USA, where wintergreen or 'winter-fresh' flavours are popular. Wintergreen is an unpopular flavour through the rest of the world – to me, its 'muscle-rub' aroma is a reminder of gym changing rooms and school PE lessons; not my personal idea of a desirable flavour. Hexyl salicylate (**81**) is used for inexpensive floral notes; benzyl salicylate (**82**) is a fixative.

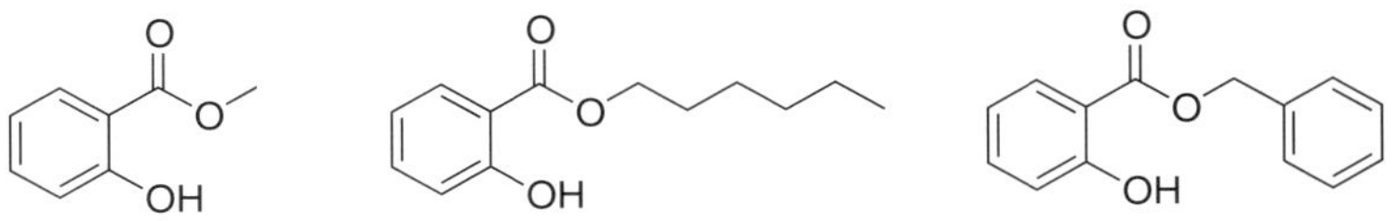

(**80**) (**81**) (**82**)

Methyl anthranilate (**83**) is unusual in that there are few amines used in the flavour and fragrance industry, where they are more commonly come across as off-notes. Some simple amines include Cadaverine (pentane-1,5-diamine), Putrescine (butane-1,4-diamine) and Spermidine (4-azaoctane-1,8-diamine) – the names are descriptive of their origins and odour characteristics. Methyl anthranilate is a component of a number of blossom aromas, and has a pleasant orange blossom odour, as does its derivative methyl *N*-methylanthranilate (**84**). It is used in its own right and in the form of Schiff's bases obtained from aldehydes, such as Lyrame® (**85**), formed from the muguet aldehyde Lyral® (*vide infra*).

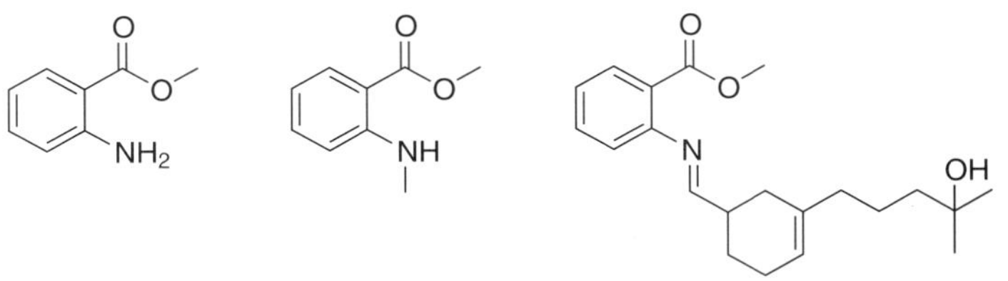

(**83**) (**84**) (**85**)

4.4.4 Lactones – gamma and delta

The two important groups here are the five-membered γ-lactones (**86**), substituted at position 5, adjacent to the ring oxygen, and the six-membered δ-lactones (**87**), substituted at position 6, again adjacent to the ring oxygen.

(**86**) (**87**)

The commonest route to γ-lactones is peroxide-catalysed addition of an alcohol to an acrylate ester [10]. Addition of (usually) ethyl acrylate and di-tertiary butyl peroxide to a hot mix of alcohol and strong acid generates the lactone, presumably by hydride abstraction

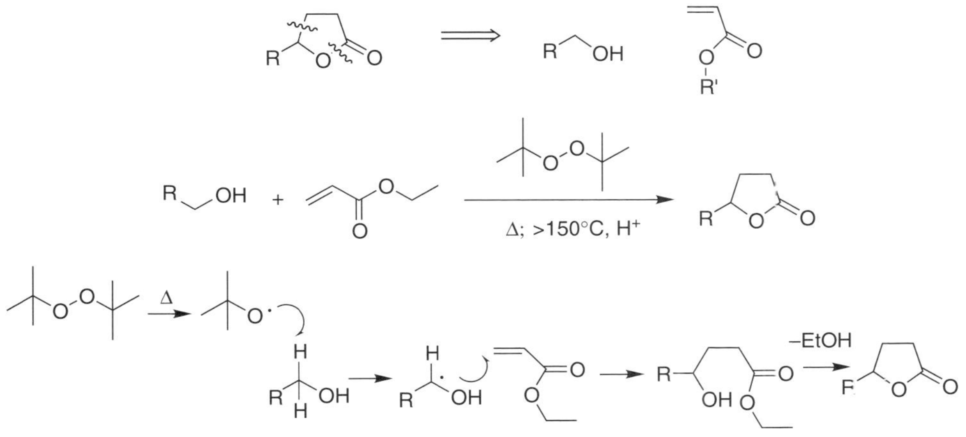

Figure 4.1 Formation of γ-lactones.

from the alcohol by the peroxide radical, followed by radical addition to the acrylate and ring closure by an acid-catalysed intra-molecular transesterification, as shown in Figure 4.1.

Bayer-Villiger oxidation of 2-alkylcyclopentanones (**88**) (*vide infra*) gives δ-lactones.

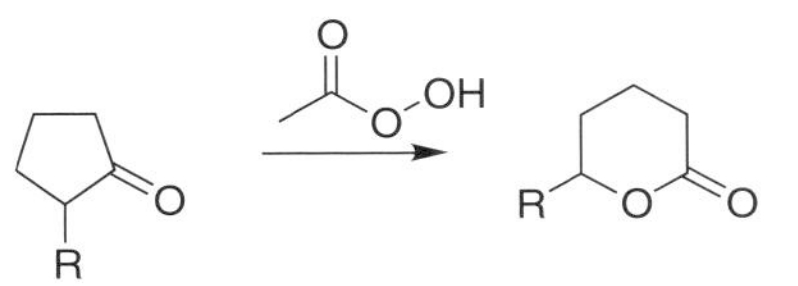

(**88**)

Since the industrial synthesis of both γ- and δ-lactones involves the ability to handle volumes of peroxides, the number of suppliers of these is much fewer than the number of suppliers of simple esters.

In common with their acyclic counterparts, both γ- and δ-lactones have generally sweet, creamy, fruity aromas; γ-octa- and γ-nonalactone have coconut aromas, and δ-deca- and γ-undecalactone (Pêche Pure®) peach-like aromas.

The lactone structure also occurs in more complex molecules. The celery odourants propylidene (**89**) and butylidene phthallide (**90**) (Ligustilide) are formed by reaction of Grignard Reagents with phthallic anhydride.

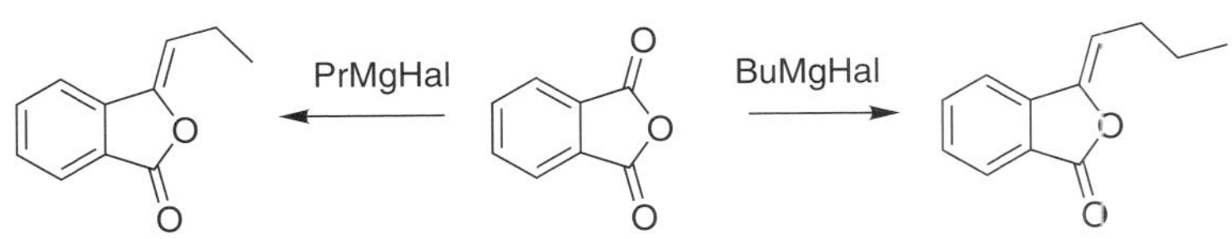

(**89**) (**90**)

Large-ring lactones, ca. sixteen members and higher, the macrolides, have musky aromas and are covered in Chapter 7, Musks.

4.4.5 Synthesis of esters

Whenever possible, esters are prepared by direct reaction between an alcohol and an acid, with a strong acid catalyst and the water removed by azeotropic distillation. With the more volatile esters, the ester itself can be used to carry the azeotrope, though care is needed with esters of water-soluble alcohols, especially ethyl esters, as the presence of too much alcohol in the distillate prevents the water from separating out. A volatile hydrocarbon can be added to help carry the azeotrope, though this lowers the volume yield and is an added cost, which can be critical with the simpler low-value esters. From *n*-butyl esters onwards this is not a problem as the alcohol is immiscible with water and separation is not an issue. Of course, as the size of the ester increases, it becomes too involatile to carry the azeotrope, and a solvent must be used.

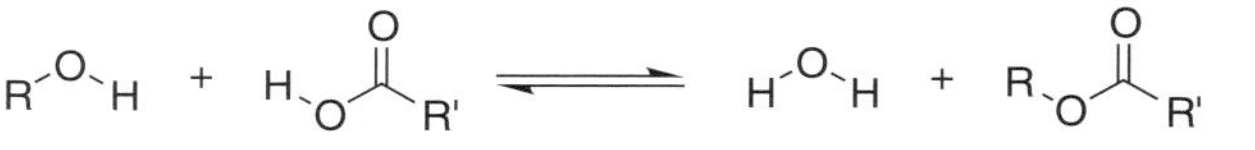

For acid-sensitive materials, a commonly used alternative is reaction of the alcohol with an anhydride, which is catalysed by tertiary amines such as pyridine, imidazole, and, in particular, 4-*N*, *N*-dimethylaminopyridine (**91**) (DMAP). For this, of course, the anhydride must be available. In practice, only the simple anhydrides such as acetic and butyric anhydrides are available on a commercial scale. Others can be prepared from the simple anhydrides by a 'transacylation' reaction driven by distilling out the more volatile acid.

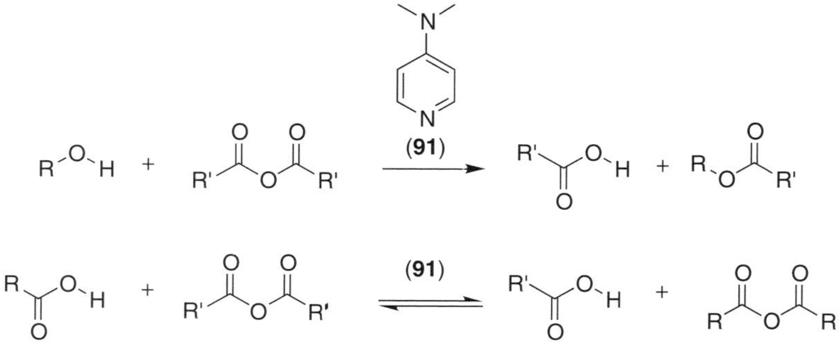

This fails with formic acid, and formic anhydride is unknown as a stable organic chemical. However, the unusual properties of formic acid means that formats can be prepared using a 'mixed anhydride', i.e. a mixture of formic acid and (usually) acetic anhydride reacts with alcohols to give the formate almost exclusively.

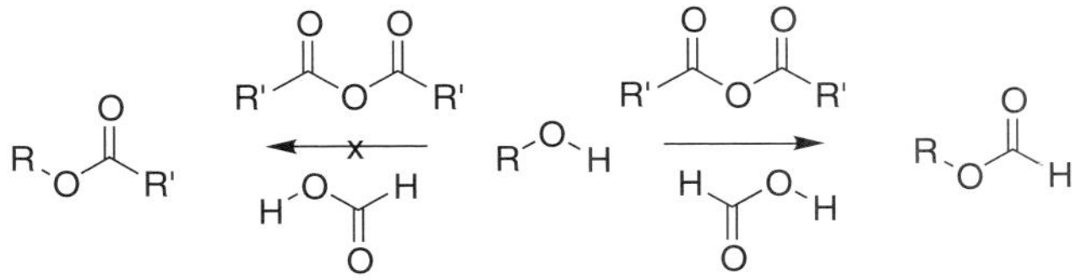

The final approach which can be taken is transesterification. Here (most commonly), a low-cost ester, such as a methyl ester, reacts with an alcohol and the reaction is driven by distilling out methanol. A similar reaction can be carried out with an acid and an ester. DMAP and alkali metal alkoxides are common and effective catalysts.

4.5 Aldehydes

4.5.1 Aliphatic aldehydes

The simplest aldehyde in use is acetaldehyde (**92**), which is ubiquitous in fruit formulations; a dilute solution has a pleasant apple taste. As the chain extends, a fatty character develops, which is exhibited by the C12 lauric aldehyde (**93**) and the synthetic material methylnonylacetaldehyde (**94**) (Aldehyde C12 MNA); this fatty, 'animalic' character has been used to effect in perfumery, most famously in the aldehyde cocktail of 'Chanel No. 5'.

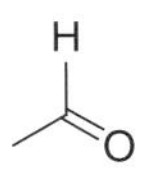

(92)

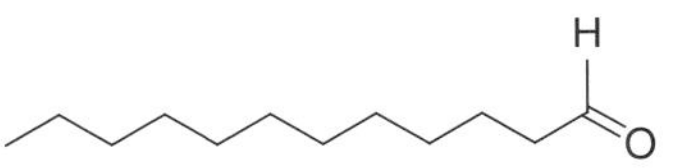

(93)

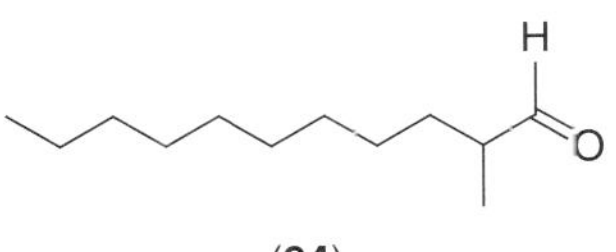

(94)

An interesting aldehyde is 12-methyltridecanal (**95**). This compound, which is FEMA GRAS, is found in beef fat and appears to originate from microorganisms in the rumen of cattle [11]. It is absorbed by the gut as plasmanogens, and released only when the beef is heated over a long period of time (e.g. stewing). Briefly roasting the meet does not release the material. Hence this gives the potential to create a boiled or stewed beef flavour, well differentiated from fried or roasted beef.

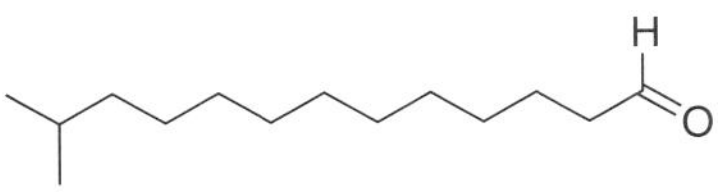

(95)

4.5.2 Unsaturated aldehydes

The most important here are the '*trans*-2-alkenals', in particular *trans*-2-hexenal (**96**) (leaf aldehyde). A Union Carbide patent [12] revealed a route using ethyl vinyl ether as 'vinylic synthon'.

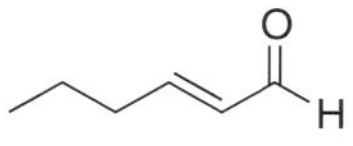

(**96**)

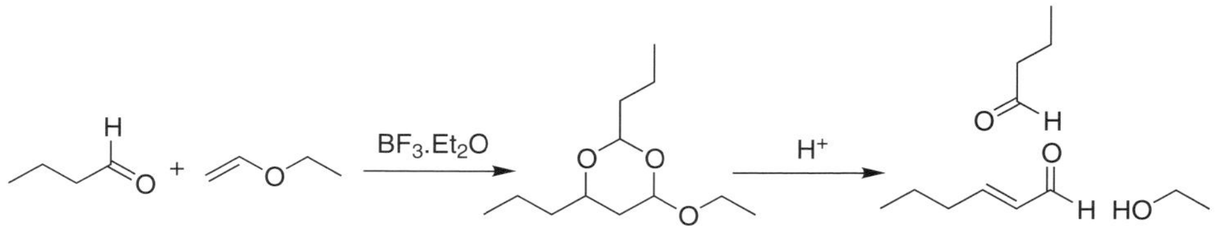

A development of this reaction uses the acetal derivative of the aldehyde as the precursor, which overcomes the need to use two equivalents of the starting aldehyde, one of which is simply regenerated. Reaction of the acetal (usually the ethyl acetal) with ethyl vinyl ether in the presence of a Lewis acid such as boron trifluoride or the clay catalyst Montorillonite K10 gives the triethoxy alkane [13]. When heated with a mixture of aqueous formic acid and an alkali metal formate this undergoes hydrolysis of the acetal function followed by elimination of the alkoxide group via the $E1_{CB}$ pathway to form the unsaturated aldehyde.

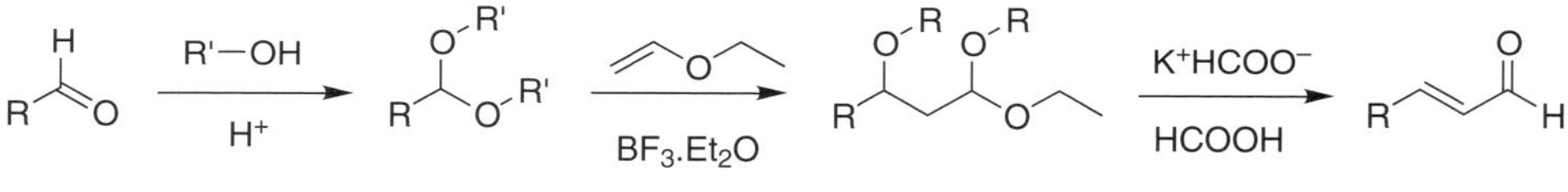

There have been many reports of synthesis of these materials via Wittig chemistry. A recent report shows that alcohols, which are in general more readily available than aldehydes, can be used as precursors via in-situ oxidation and Wittig coupling [14].

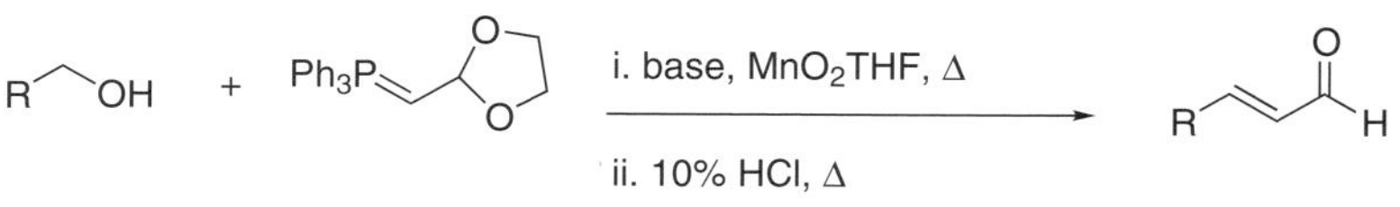

The character of the aldehydes changes significantly as the chain extends, and in general the odour threshold falls, as shown in Table 4.1. In particular, *trans*-2-dodecenal has an intense, persistent coriander aroma, which tends to live with anyone who has handled it for several days!

Table 4.1 Organoleptic properties of the *trans*-2-alkenals

Aldehyde	Odour threshold	Character
trans-2-Pentenal	1500	Sharp, acrid
trans-2-Hexenal	17	Sharp, green, almond on dilution
trans-2-Heptenal	13	Green
trans-2-Octenal	3	Fatty, citrus – *fatty lemon*
trans-2-Nonenal	0.1	Intensly fatty, cucumber on dilution
trans-2-Decenal	0.3	Oily, citrus – *orange peel*
trans-2-Undecenal		Citrus, herbaceous
trans-2-Dodecenal		Herbaceouos, intense coriander

Two non-conjugated aldehydes of importance are 10-undecanal (**97**) and *cis*-4-decenal (**98**). The former is derived from the 'cracking' of castor oil; it is claimed that commercially available materials which contain isomers other than simply the 10-isomer are more powerful.

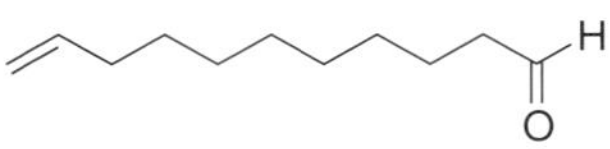

(**97**)

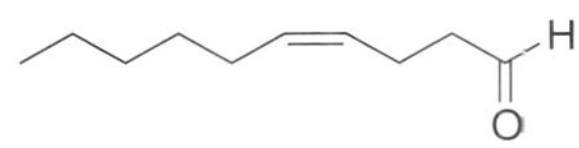

(**98**)

The most important polyunsaturated aldehyde is *trans*-2-*trans*-4-decadienal (99), which has a powerful, fatty-citrus, chicken aroma. *trans*-2-*cis*-6-Nonadienal (**100**) (violet-leaf aldehyde) has been prepared by 'one-pot' homologation using the phosphonium salt [14].

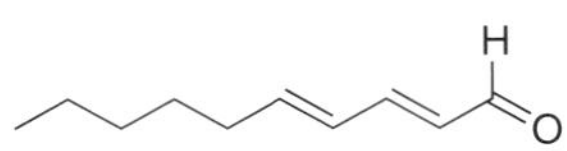

(**99**)

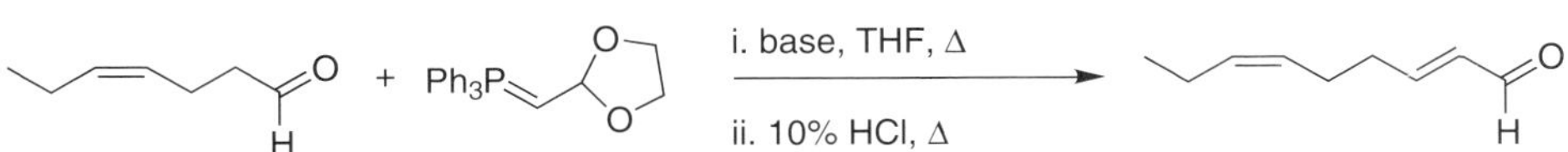

(**100**)

Amongst the terpenoids, aldehydes are less important than the alcohols. Most significant is citral, which is actually a mixture of the two isomers geranial (101) and neral (102), and citronellal (103).

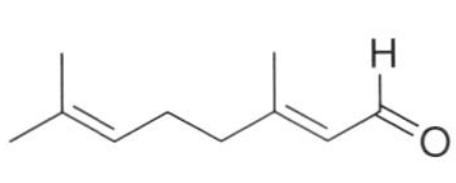

(**101**)

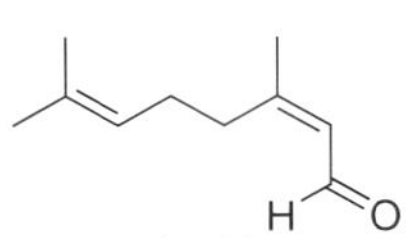

(**102**)

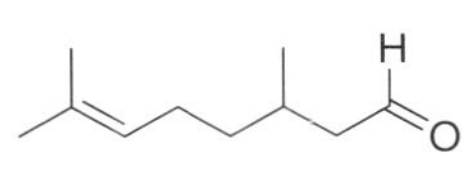

(**103**)

There are more compounds which may be considered as relatives of the terpenes which are important. 2,6-Dimethyl-5-heptenal (**104**) (Melonal) contributes a fresh, green, melon note to both flavours and fragrances. Several such as the citronellol derivative (**105**) (Muguet Aldehyde 50®) and the cyclic aldehydes (**106**) (Melafleur®) and especially (**107**) (Lyral®) are important for floral, muguet ('lily-of-the-valley') notes.

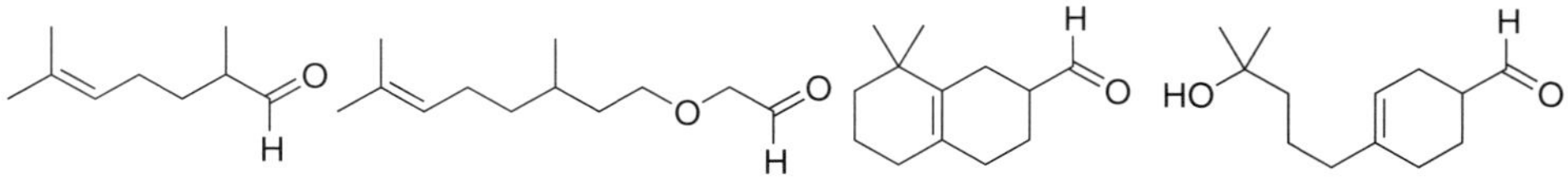

(**104**) (**105**) (**106**) (**107**)

4.5.3 Acetals

One problem that occurs in using aldehydes, especially in some demanding fragrance media, is that they are prone to oxidation, the formation of Schiff's bases with amines, and, if they have an *alpha* hydrogen, condensation reactions such as the Aldol reaction. Perhaps surprisingly, the tendency towards oxidation decreases with conjugation. Whereas simple aliphatic alcohols such as hexanal are commonly 'stabilised' by the addition of an anti-oxidant such as butylated hydroxytoluene (BHT) or butylated hydroxyanisole (BHA), this is not necessary with unsaturates such as *trans*-2-hexenal, and *trans*-2-*trans*-4-decadienal shows little tendency towards oxidation. Oxidation products are often detected by the acid value of the aldehyde; this is a simple test involving titration with potassium hydroxide solution, and is defined as the number of milligrams of potassium hydroxide needed to neutralise one gram of the analyte. Typically, the maximum value of an aldehyde would be ten. The test is also used for esters, as any hydrolysis will lead to the formation of an acid and hence a rise in acid value.

These reactions can be prevented if the aldehyde function is 'protected' by the formation of an acetal:

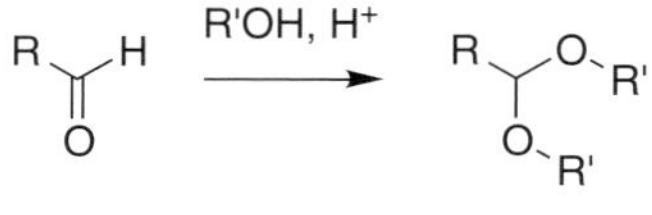

Many acetals have similar odours to their aldehyde precursors and can be used in their place; examples include citral diethyl acetal (**108**) and 2-methylundecanal dimethyl acetal (**109**) (Aldehyde C12 MNA DMA). Several acetals are used 'in their own right', including the wonderfully named grapefruit fragrance ingredient (**110**), Methyl Pamplemousse®.

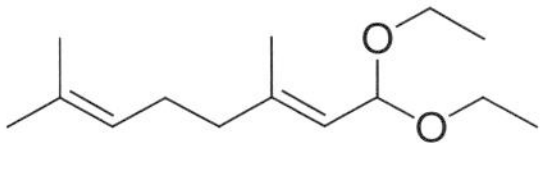

(108)

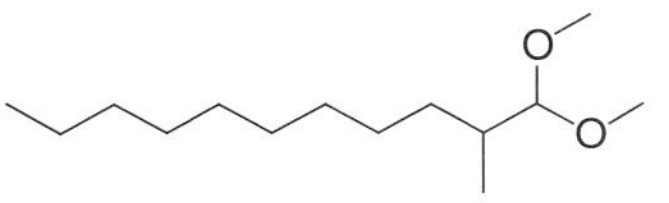

(109)

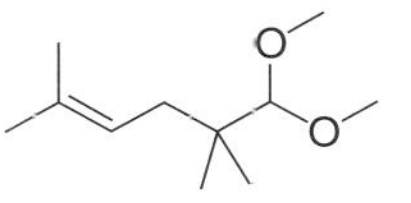

(110)

Acetals include a number of 'amber' chemicals. These materials are so-called due to their resemblance to *ambergris*, a material formed in the stomach of whales probably as a protection against intestinal damage by the 'shelly' parts of plankton. The material is occasionally found washed up on beaches, but the major source was the whaling industry. Unsurprisingly, this is now a rare and expensive material, driving the search for synthetic alternatives. Ambergris mostly consists of steroidal materials, and hence this structure was the first to be looked at. Amberketal® (**111**) resembles the ABC ring fragment of the steroids, and (**112**) (Amberconide®) could be considered a distorted version of this. More recently, molecules have been developed which have the amber tonality but do not have the steroidal character, e.g. (**113**) Karanal®, (**114**) Belambre®, (**115**) Spirambrene® and (**116**) Ysamber K®. This remains a very active area of research [15], and for the synthesis of Ambroxan, see Chapter 5.

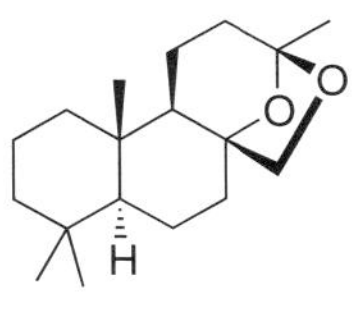

(111)

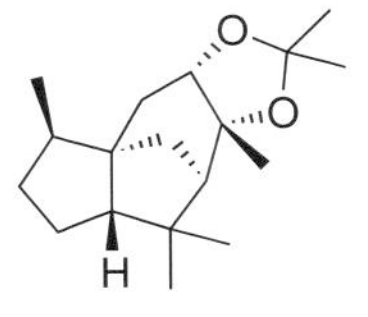

(112)

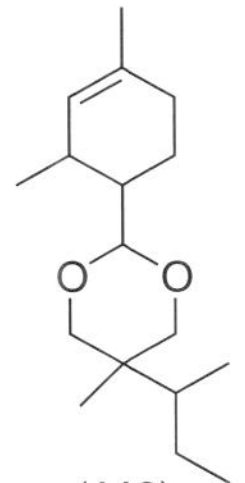

(113)

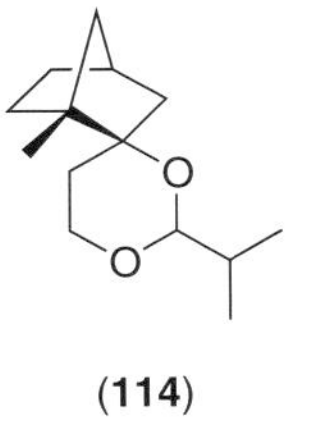

(114)

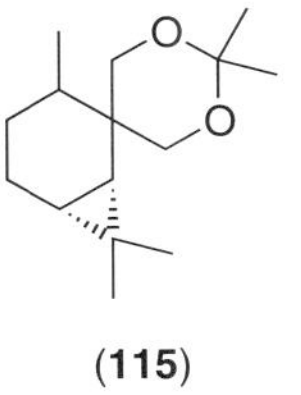

(115)

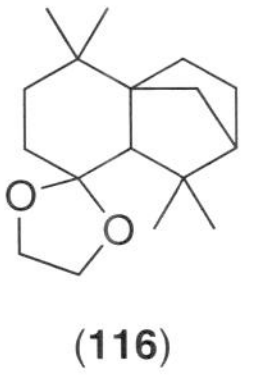

(116)

4.5.4 Aromatics

The identification of benzaldehyde (**117**) in oil of bitter almonds is one of the milestones in aroma chemistry. It has a powerful cherry-marzipan odour and is widely used in cherry flavours, even though it is not actually a significant aroma chemical in cherries. Phenylacetaldehyde (**118**) has a floral aroma common to all the 'phenylC2' compounds. Cinnamaldehyde (**119**), found in many essential oils, was another 'milestone' aroma chemical.

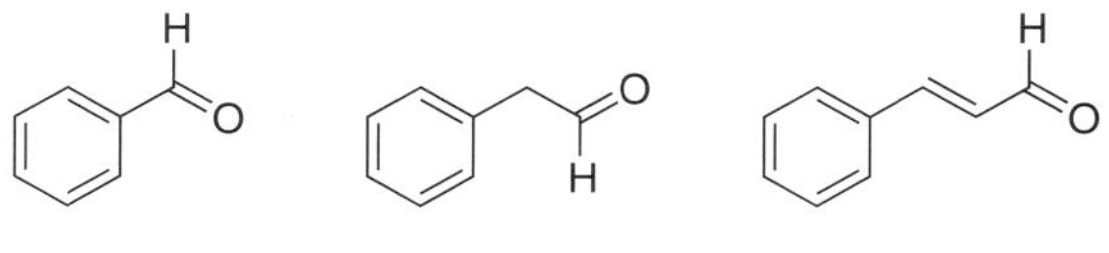

(**117**) (**118**) (**119**)

3-Phenylpropionaldehyde (**120**) (dihydrocinnamaldehyde), with a powerful hyacinth-like aroma, is obtained by the selective reduction of the alkene bond in cinnamaldehyde. It is also a by-product in the synthesis of its isomer 2-phenylpropionaldehyde (**121**) (hydratropaldehyde), which is obtained from styrene via hydroformylation (the oxo process); a recent patent claims less than 2% linear isomer when a rhodium catalyst is used [16]. This also has a powerful hyacinth aroma for floral fragrances.

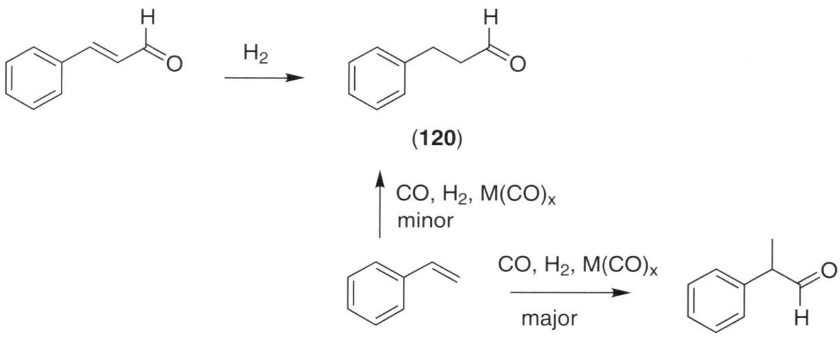

A large number of aromatic aldehydes can be considered to be cinnamaldehyde derivatives. α-Amylcinnamaldehyde (**122**) and α-hexylcinnamaldehyde (**123**) have sweet, floral jasmine notes and are widely used for domestic fragrances, especially as inexpensive alternatives to Hedione® (*vide infra*), which they resemble. They can be prepared by the condensation of benzaldehyde with the appropriate aldehyde.

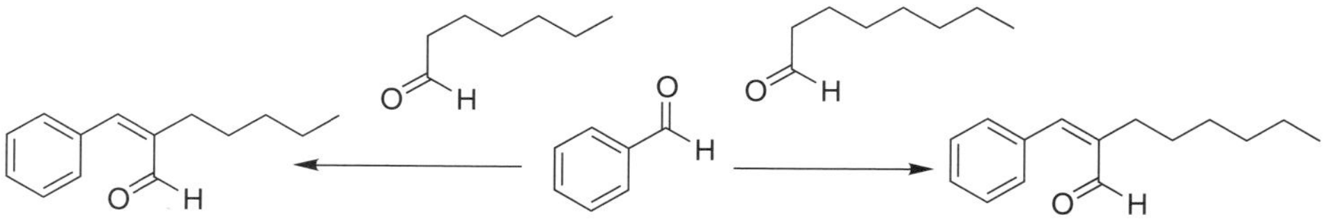

(**122**) (**123**)

Similarly, condensation of phenylacetaldehyde with acetaldehyde, isobutyraldehdye and isovaleraldehyde gives, respectively, 2-phenyl-2-butenal (**124**), 4-methyl-2-phenyl-2-pentenal (**125**) and 5-methyl-2-phenyl-2-hexenal (**126**) (Cocal®), which have sweet, vanilla-like aromas making them valuable in chocolate and sweet formulations for both flavour and fragrance.

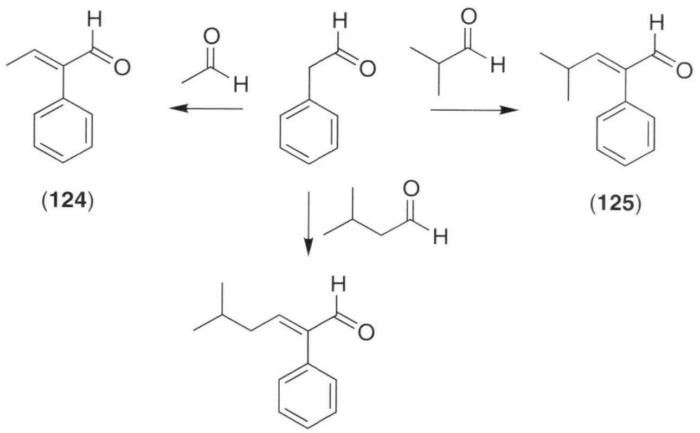

(124) (125) (126)

There are several important materials used in fragrance, especially for muguet notes which have similar structures; an aromatic ring with a methylpropanal chain and an C3 or C4 alkyl group *meta*- or *para*- to it; examples are (**127**) (Florhydral®), (**128**) (Lilial®), (**129**) (Sylvial®) and (**130**) (Cyclamen Aldehdye®); (**131**) (Canthoxal®) and (**132**) (Helional®, Tropional®) are closely related with an alkoxy group replacing the alkyl.

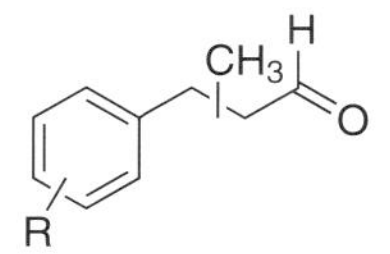

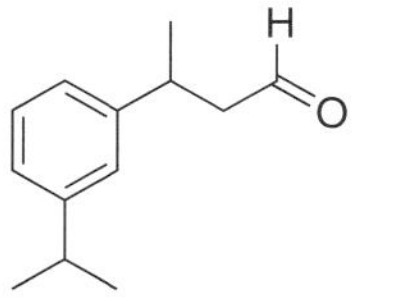

(127)

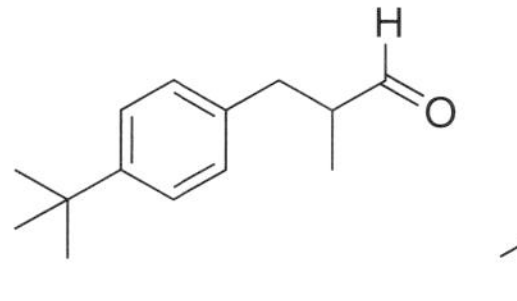

(128)

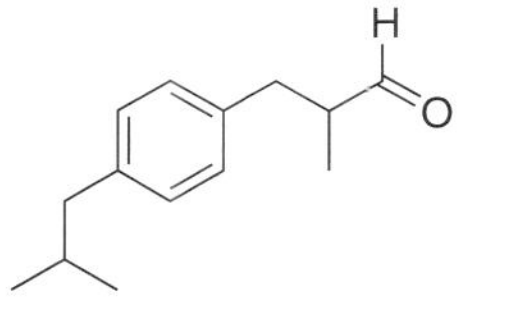

(129)

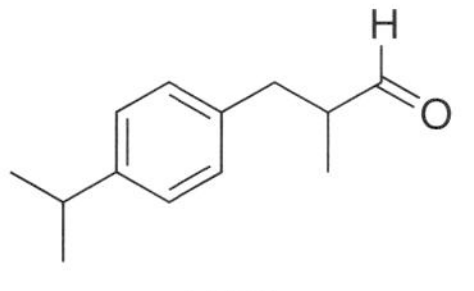

(130)

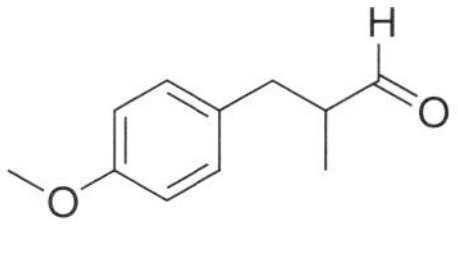

(131)

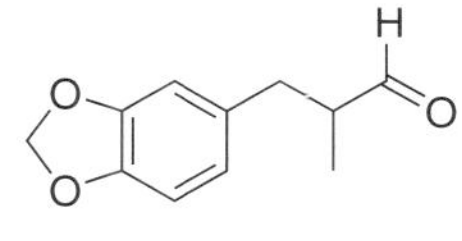

(132)

All three possible retrosynthetic disconnections have been used in the synthesis of these compounds; note the methyl group on the propanal chain is omitted for clarity, and the ring alkyl is shown *para* for the same reason:

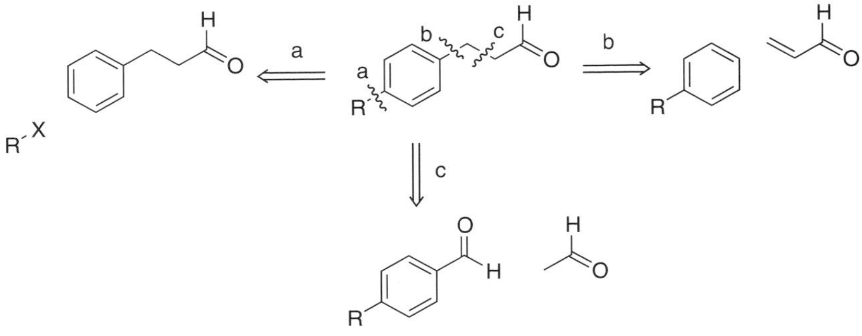

(1) Disconnection a, at the alkyl group, is illustrated in a synthesis of Lilial® (**128**). Alkylation of methyldihydrocinnamic alcohol (**133**) (available from the reduction of α-methylcinnamaldehyde, which in turn is prepared from the Aldol condensation of benzaldehyde with propionaldehyde, c.f. α-hexylcinnamaldehyde) gives the *tert*-butyl derivative (**134**), which is oxidised to the aldehyde [17, 18].

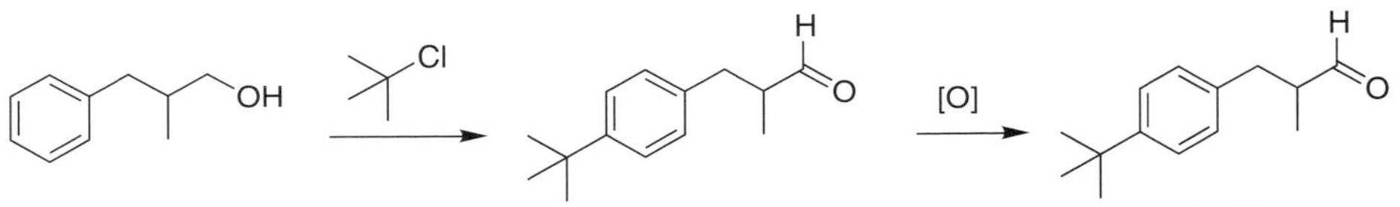

(133) **(134)** **(128)**

(2) Disconnection b, at the ring-aldehyde chain bond, is illustrated in a synthesis of Cyclamen Aldehyde® (**130**). Isopropylbenzene (**135**) undergoes a Friedel–Crafts reaction with a protected form of methacrolein, the diacetate, to form an enol acetate (**136**) which is readily hydrolysed to the aldehyde [19].

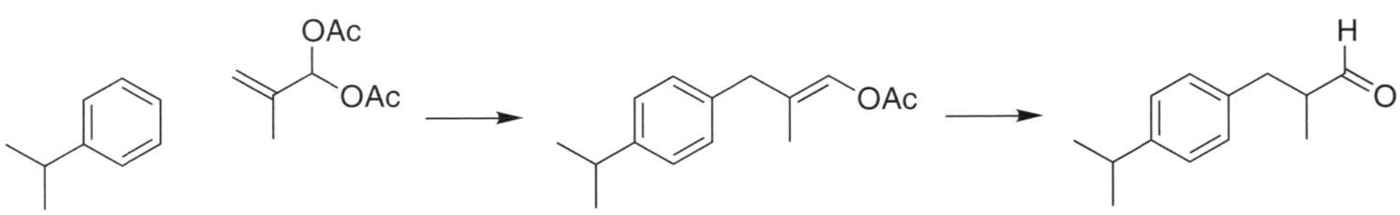

(135) **(136)** **(130)**

(3) Disconnection c, α, β to the carbonyl, is the characteristic Aldol disconnection, and is illustrated by another approach to Cyclamen Aldehyde® (**130**). Cumene aldehyde (**137**) condenses with propionaldehyde to give the unsaturated Aldol product (**138**), which is hydrogenated to the saturated aldehyde [20].

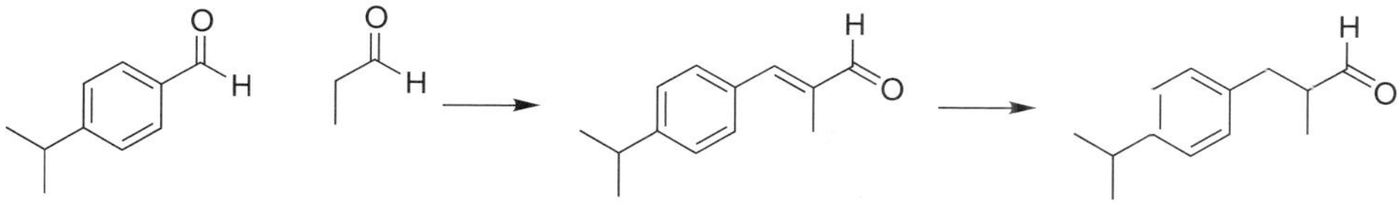

(137) **(138)** **(130)**

The most important aromatic aldehyde must be Vanillin (**139**). Its synthesis by Wilhelm Haarmann and Ferdinand Tiemann was the foundation stone of Haarmann and Reimer, now part of Symrise. It is the single most important component of vanilla, though not the only material of importance; vanilla extract has been estimated to be some ten times stronger than the Vanillin content would indicate. The synthesis of Vanillin by many routes has been reported, including carbonylation of guaiacol, and by the degradation of lignins obtained as waste from paper production; since the latter is a natural material, that is also a source for natural Vanillin. Vanillin is widely used in both flavours and fragrances; in general, it is stable, but there can be a problem with discolouration. As a phenol, it reacts with traces of iron to give a purple colouration, and indeed this could be used as a test for the metal. The isobutyl ester (**140**) (Isobutavan®) overcomes this by 'protecting' the phenolic hydroxyl. The Vanillin homologue (**141**), 'Ethyl Vanillin', has a more powerful aroma than Vanillin, though it is not found in nature.

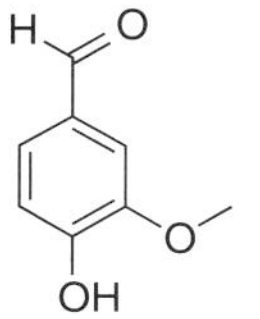

(139)

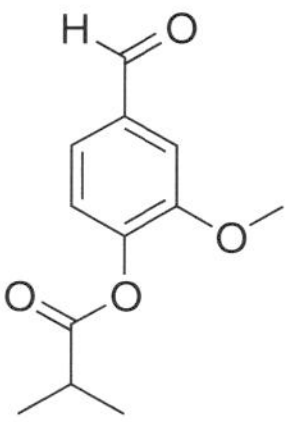

(140)

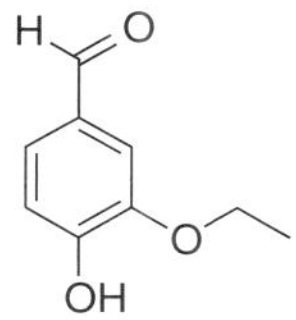

(141)

Methylenedioxybenzaldehyde (**142**) (Heliotropin, Piperonal) is found in a number of essential oils and is used for spicy, floral notes in both flavours and fragrances. Aldol condensation of Heliotropin with propionaldehyde followed by hydrogenation gives the synthetic derivative (**143**) (Helional®, Heliofolal®) which has a greener, hay-like odour.

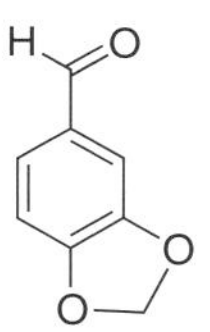

(142)

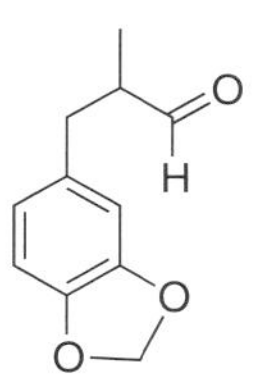

(143)

4.5.5 Nitriles

Whilst these are clearly not C, H, O compounds, it is appropriate to mention them here as they commonly have similar odours to the analogous aldehyde and are used in demanding fragrances where aldehydes cannot be used; they are more acid-stable than aldehydes, and are not susceptible to oxidation. They are frequently prepared from aldehydes via dehydration of the aldehyde oxime. Examples include geranonitrile (**144**), citronellonitrile (**145**) and cinnamanitrile (**146**).

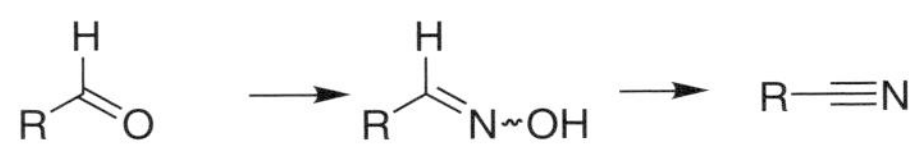

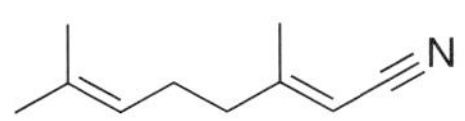

(**144**)

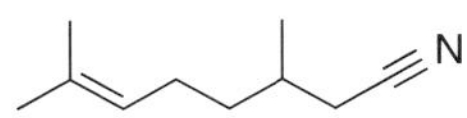

(**145**)

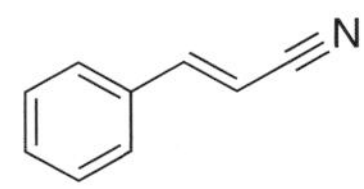

(**146**)

4.6 Ketones

Ketones are another important group of compounds in the flavour and fragrance industry, though the simple short-chain molecules, unlike the simple aldehydes, are not used. The vicinal diketone 2,3-butanedione (**147**) (diacetyl) is very important for butter flavours, and hence for flavouring margarine, as is the reduced derivative 3-hydroxybutanone (**148**) (acetoin, acetyl methyl carbinol, AMC). The series of diketones to ca. C8, with both straight and branched chains, are used for their candy, caramel notes.

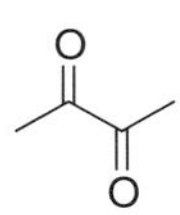

(**147**)

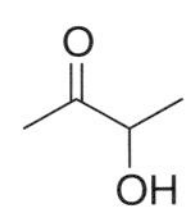

(**148**)

The cyclic diketone, 1-methylcyclopentane-2,3-dione (**149**), usually named as its enolic form 1-methylcyclopent-2-enol-3-one (**150**), is the so-called Maple Lactone or Corrylone, and has the sweet, caramel, almost burnt, aroma characteristic of maple syrup, of which

(**149**) (**150**)

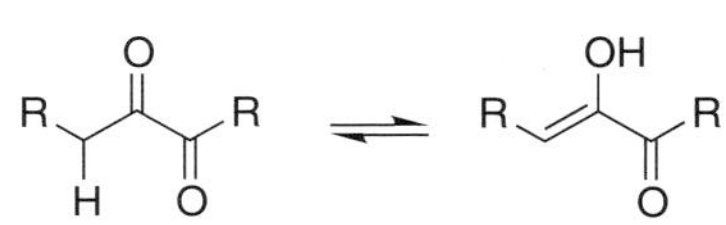

it is the major aroma component. All of the diones exist in both the keto and enol forms, which can lead to confusion in analysis, if this is not taken into account [21].

Some simple monoketones such as octan-3-one (**151**) are used, and alkan-2-ones are found in blue cheese, but more interesting is the unsaturated compound 1-octen-3-one (**152**) (Amyl vinyl ketone, AVK). This is found along with the alcohol in mushrooms, but has a much more intense odour, with a reported odour threshold of 0.005 ppb, compared to 1 ppb for the alcohol. It has also been reported as an important, and unexpected, component of other aromas, including that of elderflower [22] and chocolate [23].

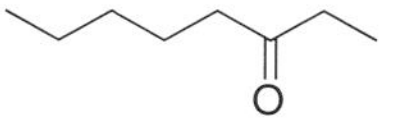

(**151**)

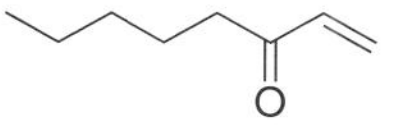

(**152**)

Amongst aromatic ketones, the most important is 4-hydroxyphenyl-2-butanone (**153**), Raspberry Ketone. As the name implies, it is a major component of raspberry aroma and is used for soft fruit flavours, as is its methylated derivative (**154**), the Bramble Ketone. Zingerone (**155**), which can be prepared from vanillin, is a component of ginger.

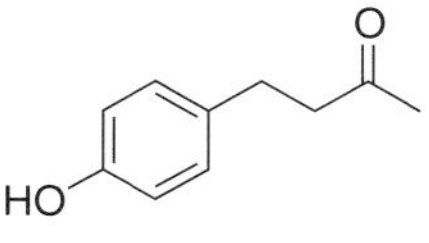

(**153**)

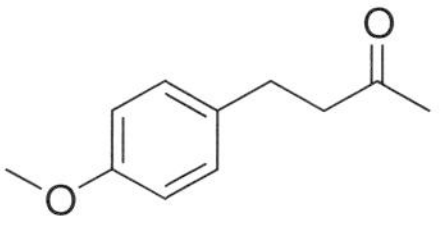

(**154**)

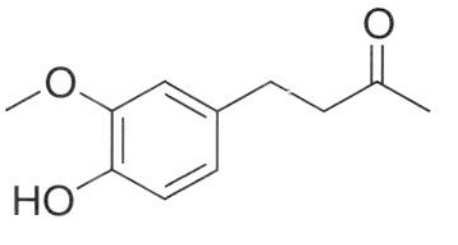

(**155**)

Derivatives of cyclopentanone are very important. A key odorant of jasmine is *cis*-jasmone (**156**); though not found in nature, dihydrojasmone (**157**) is more easily synthesised and is often used for jasmine notes.

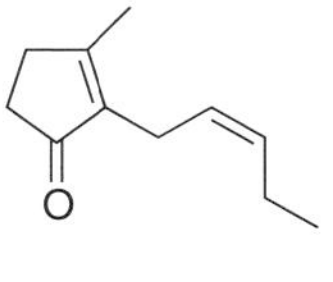

(**156**)

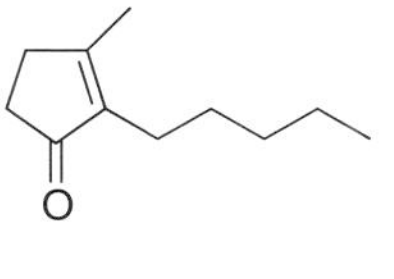

(**157**)

Analogues of this are also important. Condensation of cyclopentanone (**158**) with aldehydes such as hexanal gives the unsaturated ketone, and hydrogenation gives the saturated molecule. Both the unsaturated and saturated derivatives of pentanal, hexanal and heptanal are used, and the saturated molecules are also important as precursors of the γ-lactones (*vide supra*).

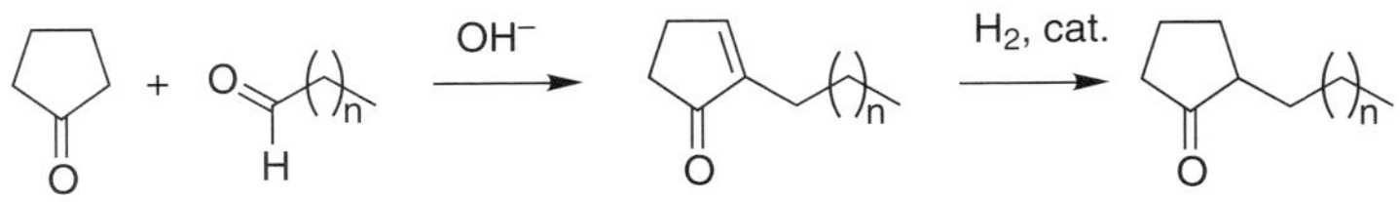

(158)

This chemistry is also used in the synthesis of the most important cyclopentanone derivative, methyl dihydroisojasmonate (**159**) (Hedione®). The introduction of this floral, jasmine molecule in the 1960s had a major impact in perfumery; first used in the famous 'Eau Savauge' by Dior, it is still widely used today. The adduct of cyclopentanone with valeraldehyde (**160**) is the precursor. Michael addition of dimethyl malonate followed by monohydrolysis and decarboxylation gives the product as a mixture of isomers, predominately *trans* (**161**). The different stereoisomers have different odour thresholds and somewhat different odours. The *cis*-isomer (**162**) is more desirable; base treatment with fractionation to remove the more volatile *cis*-form leads to enrichment, and there are also a number of products where the isomer ratios are different – Cepionate®, Kharismal®, Super Cepionate® and Hedione HC® are all forms of methyl dihydrojasmonate, with latter having up to 75% *cis*-form.

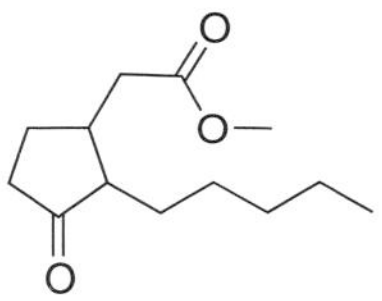

(159)

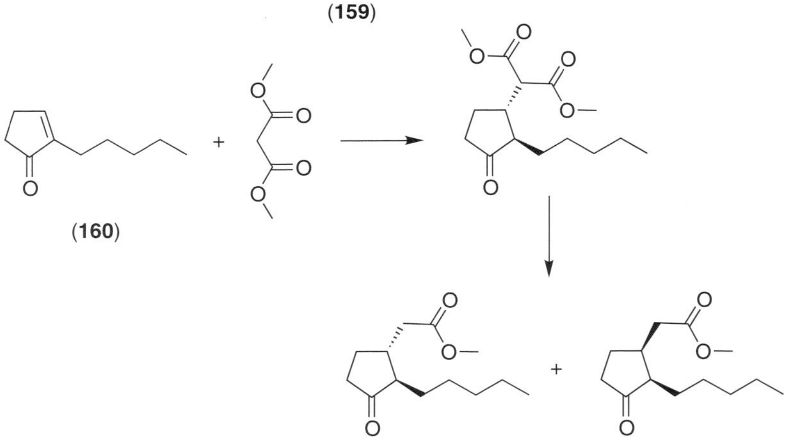

(160) **(161)** **(162)**

4.6.1 *Carotenoids, ionones, irones, damascones and related compounds*

The ionones (α-ionone (**163**), β-ionone (**164**)) are one of the oldest groups of aroma chemicals, first synthesised in the late nineteenth century, nearly a century before they were identified in nature. Here, carotenoids is an appropriate term as they are believed to be metabolites of α-carotene, possibly via peroxide cleavage, in a manner similar to the formation of *cis*-3-hexenal (*vide supra*), and indeed the main use of β-ionone is as an intermediate to vitamin A. The ionones have a powerful, sweet, 'powdery', violet aroma, and were amongst the first 'synthetics' to be used in perfumery, and research in this area is still very active today [24]. 8-Methylionone (**165**) is even more powerful than either (**163**) or (**164**), and also has a woody character. The irones such as α-irone (**166**) are methylated alpha to the geminal methyl groups.

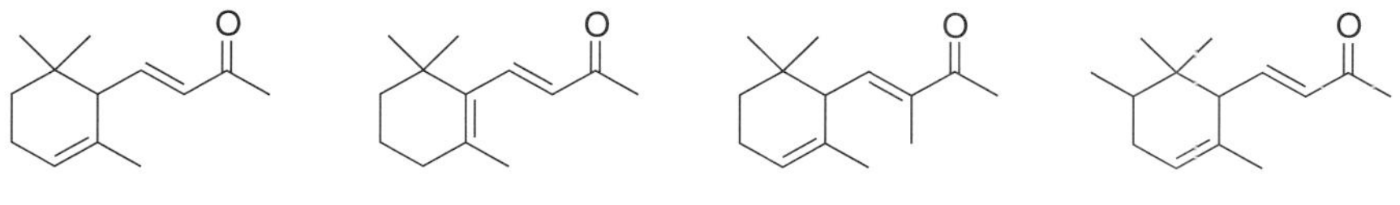

(**163**) (**164**) (**165**) (**166**)

The damascones such as α-damascone (**167**) are structural isomers of the ionones where the carbonyl is now *alpha* to the ring; the damascenones such as (**168**) have an extra double bond. β-Damscenone was first identified in Bulgarian rose oil, *Rosa damascena*, and later β-damascone was also found. Dynascone® (**169**) is a further isomeric form where the carbonyl bearing chain replaces the ring methyl; it has a green, galbanum, pineapple aroma, and is used in the popular 'Cool Water' by Davidoff.

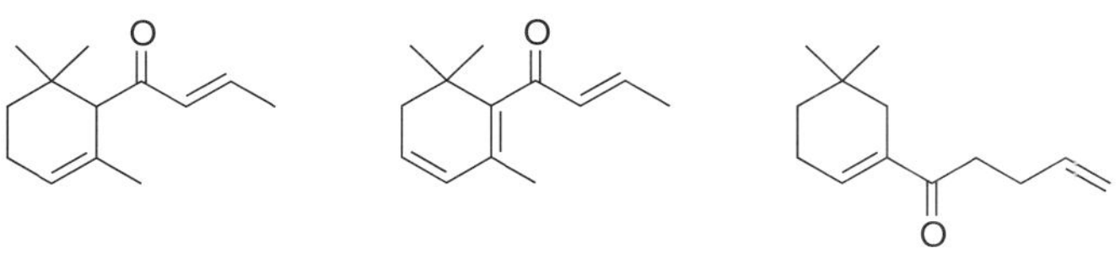

(**167**) (**168**) (**169**)

The ionones, irones, damascones and related compounds all have three main isomers, α, β or γ depending on the position of the double bond.

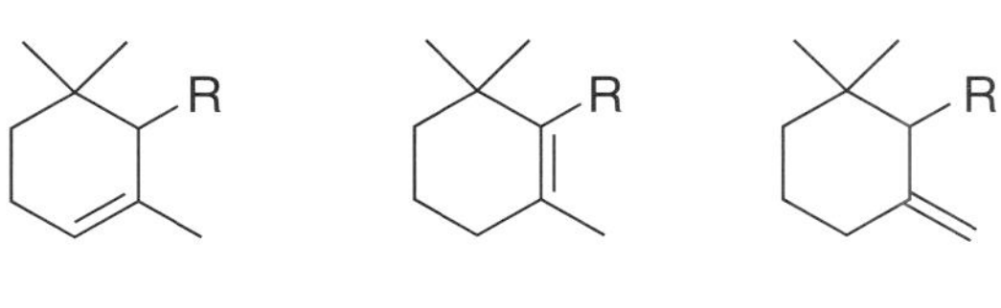

α β γ

Further research has developed new molecules loosely based on these. Diels–Alder reaction of myrcene (**170**) with 3-methyl-3-penten-2-one followed by acid-catalysed cyclisation gives the intensely *woody* molecule (**171**), Iso E Super®; the isomer (**172**), which is a minor product of the reaction, has an odour threshold some 1000 times lower and has been termed 'Iso E Super Plus'. Whilst it has not been possible to develop a commercial synthesis of this, the closely related molecule (**173**) has been introduced as Georgywood®. This is a very active field (presumably we will see Iso E Supercalifragalisticexpialidocius?) and the logic of odorant design in this area has been discussed [25].

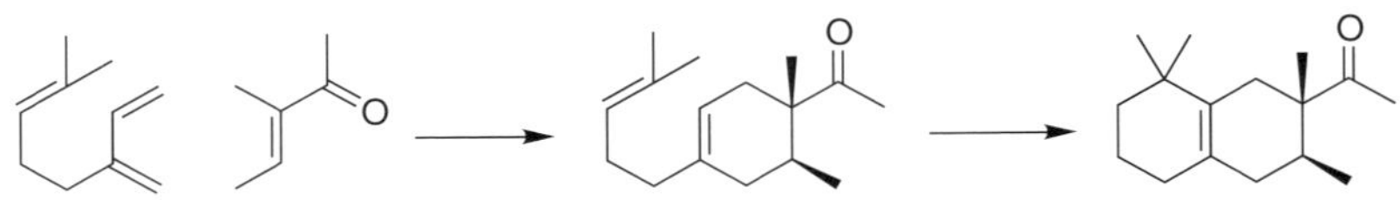

(**170**) (**171**)

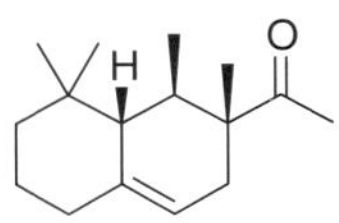

(**172**)

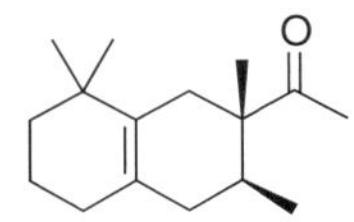

(**173**)

One of the more unusual ketones is the macrocycle (**174**) (Trimofix O®). This is a woody, amber material; other macrocylic ketones are found among the musks, and these are covered in Chapter 7.

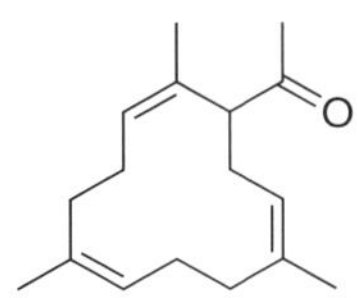

(**174**)

4.7 Hydrocarbons

Here we may stretch the definition of the compounds in this chapter to the breaking point by defining these as C, H, O compounds without the O. Since most hydrocarbons are virtually odourless, only a few have any importance as aroma chemicals. D-Limonene (**175**) is the main component of most citrus oils, and is readily available as a by-product of the orange juice industry. It has a small odour in its own right (though it remains possible that this odour is due to trace components below the level of GC detection) but is more commonly used as a solvent and as a precursor to other aroma chemicals. The pinenes, especially α-pinenes (**176**) and β-pinenes (**177**), are used for domestic fragrances, but again their most important contribution is as precursors.

An unusual group of hydrocarbons are the so-called Galbanolides. *E,Z*-1,3,5-undecatriene (**178**) is a component of galbanum oil, and mixtures of the isomers are commercially available.

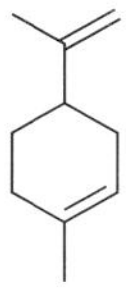

(175)

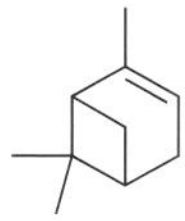

(176)

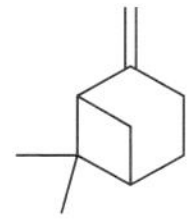

(177)

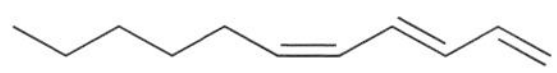

(178)

The epoxides of some terpene hydrocarbons are also used, including the woody-amber material caryophellene oxide (**179**); epoxides are also dicussed in Chapter 5.

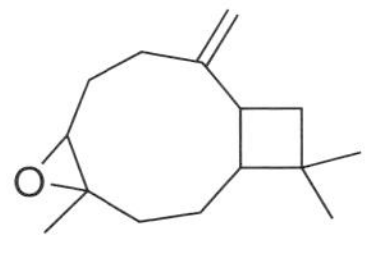

(179)

Acknowledgements

The comments made in the Acknowledgements of the Introduction are still appropriate here, though special thanks should go to Mark Dewis and Philip Kraft, who apart from working on their own chapters, made time to make valuable and constructive comments here.

References

1. Bauer, K., Garbe, D. and Surburg, H. (1997) *Common Fragrance and Flavour Materials.* Wiley-VCH, Weinheim.
2. Ohloff, G. (1994) *Scent and Fragrances.* Springer-Verlag, Berlin.
3. Teisseire, P.J. (1994) *Chemistry of Fragrant Substances.* VCH, NY.
4. Frater, G., Bajgrowicz, J.A. and Kraft, P. (1998) Fragrance chemistry. *Tetrahedron,* **54**, 7633–7703.
5. Kraft, P., Bajgrowicz, J.A., Denis, C. and Frater, G. (2000) Odds and trends: recent developments in the chemistry of odorants. *Angew. Chem. Int. Ed.*, **39**, 2980–3010.
6. Sell, C. (1999) Ingredients for the modern perfume industry. In D. Pybus and C. Sell (eds), *The Chemistry of Fragrances.* Royal Society of Chemistry, Cambridge, pp. 51–124.
7. Kondo, K., Komija, K. and Iwamoto, H. (Toyo Soda) (1976) Jap. Pat. 7,831,635.

8. Czerny, M., Mayer, F. and Grosch, W. (1999) Sensory study on the character impact odorants of roasted arabica coffee. *J. Agric. Food Chem.*, **47**, 695–699.
9. Kaiser, R. (Givaudan-Roure) (1995) Unsaturated Esters, European Pat. 773209.
10. Yamasaki, T., Kuroki, Y., Nakai, M., Oda, M., Kawaguchi, Y. and Kobayashi, M. (Ube Industrie) (1976) Jap. Pat. 7,695,058.
11. Kerscher, R., Nurnburg, K., Voigt, J., Schieberle, P. and Grosch, W. (2000) Occurrence of 12-methyltridecanal in microorganisms and physiological samples isolated from beef. *J. Agric. Food Chem.*, **48**, 2387–2390.
12. Hoaglin, R. and Hirsh, D.H. (Union Carbide) (1953) Production of unsaturated aldehydes, US Pat. 2,628,257.
13. Fishman, D., Klug, J.T. and Shani, A. (1981) α,β-Unsaturated aldehydes: Montmorillonite clay K10, an effective catalyst for the preparation of unsaturated aldehydes via condensation of acetals with enol ethers. *Synthesis*, 137–138.
14. Reid, M., Rowe, D.J. and Taylor, R.J.K. (2003) Two carbon homologated α,β-unsaturated aldehydes from alcohols using the *in situ* oxidation-Wittig reaction. *Chem. Commun.*, 2284–2285.
15. Frater, G., Helmlinger, D. and Kraft, P. (2003) Synthesis of [4.3.3]propellanes by carbenium-ion rearrangment and their olfactory characterisation. *Helv. Chim. Acta*, **86**, 678–696.
16. Mercier, F., Lelievre, S., Mathey, F. and Spagnol, M. (Rhodia Chimie) (2003) Resolution method for a racemic mixture of aldehydes, US Pat. 6,613,945.
17. Goetz, N. and Hoffmann, W. (BASF) (1980) Process for the preparation of p-substituted phenyl propanols and their esters and of p-substituted 3-phenyl-2-methylpropanols, European Pat. 32,576.
18. Webb, D. (Bush Boake Allen Ltd.) (1982) Process for the production of phenyl substituted aldehydes and of phenyl substituted alcohols, European Pat. 45,571.
19. Sciabine, I. (Rhone-Poulenc) (1958) Alkylaromatic aldehydes and their preparation, Ger. Pat. 1,145,161.
20. Dunkel, W., Eckardt, D.J. and Stern, A. (Universal Oil Products) (1968) Hydrogenation of cinnamaldehyde, US Pat. 3,520,934.
21. Zviely, M., Giger, R., Abushkara, E., Kern, A. *et al.* (2002) Application of chromatographic and spectroscopic methods for solving quality problems in several flavour aroma chemicals. In K. Swift (ed.), *Advances in Flavours and Fragrances: From the Sensation to the Synthesis*. Royal Society of Chemistry, Cambridge, pp. 39–53.
22. Jorgensen, U., Hansen, M., Christensen, L.P., Jensen, K. and Kaack, K. (2000) olfactory and quantitative analysis of aroma compounds in elderflower (*Sambucus nigra* L.) drink processed from five cultivars. *J. Agric. Food Chem.*, **48**, 2376–2383.
23. Schnermann, P. and Schieberle, P. (1997) Evaluation of key odorants in milk chocolate and cocoa mass by aroma extract dilution analysis. *J. Agric. Food Chem.*, **45**, 867–872.
24. Kraft, P. (1999) Design and synthesis of violet odorants with bicyclo[6.4.0]dodecene and bicyclo[5.4.0]undecene skeletons. *Synthesis*, 695–703.
25. Kraft, P. (2002) Designing damscone- and ionone-like odourants. In K. Swift (ed.), *Advances in Flavours and Fragrances: From the Sensation to the Synthesis.* Royal Society of Chemistry, Cambridge, pp. 138–146.

Chapter 5
Aroma Chemicals II: Heterocycles

Michael Zviely

5.1 Introduction

Of the almost 20 million chemical compounds presently characterized, almost half are heterocyclic molecules. Heterocyclic molecules are significant due to their high abundance in nature and their chemical and biological importance. In the flavour and fragrance industry heterocyclic compounds are being seen with high interest due to their varied occurrence in food flavours and due to their valuable organoleptic characteristics. Even though heterocyclic aroma chemicals are found only in minute amounts in foods, their powerful strong odours and low odour thresholds, as expressed by high φ values, make them play an important role in the boosting of flavours and fragrances.

The main heterocyclic aroma chemicals are oxygen-, sulfur- and nitrogen-containing rings. The oxygen-containing heterocyclic aroma chemicals belong to the oxirane, furan, pyran and oxepine groups. The sulfur-containing aroma chemicals belong to the thiophene family, and together with nitrogen, to the thiazole and dithiazine systems. Nitrogen-containing aroma chemicals belong to pyrrole, indole, pyridine, quinoline, pyrazine and quinoxaline systems, and together with sulfur, the thiazole and the dithiazine families.

5.2 Introduction to heterocyclic compounds

Heterocyclic compounds are cyclic molecules where one atom or more in the ring is not a carbon atom.

The heterocyclic compounds are divided into two groups – aromatic and non-aromatic. The non-aromatic heterocyclic compounds have similar characteristics to their non-cyclic counterparts; tetrahydrofuran (THF) (1) and 1,4-dioxane (2) are typical ethers, and 1,3,5-trioxane (3) behaves like an acetal.

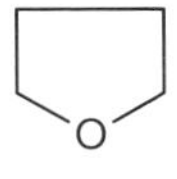

(1)

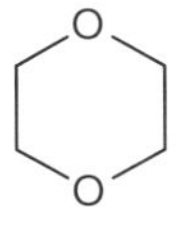

(2)

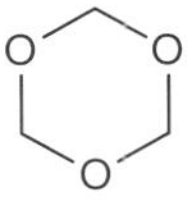

(3)

Pyrrolidine (**4**) and piperidine (**5**) are typical secondary amines, and quinuclidine (**6**) a tertiary amine.

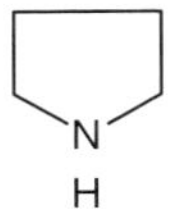

(**4**)

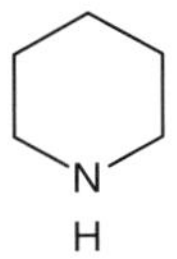

(**5**)

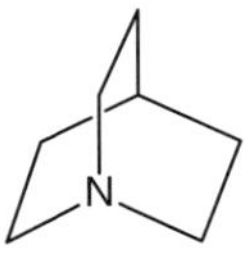

(**6**)

The aromatic heterocyclic group includes furan (**7**), pyrrole (**8**) and thiophene (**9**).

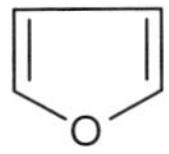

(**7**)

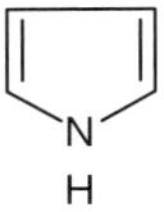

(**8**)

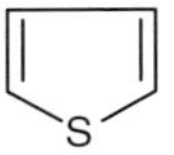

(**9**)

Other compounds contain more than one heteroatom, or are polycyclic molecules – pyrazine (1,4-diazine) (**10**), thiazole (**11**) and indole (**12**).

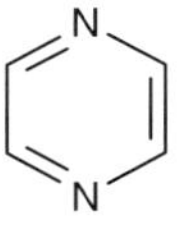

(**10**)

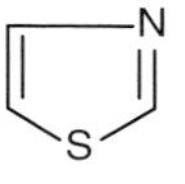

(**11**)

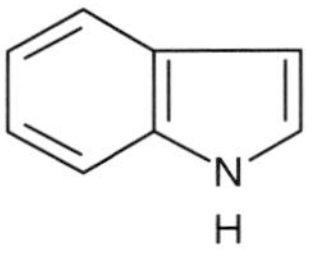

(**12**)

5.2.1 Terminology of heterocycles

The name of the heterocyclic ring is based on the carbon ring with the following prefixes:

Aza – for nitrogen;
Oxa – for oxygen;
Thia – for sulfur, e.g. 1,3-oxathiane (**13**).

The ring size is noted by the following suffixes:

-irane – three-membered ring;
-etane – four-membered ring;
-olane – five-membered ring, e.g. oxolane (**14**) and 1,3-dioxolane (**15**);
-ane – six-membered ring;
-epam – seven-membered ring.

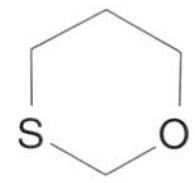

1,3-Oxathiane
(**13**)

Oxolane
(**14**)

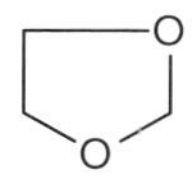

1,3-Dioxolane
(**15**)

5.2.2 Non-aromatic heterocyclic compounds

Heterocyclic molecules containing one heteroatom:

Oxirane (Epoxide)
(Ethylene oxide)

Thiirane (Thioepoxide)
(Ethylene sulfide)

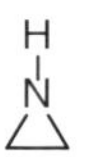

Aziridine
(Ethyleneimine)

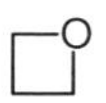

Oxetane

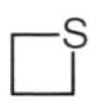

Thietane

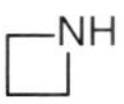

Azetidine

Oxolane
(Tetrahydrofuran)

Thiolane
(Tetrahydrothiophene)

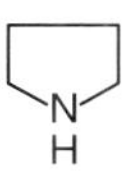

Pyrrolidine

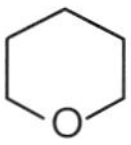

Oxane
(Tetrahydropyran)

Thiane

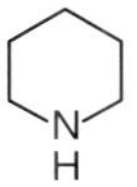

Piperidine

Additional heteroatoms are numbered, e.g. 1,2-di or 1,3,5-tri, etc. The numbering of heterocyclic molecules starts at the heteroatom.

Due to the highly potent odour character of a great part of the heterocyclic compounds used in flavours and fragrances, the term φ value will be used for some of the materials described.

A relatively new concept, φ value gives a better apprehension of the odour intensity of a single molecule, taking into consideration its molecular mass (MM), analogous to ε value in UV/VIS data of particular molecules.

φ Value of a molecule is defined as following [1]:

$$\varphi = \frac{\text{MM} \times 10^3}{\text{Threshold (ppm)}}$$

5.3 Oxygen-containing heterocyclic aroma chemicals

The oxygen-containing heterocyclic aroma chemicals belong to the following groups:

I. Oxiranes (epoxides);
II. Furans and hydrofurans;
III. Pyrans;
IV. Oxepines.

I. Oxiranes
The epoxide-containing aroma chemicals include glycidate esters. Ethyl 3-phenylglycidate (**16**) is the so-called Aldehyde C-16 Special, which is an artificial aroma chemical, having a fruity, strawberry, fermented, honey odour. It is mainly applied in fruit flavours (strawberry) and ice cream flavourings [2]. Ethyl 3-methyl-3-phenylglycidate (**17**), the so-called Aldehyde C-16 (strawberry aldehyde), which is also an artificial aroma chemical with fruity strawberry, ferment, honey odour, is applied to fruit flavours, ice cream flavourings; strawberry, raspberry, cherry, grape, apricot and peach.

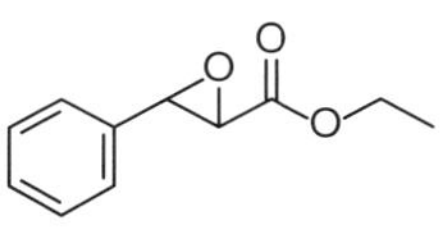

Ethyl 3-phenylglycidate (**16**) (Aldehyde C16 Special, so-called)

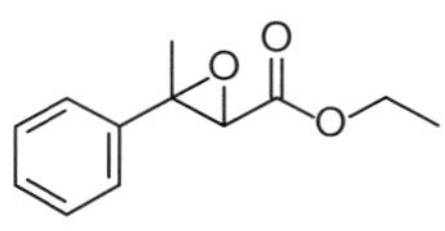

Ethyl 3-methyl-3-phenylglycidate (**17**) (Aldehyde C16, so-called)

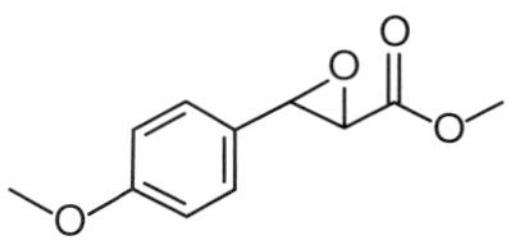

Methyl anisylglycidate Anoxirane® (**18**)

Methyl anisylglycidate (**18**), or methyl (4-methoxyphenyl) oxiranecarboxylate, has a fresh aromatic odour, fruity, somewhat reminiscent of berries.

The synthesis of glycidates is done generally using the Darzens condensation (Darzens–Claisen reaction; Glycidic Ester condensation) [3].

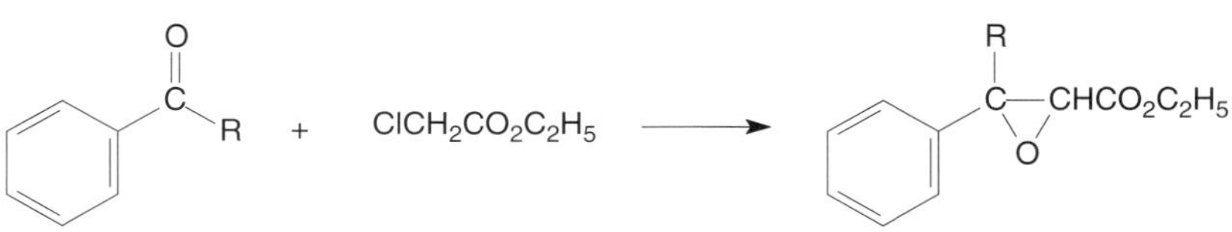

Other epoxides include α-cedrene epoxide (**19**), which has a precious woody odour, reminiscent of ambergris, tobacco and sandalwood. This epoxide, which is prepared from α-cedrene, is applied in soap compound and is useful in artificial patchouli.

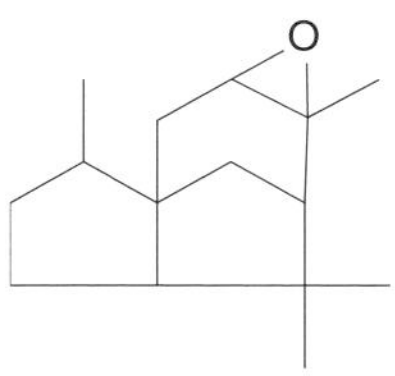

α-Cedrene epoxide
(**19**)
Andrane®

II. Furans and hydrofurans

One of the most important molecules is 4-hydroxy-2,5-dimethyl-3(2*H*)-furanone (**20**), known also by the name of Furaneol®. This material occurs in strawberry, pineapple and in beef. It has a sweet, caramel-fruity (pineapple-like) odour and flavour with fried meat aspects. 4-Hydroxy-2,5-dimethyl-3(2*H*)-furanone is applied mainly to fruity (strawberry), meat flavourings and ice cream formulations.

Another important molecule is furfural (furan-2-carboxaldehyde) (**21**), occurring in almost every type of food flavour, including dairy, cereals, roasted products and meats. Furfural has a sweet caramel-like, nutty, baked bread, almond odour and flavour, and it is applied in cereal, roasted and meat flavourings.

2-Methylfuran-3-thiol (**22**), which occurs in meat and coffee, has a sulphurous, coffee- and meat-like odour and flavour on extreme dilution. 2-Methylfuran-3-thiol is applied mainly in roasted and meat flavourings.

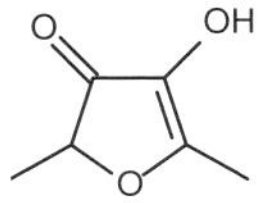

4-Hydroxy-2,5-dimethyl-3(2*H*)-furanone
(**20**)
Furaneol®
Fraision®

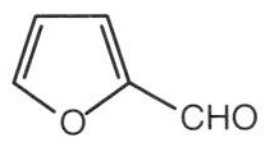

Furfural
(**21**)
φ 3.2 × 10^4

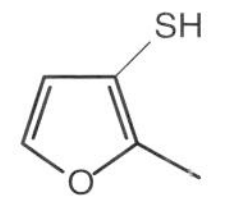

2-Methylfuran-3-thiol
(**22**)
Oxycyclothione®

4-Hydroxy-2,5-dimethyl-3(2*H*)-furanone (**20**) can be synthesized by dimerization of methylglyoxal and cyclization of the diol-dione intermediate [4].

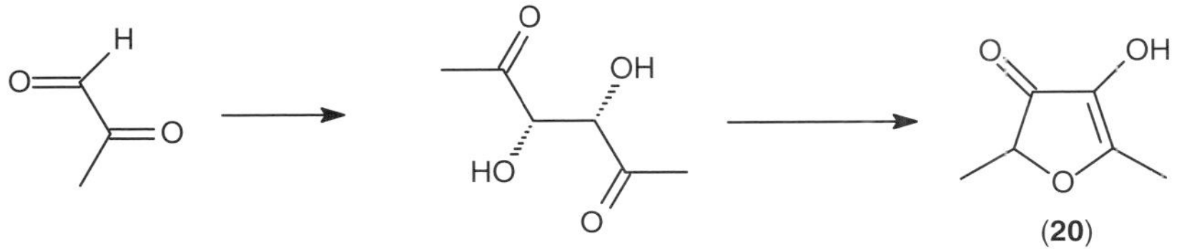

(**20**)

Another important route to 4-hydroxy-2,5-dimethyl-3(2*H*)-furanone (**20**) is shown herein [5]:

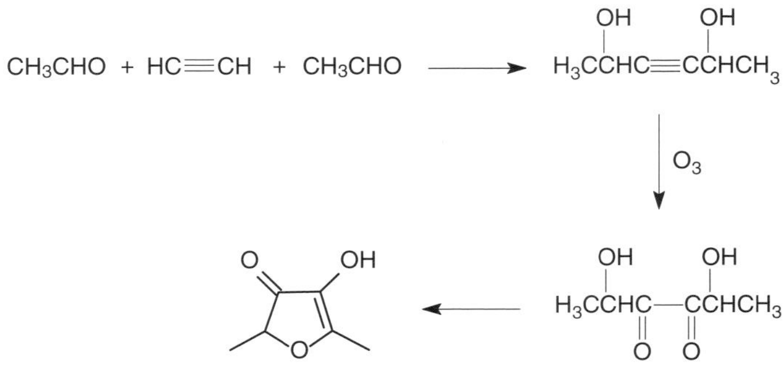

(20)

Furfural (2-furaldehyde (**21**)) is produced from plant residues which are rich in pentoses, e.g. bran, by treatment with dilute sulfuric acid followed by steam distillation [6].

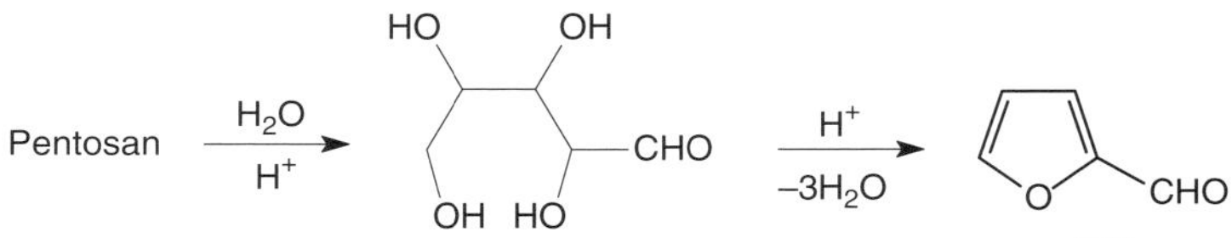

(21)

2-Substituted furans are available also by direct reaction, e.g. 2-acetylfuran, and this is also true for pyrroles and thiophenes.

2-Methyltetrahydrofuran-3-one (dihydro-2-methyl-3(2*H*)-furanone (**23**)) occurs in coffee, roasted hazelnuts, beer, rum, roasted almond and potato chips. It has a bread-like, buttery top-note, and a nutty and astringent flavour with a slight creamy almond nuance. 2-Methyltetrahydrofuran-3-one is applied in roasted flavourings, and for almond, rum, cocoa, brandy and caramel.

2-Pentylfuran (**24**) occurs in many foodstuffs, e.g. fruits, vegetables, meat, roasted products, coffee, cocoa, tea and fish. The material has a green, waxy aroma with musty, cooked caramellic nuances, and is applied in fruit, vegetable, coffee, nut and bread flavourings. Furfuryl mercaptan (coffee mercaptan, 2-furanmethanethiol (**25**)) is found in coffee, chicken, beef, grilled and roasted pork, and has, on dilution, a strong coffee-like, sulphurous, roasted, onion, garlic odour and flavour. Furfuryl mercaptan is applied mostly to coffee, mocca and ice cream flavourings.

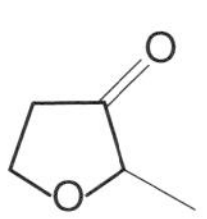

2-Methyltetrahydrofuran-3-one
(**23**)
Coffee Furanone®

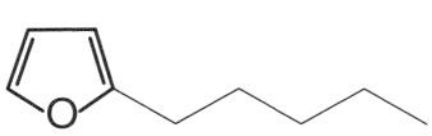

2-Pentylfuran
(**24**)
φ 2.3 × 10^7

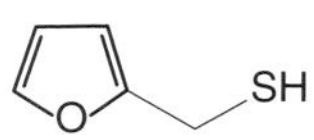

Furfuryl mercaptan
(**25**)
φ 2.6 × 10^9
Coffee mercaptan®

Furfuryl mercaptan (**25**) can be synthesized from furfuryl alcohol, by reaction with thiourea in the presence of hydrogen chloride. The resulting *S*-furfurylisothioronium chloride is cleaved with sodium hydroxide to give furfuryl mercaptan (**25**) [7]:

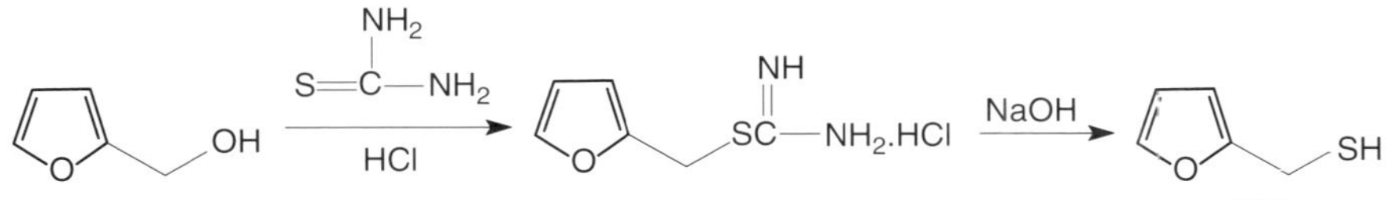

Furfuryl alcohol (**25**)

Some possible synthetic pathways towards 2-methyltetrahydrofuran-3-one (**23**) [8] are shown below:

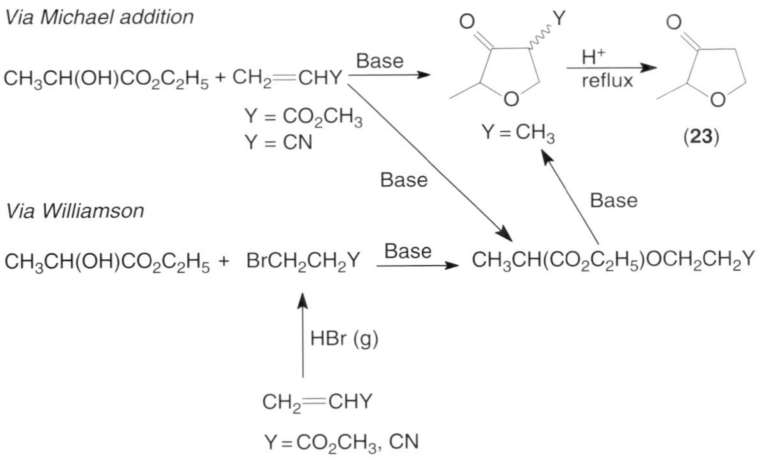

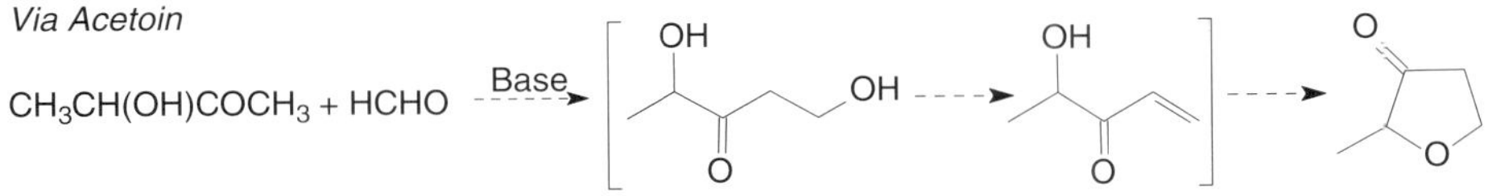

(**23**)

1,4-Cineole (1,4-epoxy-*p*-menthane (**26**)) is a tetrahydrofuran derivative having fresh, cooling, camphoraceous, minty and eucalyptus-like odour, with terpene-like and green nuances.

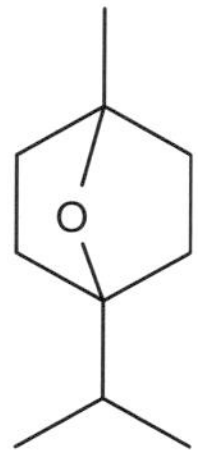

1,4-Cineole
(26)

An interesting heterocyclic oxygen-containing molecule is dodecahydro-3a,6,6,9a-tetramethylnaphtho[2,1-b]furan, known also as Ambroxan®, Cetalox®, L'ambroxine®, Ambermor® (27).

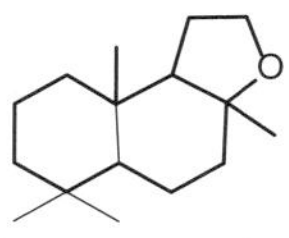

(+) isomer φ 7.88×10^8
(–) isomer φ 9.09×10^7
(27)
Ambroxan®, Cetalox®, L'ambroxine®, Ambermor®

This molecule, having a tetranorlabdanolic system, has a strongly amber odour, enriched by dry and woody nuances. This compound is very useful as a modifier for all floral fragrance formulations. Its synthesis is shown herein [9]:

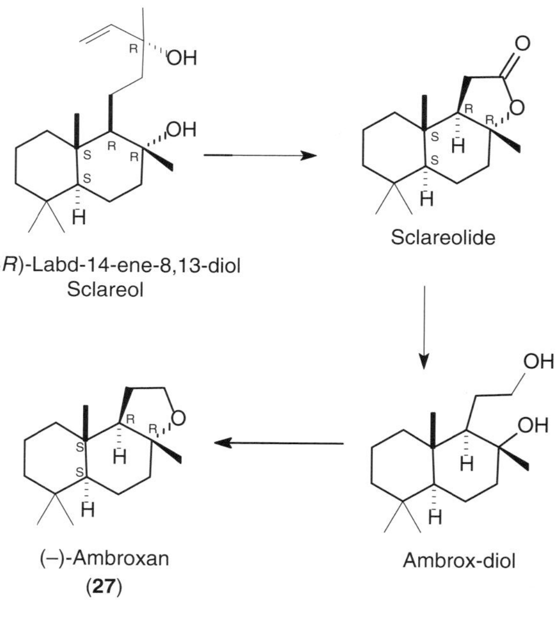

Another tetrahydrofuran worth mentioning is theaspiran (2,6,10,10-tetramethyl-1-oxa-spiro[4,5]decene) (**28**). Theaspirane occurs in blackberry, raspberry, grape, tea, passion fruit and guava and has a cooling minty, camphoraceous flavour, with a herbal and piney nuance.

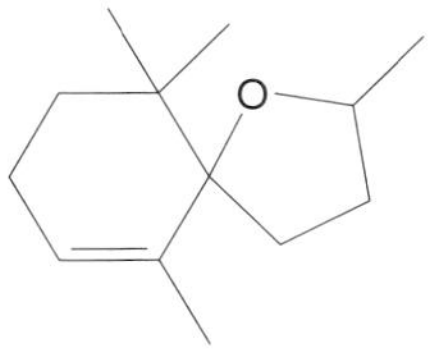

Theaspiran
(**28**)
Spiroxide®

Theaspirane is applied to tea, guava, mint, berry, citrus and grapefruit flavour formulations and in minty, herbal and woody fragrance compounds.

III. Pyrans

Three examples will be shown of this group of oxygen-containing heterocycles. 1,8-Cineole, or Eucalyptol (**29**), is a molecule occurring in many essential oils such as *Eucalyptus globulus*. Eucalyptol has a fresh, strong eucalyptus-like, camphoraceous, minty odour and a cooling taste. It is applied to candies, toothpaste and chewing gum.

Eucalyptol
1,8-cineol
(**29**)

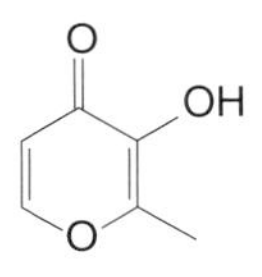

Maltol
(**30**)

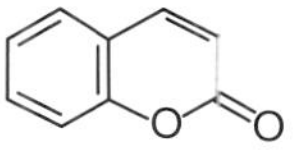

Coumarin
(**31**)

Maltol (**30**), or 3-hydroxy-2-methyl-4-pyrone, can be found in strawberry, bread, dairy products, cocoa, coffee, barley, filbert and peanut. It has a sweet-aromatic, caramellic, cooked fruit-like odour, with fruit and berry notes. Maltol is applied in fruit flavours and roasted food formulations such as strawberry, pineapple, caramel, brown sugar, brandy and butterscotch.

Ethyl maltol, which is not found in nature, is more powerful than Maltol; its synthetic pathway is shown below:

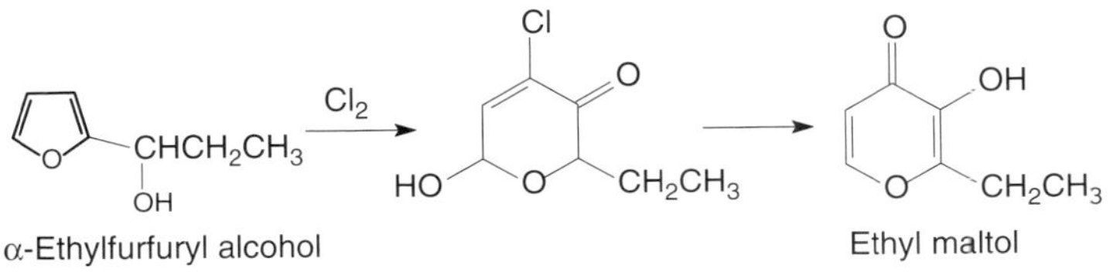

α-Ethylfurfuryl alcohol

Ethyl maltol

This one-pot process starts from α-ethylfurfuryl alcohol, which is treated with chlorine to give 4-chloro-6-hydroxy-2-ethyl-2*H*-pyran-3(6*H*)-one, which need not to be isolated and can be converted to ethyl maltol by aqueous hydrolysis [10].

Coumarin (2*H*-1-benzopyran-2-one) (**31**) occurs in bilberry, raspberry, mint oil, tea, soybean, cloudberry, matsutake and chicory. It has a sweet-aromatic, somewhat coconut-like odour and flavour and it is applied in sweet-aromatic compounds for functional perfumery and in sweet-aromatic, spicy formulations, for candies and cookies. Coumarin is prepared from salicylic aldehyde and acetic anhydride, in presence of sodium acetate [11].

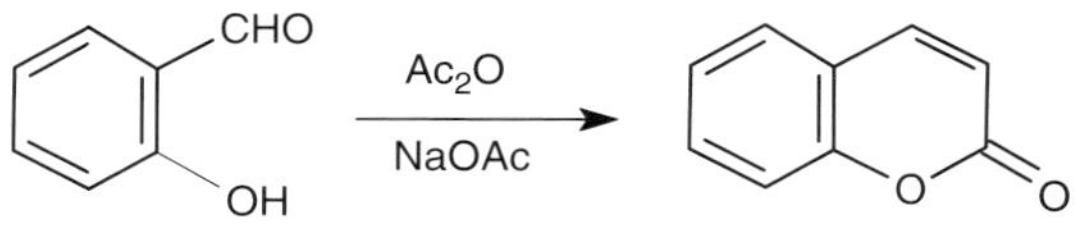

(**31**)

2-Isobutenyl-4-methyltetrahydropyran (rose oxide) (**32**) is found in essential oils such as rose and geranium. It is described as floral, rose-like, fruity, green, vegetative and herbal, and is applied in floral, rose-like fragrances.

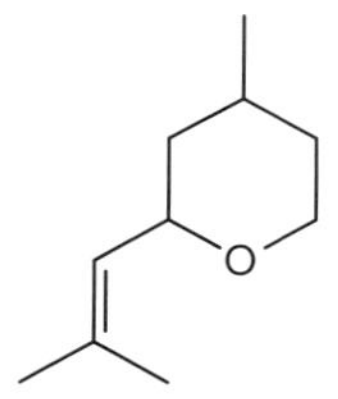

Rose oxide
(**32**)

One of the syntheses of rose oxide is shown herein. This route begins by acetylation of citronellol, followed by allylic bromination of citronellyl acetate. Elimination of hydrobromic acid gives two acetylated dienes. Hydrolysis of these dienes gives the alcoholic dienes, which undergo cyclization to yield *cis* and *trans* rose oxide isomers [12].

The stereoselective synthesis of all four stereoisomers of rose oxide has been achieved starting from optically active citronellol [13]. Stereochemistry at the 4-methyl group was controlled by the chirality of the 3-methyl group in citronellol. For example, (4*R*)-rose oxide is obtained from (*S*)-citronellol and the (4*S*)-enantiomer from (*R*)-citronellol, respectively. Relative stereochemistry at the 2-position is controlled by the choice of catalyst for the cyclization in the last stage. Acid-catalyzed cyclization of the (S)-diene gives (2*S*,4*R*)-rose oxide (**33**) with 95% *cis*-selectivity, whereas Pd-(S)-BINAP gives a 1:1 mixture of (2*R*,4*R*) (**34**) and (2*S*,4*R*) (**33**) isomers. The (2*R*,4*R*)-rose oxide (**34**) is

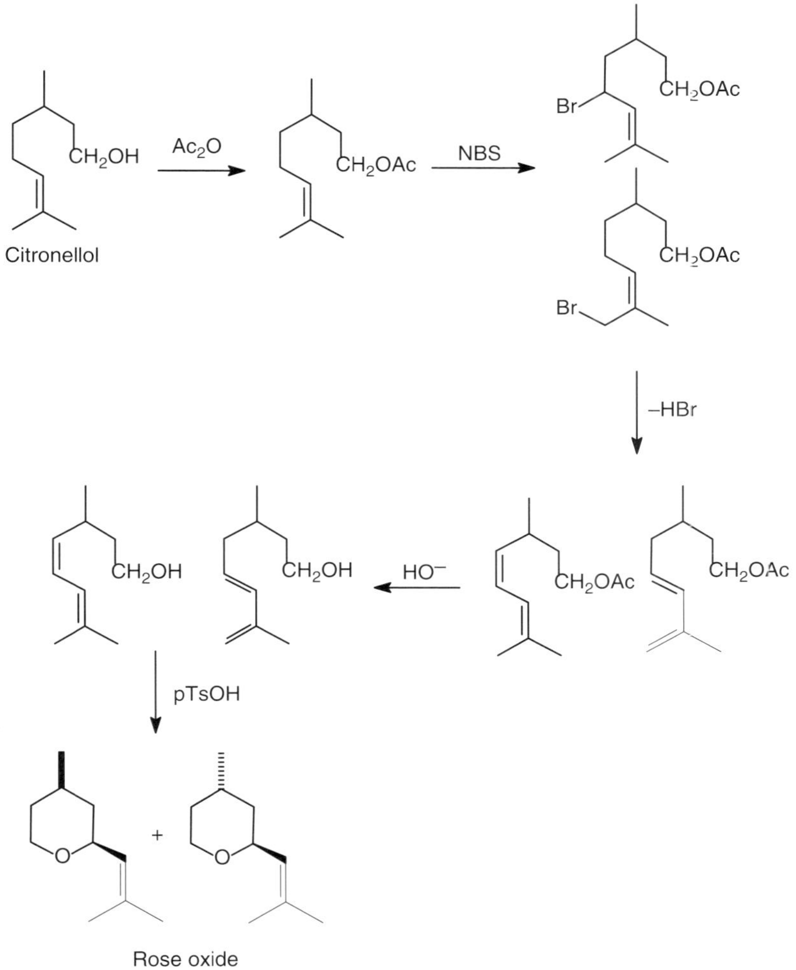

obtained by column chromatographic separation of this mixture. Similarly, (2*R*,4*S*)- (**35**) and (2*S*,4*S*)- (**36**) rose oxides, have been prepared from (*R*)-citronellol

(2*S*,4*R*)-rose oxide (**33**) has the higher φ value of all isomers.

The odour character of (2*S*,4*R*)-rose oxide (**33**) is greatly superior to that of the other isomers in terms of both quality and quantity [14], which can be seen in Table 5.1.

Another important pyran aroma chemical is the polycyclic indane musk 4,6,6,7,8,8-hexamethyl-1,3,4,6,7,8-hexahydrocyclopenta[g]benzopyran (**37**).

This molecule has the powerful and clean musk odour characteristic resembling that of macrocyclic musks. This benzopyran is used in a wide range of perfume compounds, e.g. in soap, perfumery, household products, cosmetics and alcoholics.

The synthesis of 4,6,6,7,8,8-hexamethyl-1,3,4,6,7,8-hexahydrocyclopenta[g]-benzopyran (**37**), starts from 1,1,2,3,3-pentamethylindane, which is prepared by cycloaddition of *tert*-amyl alcohol to α-methylstyrene. The pentamethylindane is hydroxylated with

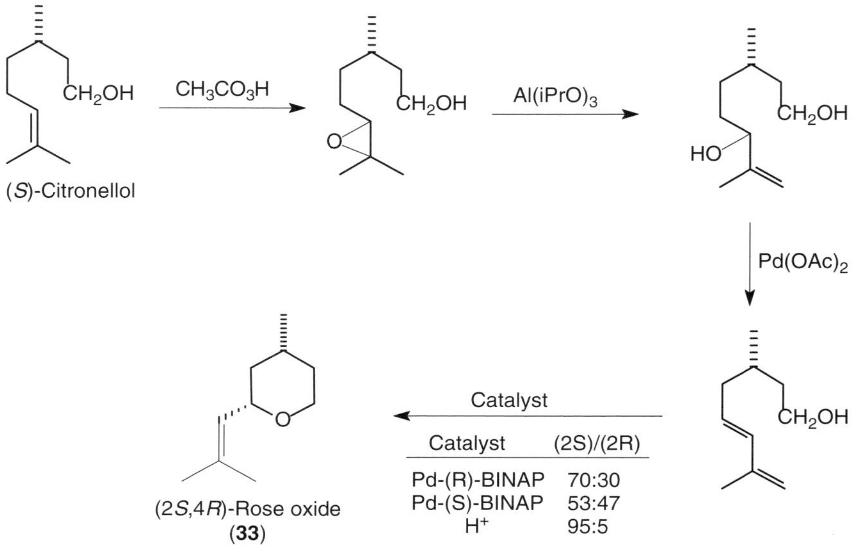

Table 5.1 Odour characteristics of rose oxide diasterioisomers

Rose oxide diasterioisomer	φ Value	Odour description
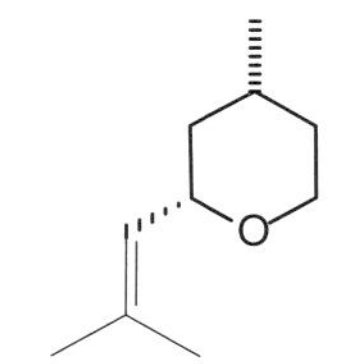 (2*S*,4*R*)-Rose oxide (**33**)	3.08 x 10^7	Floral-green; clean, light, rose, green, diffusive, strong
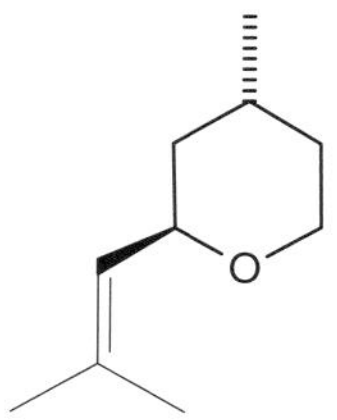 (2*R*,4*R*)-Rose oxide (**34**)	9.64 x 10^5	Floral-green; green herbal (minty), fruity

Table 5.1 (Continued)

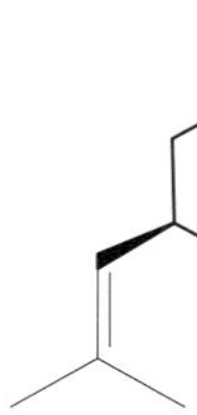 (2*R*,4*S*)-Rose oxide (**35**)	3.08×10^6	Herbal-green-floral; hay green, earthy, heavy
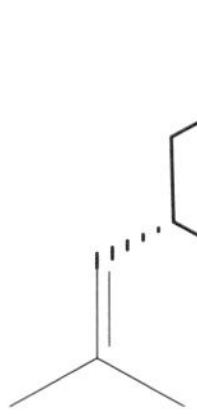 (2*S*,4*S*)-Rose oxide (**36**)	1.93×10^6	Herbal-green-floral; fruity, herbal rose, citrus (bitter peel)

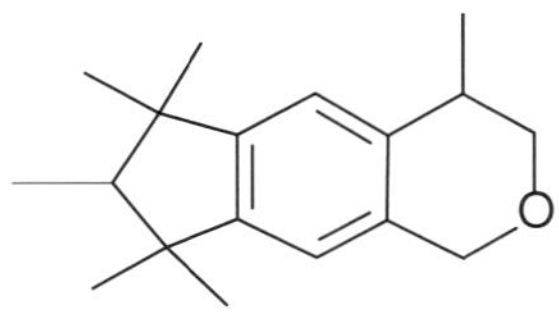

(**37**)

Galaxolide 50®, Abbalide®, Musk 50®

propylene oxide in a Friedel–Crafts reaction, using aluminum chloride as a catalyst. Ring closure of the resulting 1,1,2,3,3-pentamethyl-5-(β-hydroxyisopropyl)indane is accomplished with paraformaldehyde and a lower aliphatic alcohol via the acetal [15], or with paraformaldehyde and a carboxylic acid anhydride via the acylate [16]:

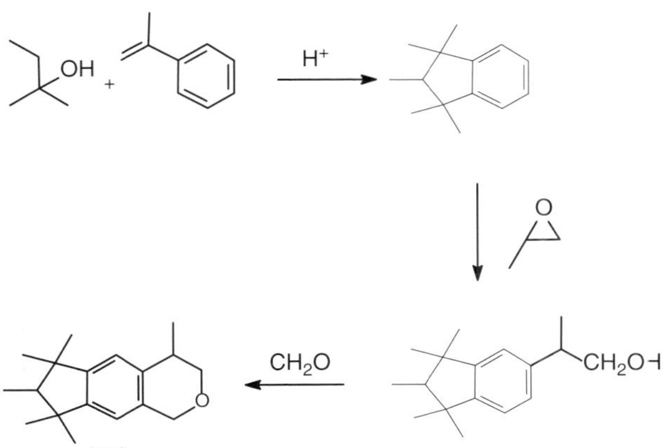

(**37**)

Tetrahydro-3-pentyl-2*H*-pyran-4-yl acetate (**38**) has an oily, plant-like, herbaceous lavender, slightly mushroom odour, with floral tea-like jasmine note.

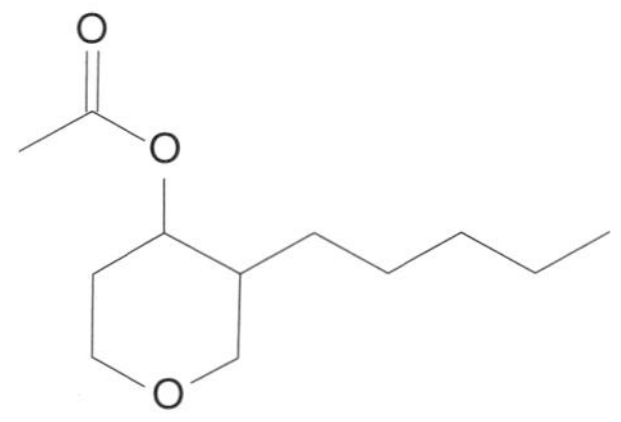

(**38**)
Jasmine acetate®

This material is prepared from octene, formaldehyde and acetic acid in the Prins reaction, and is obtained among other components. Jasmine acetate is an economic substitute for jasmonates in floral compounds.

IV. Oxepines

An important seven-membered ring is 7-methyl-3,4-dihydro-2*H*-benzodioxepin-3-one (**39**), a fragrance ingredient having a fresh 'oceanic', floral odour, which is also diffusive with watermelon notes. This molecule is used in fragrances as an extremely valuable component for marine notes in compounds for alcoholic and cosmetic perfumery.

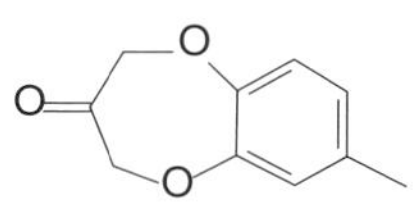

7-Methyl-3,4-dihydro-2*H*-benzodioxepin-3-one
(**39**)
Calone®, Oxalone®, Ganone®, Aquamor®

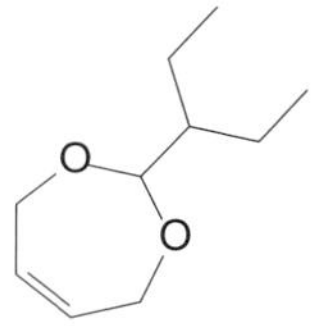

4,7-Dihydro-2-isopentyl-2-methyl-1,3-dioxepin
(**40**)
Karotin®

4,7-Dihydro-2-isopentyl-2-methyl-1,3-dioxepin (**40**) has a citrus, herbal-floral odour, with notes of lemon, neroli and jasmine. It is applied in a range of fragrances for cosmetic and functional perfumery for its herbal-floral note.

7-Methyl-3,4-dihydro-2*H*-benzodioxepin-3-one (**39**) is synthesized from 4-methylcatechol as following [17]:

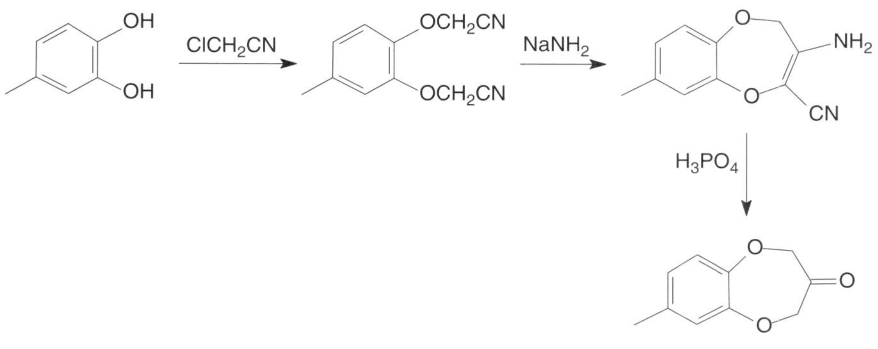

(**39**)

Condensation of 4-methylcatechol with two moles of chloroacetonitrile gives the double cyanomethyl ether, which cyclizes using sodium amide to obtain the α-aminocyano unsaturated dioxepine ring. Hydrolysis of this disubstituted ring in presence of phosphoric acid yields the desired 7-methyl-3,4-dihydro-2*H*-benzodioxepin-3-one (**39**).

5.4 Heterocyclic compounds containing nitrogen and/or sulfur

The nitrogen and sulfur heterocyclic aroma chemicals can be divided into the following main groups:

I. Non-aromatic molecules;
II. Aromatic molecules.

I. Non-aromatic molecules

2-Acetyl-3,4-dihydro-5*H*-pyrrole (**41**) has the characteristic odour of white bread crust, but a short half-life restricts its use. 2-Isobutyl-4,6-dimethyldihydro-1,3,5-dithiazine (**43**) (R = iPr) occurs in yeast extract, having a fruity, tropical fruit note, green roasted, cocoa odour and flavour. This and related molecules are applied in fruity and roasted formulations.

2-Acetyl-3,4-dihydro-5*H*-pyrrole (**41**)

1-Pyrroline (**42**)

1-Pyrroline (3,4-dihydro-[2*H*]-pyrroline (**42**)) occurs in sea fruit, clam and squid. It has a fishy flavour and is applied in seafood formulations.

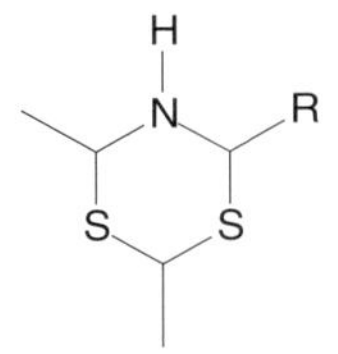

R = Me, iPr, iBu, 2-Bu
(**43**)

Mixtures containing dithiazines are formed when aldehydes are treated with a mixture of ammonia and hydrogen sulfide. A more directed synthesis of 2,4,6-trimethyldihydro-1,3,5-dithiazine is shown below:

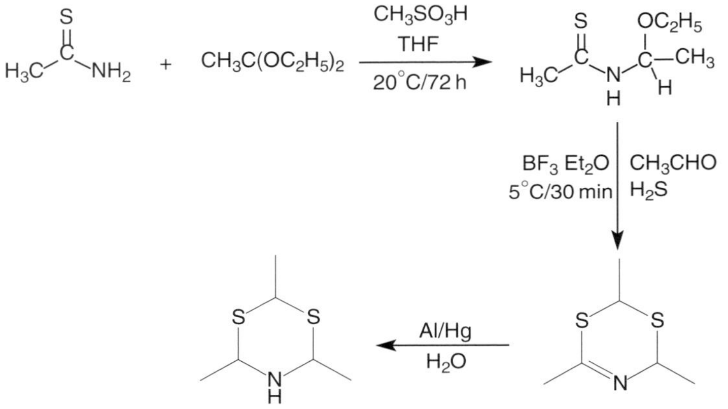

($R = CH_3$)
(**43**)

Thioacetamide is reacted with acetaldehyde diethylacetal in presence of methane-sulphonic acid to yield *N*-(1-ethoxyethyl)thioamide as an intermediate. The thioamide cyclizes with acetaldehyde and hydrogen sulfide, in presence of boron trifluoride etherate as catalyst, to give 2,4,6-trimethyl-4*H*-1,3,5-dithiazine. Reducing the double bond with aluminum amalgam yields 2,4,6-trimethyldihydro-1,3,5-dithiazine (**43**) ($R = CH_3$).

II. Aromatic molecules

Aromatic molecules containing nitrogen and/or sulfur belong to the following groups:

(i) Thiophene derivatives;
(ii) Pyrrole and indole derivatives;
(iii) Pyridine and quinoline derivatives;
(iv) Pyrazine and quinoxaline derivatives;
(v) Thiazole derivatives.

(i) Thiophene derivatives

3-Acetyl-2,5-dimethylthiophene (**44**) occurs in boiled and in cooked beef. It has a burnt, roast flavour and it is applied to roasted and smoked flavourings, e.g. boiled and fried meat.

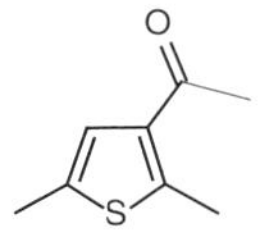

3-Acetyl-2,5-dimethylthiophene
(**44**)

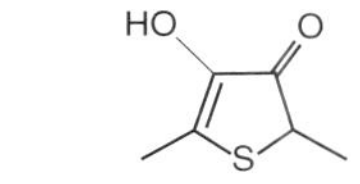

4-Hydroxy-2,5-dimethyl-2,3-dihydrothiophen-3-one
(**45**)

4-Hydroxy-2,5-dimethyl-2,3-dihydrothiophen-3-one (**45**) is not yet reported as being found in flavours, but is probably present in prepared meat. It has a sulphurous aroma, and on dilution, a fried and prepared meat-like flavour, and is used in meat and vegetable formulations.

3-Acetyl-2,5-dimethylthiophene (**44**) can be synthesized by cyclization of 1,4-dicarbonyl compounds analogues in the presence of phosphorous pentasulfide or H_2S [18], followed by acetylation of the 2,5-dimethylthiophene intermediate:

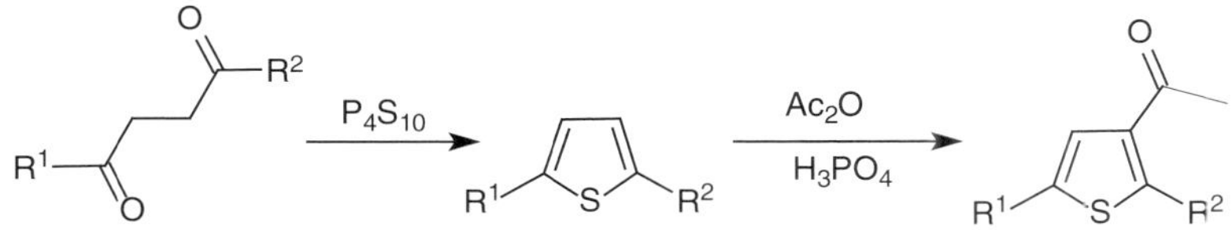

(**44**)

$R^1 = R^2 = CH_3$

(ii) Pyrrole and indole derivatives

2-Acetylpyrrole (**46**) occurs in vegetables, bread, meat, alcoholic and roasted products. It has a sweet musty, nutty and tea-like odour and flavour. 2-Acetylpyrrole is mainly applied in baked, fried and roasted flavourings. Indole (2,3-benzopyrrole) can be found in bread, cheese, egg, beef, wine, beer, brandy, cocoa, coffee, tea, etc. It has an erogenic-floral, animal, slightly musk, and slightly faecal odour on dilution. Indole is applied as a trace component in erogenic-floral fragrance compounds.

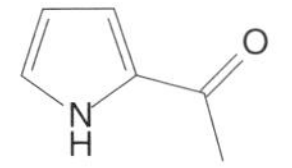

2-Acetylpyrrole
(**46**)

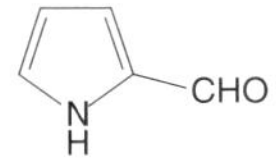

Pyrrole 2-carboxaldehyde
(**47**)

Pyrrole 2-carboxaldehyde (**47**) can be found asparagus, onion (roasted), wheaten bread, beef, beer, white wine, cocoa, coffee, tea and barley. This material is applied in roasted flavourings, like nuts and meat. It can be prepared from pyrrole via the [2 + 2] cycloaddition with dichlorocarbene, which is in competition with the Reimer–Tiemann formylation [19]:

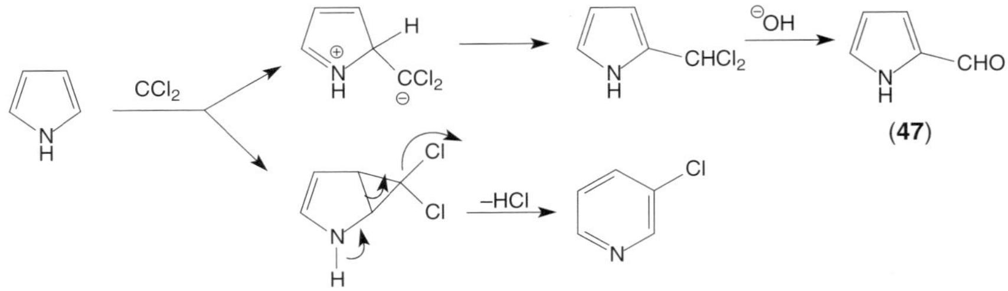

Under strongly basic conditions, i.e. when generating dichlorocarbene from chloroform and potassium hydroxide, electrophilic substitution of pyrrole by dichlorocarbene dominates, leading eventually to pyrrole-2-carboxaldehyde (**47**). In a weakly basic medium (e.g. when generating dichlorocarbene by heating sodium trichloroacetate), the [2 + 1] cycloaddition prevails. The primary product eliminates hydrogen chloride to give 3-chloropyridine.

Indole (1-*H*-benzo[b]pyrrole (**48**)) occurs in bread, cheese, egg, beef, wine, beer, brandy, cocoa, coffee, tea and cauliflower. Indole has erogenic-floral, animalic, cheese odour and flavour, which becomes slightly faecal on dilution. Indole is applied in some cheese modifications and in erogenic-floral fragrances. Its homologue skatole (3-methylindole (**49**)) can be detected in jasmine, in dairy products (cheese), coffee, tea and dried bonito fish. It has a putrid, sickening, mothballs, decayed, faecal odour; on extreme dilution it has a jasmine-like odour. It is a degradation product of tryptophan, and as the name implies, it is found in faecal material. Skatole is applied in very low concentration in floral fragrances.

Indole (**48**) φ 1.17×10^9

Skatole (**49**) φ 1.31×10^8

1-Dimethylamino-2-(*o*-nitrophenyl)ethenes, obtained from *o*-nitrotoluene and *N,N*-dimethylformamide dimethyl acetal, yield indoles on reductive cyclization of the corresponding *o*-aminophenyl derivative [20]:

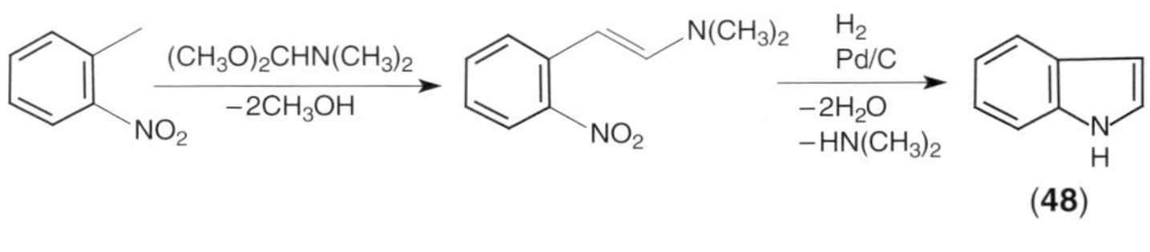

(**48**)

(iii) Pyridine and quinoline derivatives

2,6-Dimethylpyridine (2,6-lutidine (**50**)) occurs in bread, cheese, beef, lamb, sherry, coffee and tea. It has a nutty, coffee, cocoa, musty, bread and meaty flavour and is used in roasted, vegetable and fried flavours.

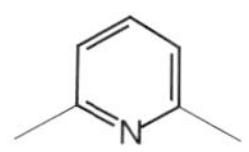

2,6-Dimethylpyridine
(**50**)

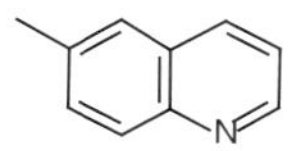

6-Meythylquinoline
(**51**)

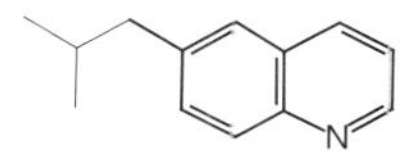

6-Isobutylquinoline
(**52**)

1,5-Dicarbonyl compounds, on treatment with ammonia, undergo cyclocondensation to give 1,4-dihydropyridines, which on dehydrogenation yield pyridines [21]:

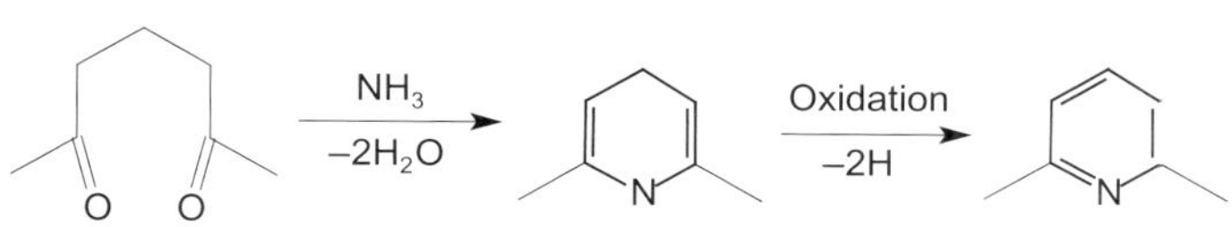

(**50**)

Quinolines, or benzopyridines, are a small group of aroma ingredients used mainly in flavours. 6-Methylquinoline (**51**) is found in tea and it has a narcotic, earthy, green odour. 6-Methylquinoline is applied into tea formulations. 6-Isobutylquinoline (**52**) has a narcotic, slightly roasted-burnt, somewhat green root-like odour and flavour, and it is used in roasted formulations.

(iv) Pyrazine and quinoxaline derivatives

This group of aroma ingredients contains a 1,4-diazine group in its structure.

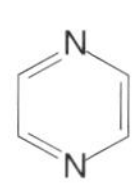

Pyrazine (1,4-diazine)

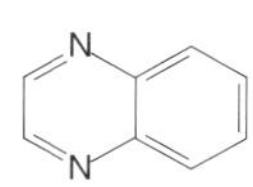

Quinoxaline (Benzopyrazine)

Pyrazines are well-known aroma substances with unique flavour properties and, usually, low odour thresholds, many of which occur in nature. Some are found in plants, e.g. 2-methoxy 3-(1-methyl-propyl)-pyrazine smells like green peas and 2-methoxy-3-(2-methylbutyl)-pyrazine is the key odour compound of bell pepper. Other pyrazines are formed in thermal reactions such as the Maillard reaction [22, 23] e.g. the alkyl substituted pyrazines which are found in roasted beef and coffee. Because of these findings, pyrazines are of high relevance for the flavour industry, especially those with acetyl-, alkyl-, alkoxy- and sulfur-containing groups. We may divide them into four main sub-groups – alkylpyrazines, acetylpyrazines, alkoxypyrazines and thioalkoxypyrazines. These are defined in Tables 5.2–5.5.

Table 5.2 Alkylpyrazines

Compound	Odour and flavour	Occurrence [2]	Structure φ Value
2-Methylpyrazine	Musty, nutty, roasted, cocoa, light, peanut	Occurs widely in nature in baked, fried and roasted products	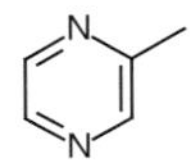 **(53)** φ 9.41×10^2
2,3-Dimethylpyrazine	Roasted meat, in higher dilution-vanilla and chocolate like	In many flavours, vegetables, dairy, meat, roasted products, cocoa, coffee, potatochips, shrimp, papaya, asparagus, cabbage and fried potato	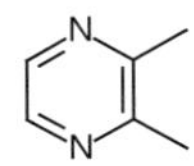 **(54)** φ 3.09×10^3
2,5-Dimethylpyrazine	Earthy, nut-like, raw potato, musty	In many food flavours, vegetable, dairy, meat and roasted products	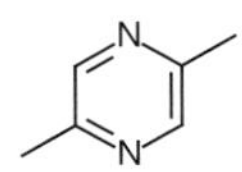 **(55)** φ 6.0×10^4
2,6-Dimethylpyrazine	Roasted, cocoa-like	In many food flavours, vegetable, dairy, meat and roasted products	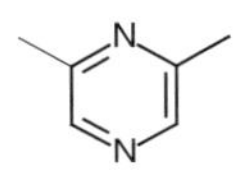 **(56)** φ 7.20×10^4
2,3,5-Trimethylpyrazine	Burnt roasted, earthy, tobacco-like	Nuts, meat, roasted products (coffee, cocoa), rum, whisky and popcorn	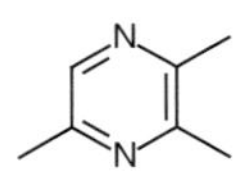 **(57)** φ 3.05×10^5
2,3,5,6-Tetramethylpyrazine	Caramellic-milky, fermented soya beans, slightly woody	In guava, boiled egg, roasted beef, cocoa, coffee, green tea, whisky, shrimp	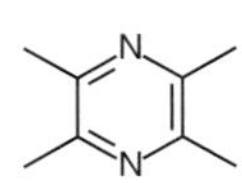 **(58)**
2-Ethylpyrazine	Musty, nutty, burnt	In many flavours, bread, beef, brandy, rum, whisky, coffee, cocoa, etc.	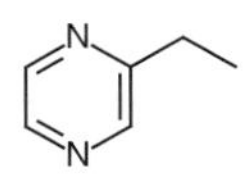 **(59)** φ1.8×10^4

Table 5.2 (Continued)

2-Ethyl-5-methyl pyrazine	Roasted, coffee-like	In many flavours, vegetables, bread, meat, whisky, coffee, cocoa	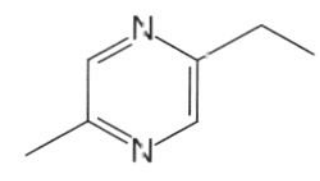 **(60)** φ 7.33×10^5
2-Ethyl-6-methylpyrazine	Roasted, cocoa and coffee-connotations	In many food flavours, fish, meat, roasted products, in alcoholic beverages	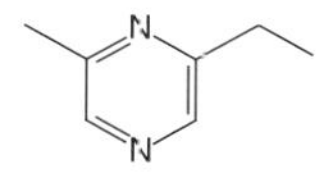 **(61)** φ 1.22×10^6
2-Isopropylpyrazine	Dusty, roasted nuts	In fried chicken, cocoa, filbert, peanut	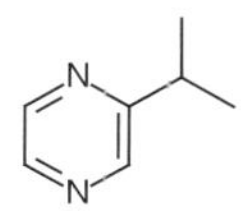 **(62)**
2-*tert*-Butylpyrazine	Green, earthy, carrot	Not yet reported as being found in nature; probably exists in roasted products	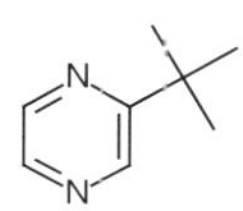 **(63)**

A newly available alkylpyrazine is the isomeric mixture of 3,5-dimethyl-2-isobutylpyrazine and 3,6-dimethyl-2-isobutylpyrazine (**64**). These isomers were isolated from several natural sources, e.g. skin and flesh of potato cultivars after baking [24]. The odour threshold of 3,5-dimethyl-2-isobutylpyrazine and 3,6-dimethyl-2-isobutylpyrazine indicates a relatively low φ value.

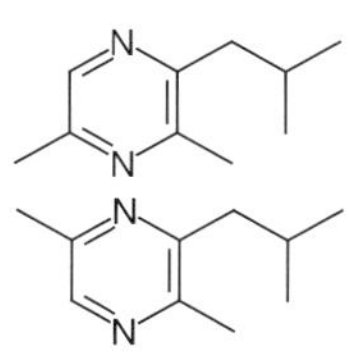

(64)
φ 8.21×10^4

The odour and flavour of (**64**) has notes of cocoa and hazelnut and it is slightly musky animalic, with patchouli and vetiver tones and a menthol note. Food applications include chocolate, cocoa, baked goods, breakfast cereals, milk products, soups, sweet sauces and mint flavours. It can also be used in fragrance compositions to impart a strong and natural, original and long-lasting effect, and to boost woody, chypre, oriental and fougere notes. The organoleptic profile of 3,5-dimethyl-2-isobutylpyrazine and 3,6-dimethyl-2-isobutylpyrazine mix is shown by the radar description in Figure 5.1.

Table 5.3 Acetylpyrazines

Compound	Odour and flavour	Occurrence [2]	Structure φ Value
2-Acetylpyrazine	Popcorn, bread-crust like, nutty	In roasted flavours, bread, popcorn, cocoa, coffee, barley, peanut, filbert	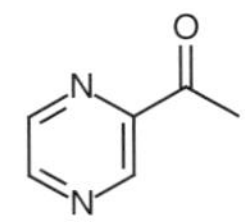 (**65**) φ 1.97×10^6
2-Acetyl-3-methylpyrazine	Roasted potatoes, nutty, vegetable and cereal	In meat, cocoa, coffee, fried potato	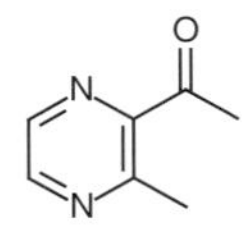 (**66**)
2-Acetyl-6-methylpyrazine	Aromatic and roasted, coffee- and cocoa-like	In coffee, roasted sesame seeds	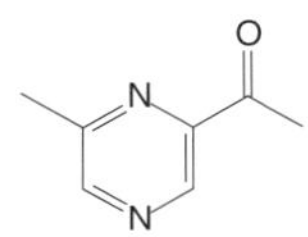 (**67**)
2-Acetyl-3,6-dimethylpyrazine	Roasted, caramellic, hazelnut, popcorn	In coffee	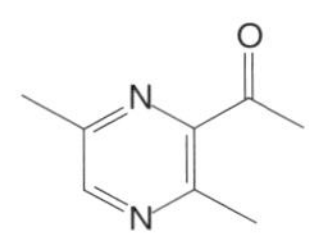 (**68**)
5-Acetyl-2,3-dimethylpyrazine	Roasted, slightly aromatic green	In coffee	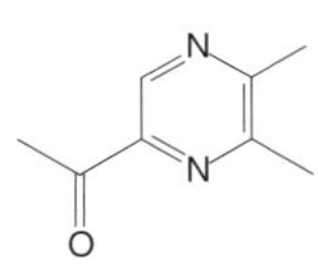 (**69**)

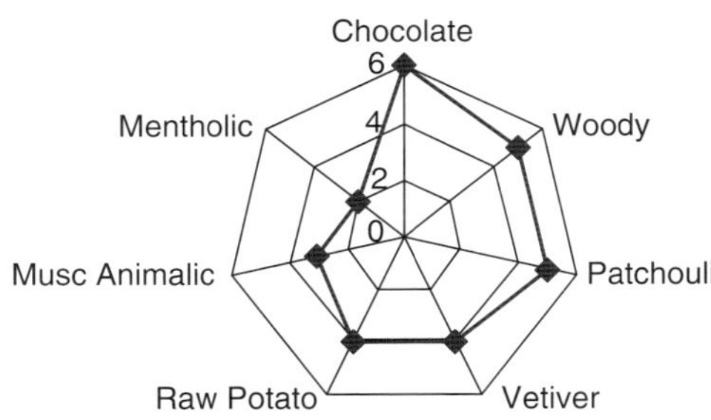

Figure 5.1 Radar organoleptic description of 3,5-dimethyl-2-isobutylpyrazine and 3,6-dimethyl-2-isobutylpyrazine mix [1].

Table 5.4 Alkoxypyrazines

Compound	Odour and flavour	Occurrence [2]	Structure φ Value
2-Methoxy-3-methylpyrazine	Roasted, peanut, somewhat green	In sherry, coffee, krill	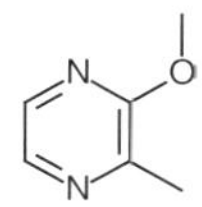 **(70)** φ 2.48×10^7
2-Methoxy-3-ethylpyrazine	Green, earthy, nutty	In sherry	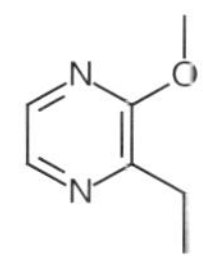 **(71)**
2-Methoxy-3-isopropylpyrazine	Bell pepper, earthy, green	In asparagus, carrot, cucumber, lettuce, garlic, peas, potato, capsicum, etc.	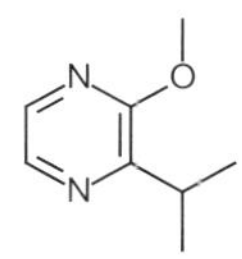 **(72)**
2-Methoxy-3-*n*-propylpyrazine	Vegetable, earthy, green	In sherry	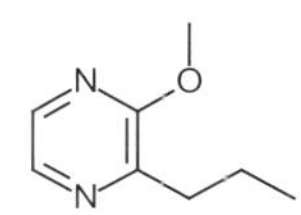 **(73)**
2-Methoxy-3-isobutyl-pyrazine	Green bell pepper, green peas	In green bell pepper, petitgrain oil	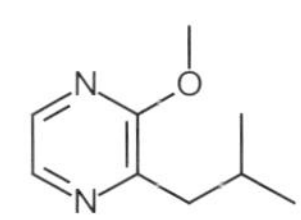 **(74)**
2-Methoxy-3-*sec*-butylpyrazine	Pea, green, earthy, potato, galbanum	In coffee and green peas	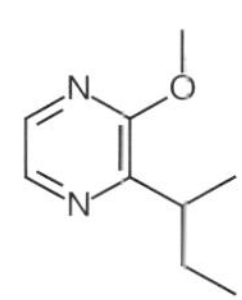 **(75)**
2-Methoxy-3-isopentylpyrazine	Green, leafy, lavender at low levels	Not yet reported as being found in nature	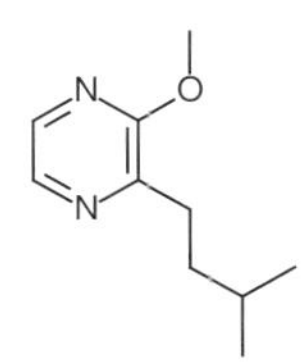 **(76)**

Table 5.4 (Continued)

Compound	Odour and flavour	Occurrence [2]	Structure φ Value
2-Methoxy-3-isohexylpyrazine	Powerful, sweet greenish, with asparagus note	Not yet reported as being found in nature	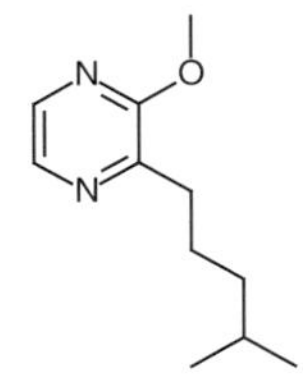 (**77**)
2-Ethoxy-3-methylpyrazine	Roasted, baked, fried, somewhat nut-like	Not yet reported as being found in nature (probably in coffee, krill)	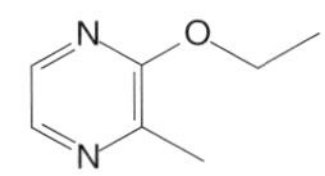 (**78**)
2-*n*-Propoxy-6-methylpyrazine	Green, woody, mint at low levels	Not yet reported as being found in nature. Probably in roasted products, e.g. coffee	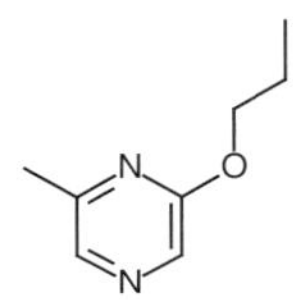 (**79**)
2-Isopropoxy-3-methylpyrazine	Dusty, green, somewhat paprika-like	Not yet reported as being found in nature	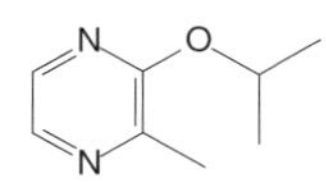 (**80**)

A convenient method for pyrazine synthesis is cyclocondensation between 1,2-dicarbonyl compounds and 1,2-diaminoethanes, with double imine formation, to afford 2,3-dihydropyrazines, which are conveniently oxidized to pyrazines by manganese dioxide or copper oxide in ethanolic potassium hydroxide [25]:

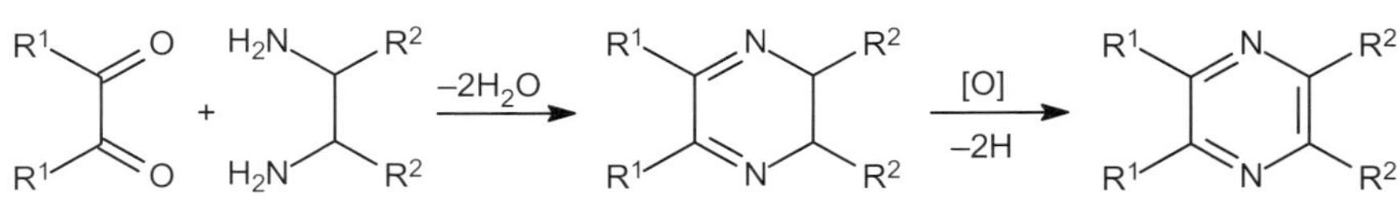

This method yields the best results with symmetrical starting compounds.

The classical pyrazine synthesis involves the self-condensation of two molecules of an α-aminocarbonyl compound furnishing 3,6-dihydropyrazine followed by oxidation, usually under mild conditions to pyrazine [26].

Table 5.5 Thioalkoxypyrazines

Compound	Odour and flavour	Occurrence [2]	Structure
2-Methylthio-3-methylpyrazine	Nutty, sweet, meaty, slightly green	Not yet reported as being found in nature	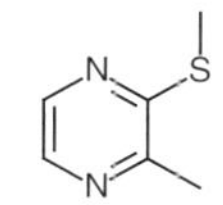 (**81**)
2-Methylthio-3-isopropylpyrazine	Nutty, burnt, vegetable	Not yet reported as being found in nature	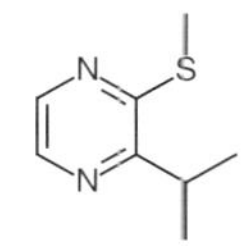 (**82**)

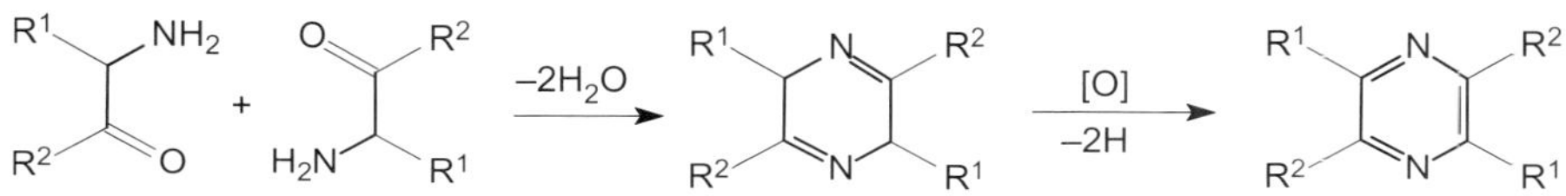

This route is restricted by the commercial non-availability of the aminocarbonyl precursor.

An alternative synthesis of the dihydropyrazine utilizes a cyclizing aza-Wittig reaction of two molecules of α-phosphazinyl ketones which are accessible from α-azido ketones and triphenylphosphane [26].

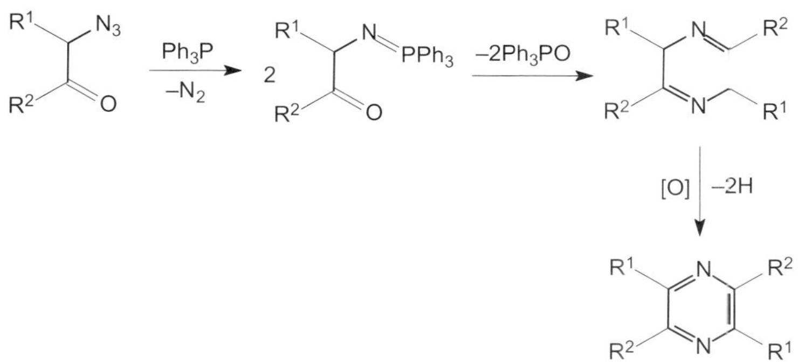

Quinoxalines are benzopyrazines, and several derivatives are also used in the flavours and fragrances industry.

2-Methylquinoxaline (2-methylbenzopyrazine) (**83**) occurs in pork, cocoa, coffee, passion fruit juice. It has a musty, roasted flavour and is used in roasted and meat flavourings.

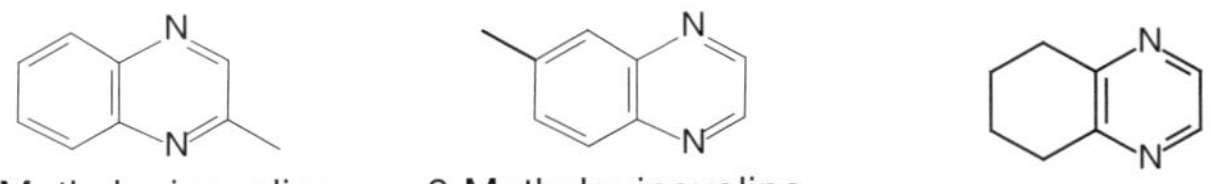

2-Methylquinoxaline (**83**) 6-Methylquinoxaline (**84**) 5,6,7,8-Tetrahydro-quinoxaline (**85**)

6-Methylquinoxaline (**84**) occurs in pork liver, has a roasted and toasted flavour and is used in roasted flavourings, such as coffee and meat preparations.

5,6,7,8-Tetrahydroquinoxaline (cyclohexapyrazine) (**85**) occurs in bread, heated beef and pork, cocoa, coffee and in roasted peanut. It is applied in fried and in roasted flavourings.

5,6,7,8-Tetrahydroquinoxaline can be prepared using either 1,2-diaminocyclohexane or 1,2-cyclohexanedione as the starting material. 1,2-Diaminocyclohexane is condensed with glyoxal to give the dihydro intermediate, which undergoes dehydrogenation to give the quinoxaline. The second alternative is a double condensation of 1,2-diaminoethane with 1,2-cyclohexanedione followed by dehydrogenation–aromatization step.

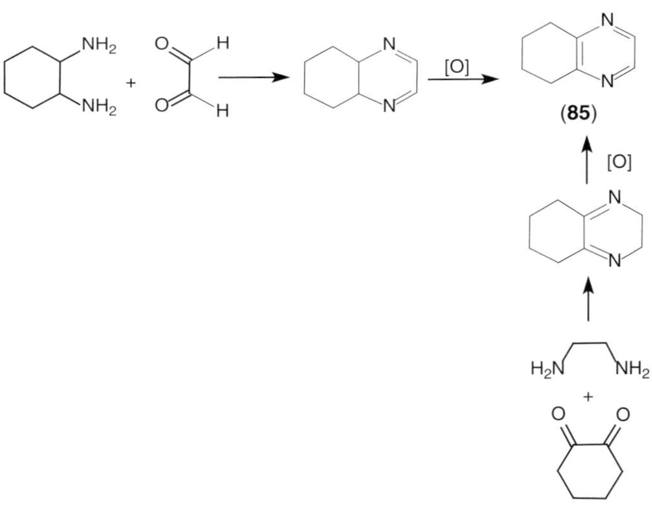

Routes for preparation of 5,6,7,8-tetrahydroquinoxaline (**85**)

(v) Thiazole derivatives

Thiazoles are sulfur- and nitrogen-containing five-membered aromatic rings. The most important types are alkylthiazoles, acetylthiazoles and hydroxyethylthiazoles.

(a) Alkylthiazoles

2,4,5-Trimethylthiazole (**86**) occurs in meat, potato, coffee, cooked beef, heated lamb, and in cooked chicken. It has a chocolaty, nutty, coffee-like aroma, with a meaty flavour and is applied in roasted flavours, beverages, sauces and ice cream.

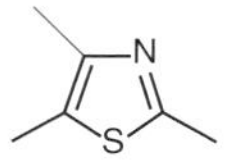

2,4,5-Trimethylthiazole
(**86**)

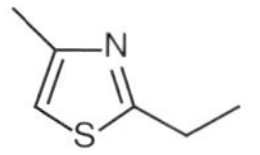

2-Ethyl-4-methylthiazole
(**87**)

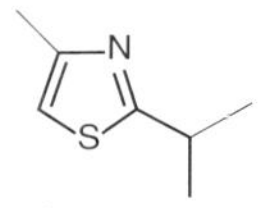

2-Isopropyl-4-methylthiazole
(**88**)

2-Ethyl-4-methylthiazole (**87**) occurs in coffee. It has a coffee, cocoa, meaty, rubbery odour and flavour and it is applied in baked goods, candy, coffee, cocoa, nut, savoury and in meat flavourings. 2-Isopropyl-4-methylthiazole (**88**) has a green, vegetable character with a tropical fruity nuance odour and flavour and it is widely used in fruit flavourings, especially peach.

Another example of thiazole of this type is 4-vinyl-5-methylthiazole (**89**) [27]. This molecule occurs in coffee, filbert and in yellow passion fruit. It has a nutty, musty, earthy, cocoa-powder flavour and is applied to nut, cocoa and chocolate, coffee, meat and potato flavourings.

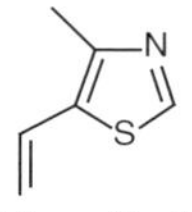

4-Vinyl-5-methylthiazole
(**89**)

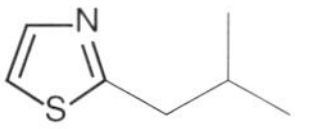

2-Isobutylthiazole
(**90**)
φ 4.7 × 10^7

2-Isobutylthiazole (**90**) occurs in tomato leaves. It has a powerful green, tomato, vine-like aroma, and it is used in tomato soup vegetable flavourings.

(b) Acetylthiazoles

2-Acetylthiazole (**91**) occurs in asparagus, potato, beef, pork, beer, whisky, coffee, beans and white bread. It has a nutty, cereal-note, hazelnut, cocoa and popcorn-like flavour. 2-Acetylthiazole is used in roasted flavourings and in nut and meat flavours. 2,4-Dimethyl-5-acetylthiazole (**92**) has not yet been reported as being found in nature, but it may be possibly found in meat. It has a meaty, roasted, nutty, boiled beef odour and flavour; savoury with woody and coffee nuances flavour.

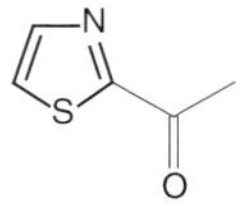

2-Acetylthiazole
(**91**)

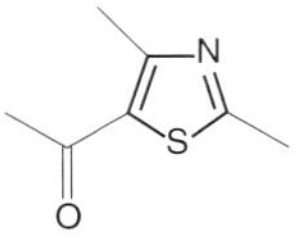

2,4-Dimethyl-5-acetylthiazole
(**92**)

Acetylthiazoles cannot be made by direct substitution of a thiazole. 2-Acetylthiazole can be prepared from the readily available 2-aminothiazole via a diazonium chemistry, using the following route:

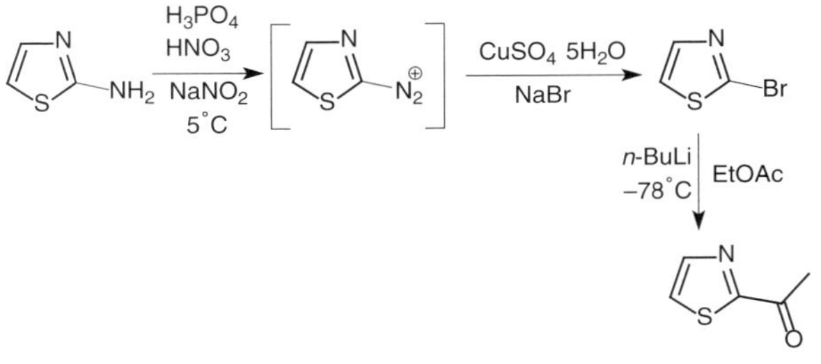

2-Acetylthiazole
(**91**)

(c) Hydroxyethylthiazoles

Hydroxyethylthiazoles are derived from Thiamine (Vitamin B1) (**93**). Thiamine occurs in yeast, in rice husk and in other cereals.

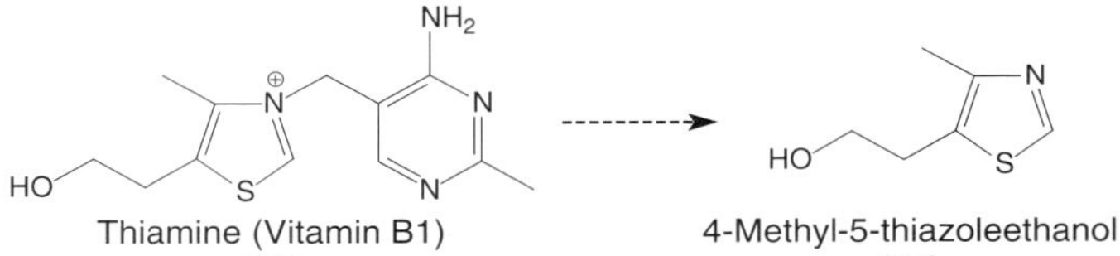

Thiamine (Vitamin B1)
(**93**)

4-Methyl-5-thiazoleethanol
(**94**)

The first and the most important derivative is 4-methyl-5-thiazoleethanol (**94**) and the second is its ester, namely, 4-methyl-5-thiazoleethyl acetate.

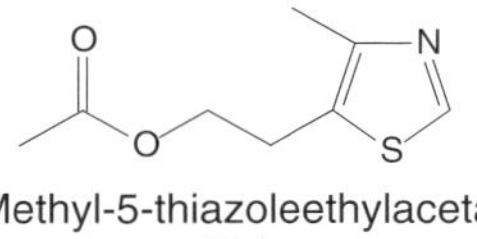

4-Methyl-5-thiazoleethylacetate
(**95**)

4-Methyl-5-thiazoleethanol (Sulfurol) (**94**) occurs in cocoa, cooked beef, roasted peanut, beer, cognac and liver. It has a beefy, somewhat nutty, meaty, broth, roasted, metallic flavour and is used in meat, savoury and dairy flavours. It has a relative low φ value of 1.43×10^4. Sulfurol acetate (**95**) has not yet been reported as being found in nature. It has a meaty, bready, beefy, bloody and chicken note flavour, and is applied in meat, nut and dairy, cocoa and bread flavourings. 4-Methyl-5-thiazoleethyl acetate is used also in fragrances to impart woody, sandal-like odour.

Thiazoles are generally prepared by the cyclocondensation of α-halocarbonyl compounds with thioamides, known as Hantzsch synthesis, as shown in the following scheme:

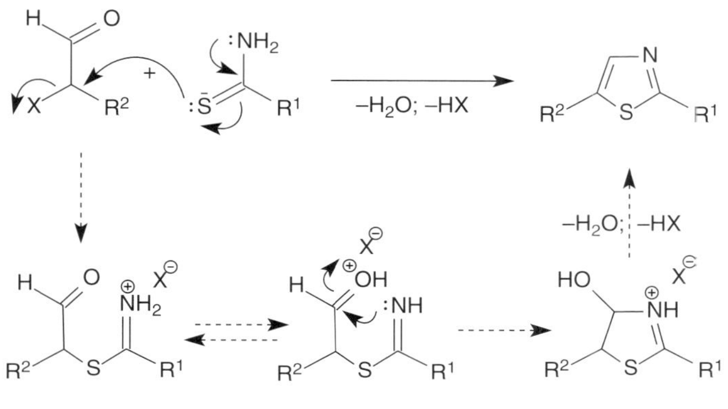

The Hantzsch synthesis mechanism

4-Methyl-5-thiazoleethanol (**94**) can be prepared from α-aceto-γ-butyrolactone, by chlorination followed by hydrolysis and decarboxylation. The obtained oxo-chloropentanol is cyclized with thioformamide in the Hantzsch synthesis, as shown in the following scheme [28].

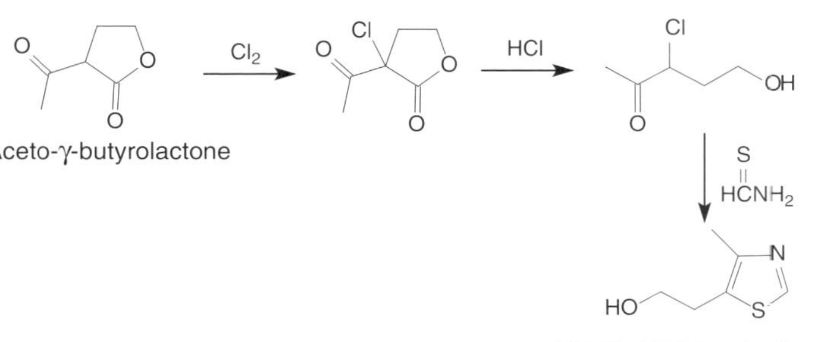

α-Aceto-γ-butyrolactone

4-Methyl-5-thiazoleethanol
(**94**)

References

1. Zviely, M., Abushkara, E. and Hodrien, D. (2003) Chocarom Pyrazine, A Pyrazine for Flavours and Fragrances. *Perfumer & Flavorist*, **28**, 36.
2. The organoleptic description of this material and others is taken mostly from FRM 2001 Database of Flavour (Raw) Materials, and PMP 1996 Database of Fragrance (Raw) Materials Boelens Aroma Chemical Information Service, The Netherlands.
3. Bauer, K., Garbe, D. and Surburg, H. (2001) *Common Fragrance and Flavor Materials*, Wiley-VCH, p. 161.
4. Buchi, G., Demole, E. and Thomas, A.F. (1973) Syntheses of 2,5-dimethyl-4-hydroxy-2,3-dihydrofuran-3-one (Furaneol), a flavor principle of pineapple and strawberry. *J. Org. Chem.*, **38**, 123.
5. Re, L. and Ohloff, G. (1967) Swiss Patent CH 474,500, assigned to Firmenich.
6. Eicher, T. and Hauptmann, S. (1995) *The Chemistry of Heterocycles*, Georg Thieme Verlag, p. 60.
7. Bauer, K., Garbe, D. and Surburg, H. (2001) *Common Fragrance and Flavor Materials*, Wiley-VCH, p. 162.
8. Gianturco, M.A., Friedel, P. and Giammarino, A.S. (1964) The Volatile Constituents of Coffee – III. *Tetrahdron Lett.*, **20**, 1763; Colovray, G. and Descotes, G. (1965) Sur les Alcoyl-2 tetrahydrofuranones-3. I. Preparation. *Bull. Soc. Chim. Fr*, 1985; Wynberg, H. (1963) Tetrahydrofuran-3-one, spirane und dithienyle. *Angew. Chem.*, 453.
9. Bauer, K., Garbe, D. and Surburg, H. (2001) *Common Fragrance and Flavor Materials*, Wiley-VCH, p. 146.
10. Brennan, T.M., Brannegan, D.P., Weeks, P.D. and Kuhla, D.E. (1981) CA Patent 1,110,254, assigned to Pfizer Inc.
11. Bauer, K., Garbe, D. and Surburg, H. (2001) *Common Fragrance and Flavor Materials*, Wiley-VCH, p. 160.
12. Eschenmoser, A., Seidel, C.F. and Felix, D. (1964) US Patent 3,161,657, assigned to Firmenich.
13. Matsuda, H., Yamamoto, H. and Kanisawa, T. (1995) *Proceedings of 13th International Congress of Flavours, Fragrances and Essential Oils*, Istanbul.
14. Yamasaki, T. (1997) *Flavours and Fragrances*, K.A.D. Swift (ed.), The Royal Society of Chemistry.
15. Sanders, J.M. and Michael, L.H. (1975) US Patent 3,910,964, assigned to International Flavors & Fragrances.
16. Finkelmeier, H. and Hopp, R. (1984) European Patent EP120,257, assigned to Haarman & Reimer.
17. Rooney, C.S., Stuart, R.S., Wasson, B.K. and Williams, H.W.R. (1975) 3,4-Dihydro-*2H*-1,5-benzodioxepines. A novel class of β-adrenergic stimulants. *Can. J. Chem.*, **53**, 2279 (The article mentions catechol not 4-methylcatechol).
18. Eicher, T. and Hauptmann, S. (1995) *The Chemistry of Heterocycles*, Georg Thieme Verlag, p. 76.
19. Eicher, T. and Hauptmann, S. (1995) *The Chemistry of Heterocycles*, Georg Thieme Verlag, p. 93.
20. Cournoyer, R., Evans, D.H., Stroud, S. and Boggs, R. (1991) Synthesis of an Unsymmetrical naphthalein indicator dye from an indole-6-sulfonamide. *J. Org. Chem.*, **58**, 4576.
21. Eicher, T. and Hauptmann, S. (1995) *The Chemistry of Heterocycles*, Georg Thieme Verlag, p. 298.
22. Ohloff, G., Flament, I. and Pickenhagen, W. (1985) Flavor chemistry. *Food Rev. Int.*, **1**, 99.
23. Rizzi, G.P. (1988) The biogenesis of food-related pyrazines. *Food Rev. Int.*, **4**, 375.
24. Oruna-Concha, M.J., Craig Duckham, S. and Ames, J.M. (2001) Comparison of volatile compounds isolated from the skin and flesh of four potato cultivars after baking. *J. Agric. Food Chem.*, **49**, 2414.

25. Eicher, T. and Hauptmann, S. (1995) *The Chemistry of Heterocycles*, Georg Thieme Verlag, p. 420.
26. Eicher, T. and Hauptmann, S. (1995) *The Chemistry of Heterocycles*, Georg Thieme Verlag, p. 421.
27. Werkhoff, P., Guntert, M., Krammer, G., Sommer, H. and Kaulen, J. (1998) Vacuum headspace method in aroma research: flavor chemistry of yellow passion fruits. *J. Agric. Food Chem.*, **46**, 1076.
28. Buchman, E.R. (1939) Reichs Pat. 673,174, assigned to E. Merck.

Chapter 6
Aroma Chemicals III: Sulfur Compounds

Simon B. Jameson

Sulfur compounds are an important family of aroma chemicals found in a wide variety of fragrances and flavourings. Although normally present in trace quantities relative to other components, their low odour thresholds and high impact nature make them of vital importance to the overall aroma and flavour of foods such as bread (3-(methylthio)propionaldehyde (methional) (**1**)), meat (2-methylfuran-3-thiol (MFT) (**2**)), coffee (furfuryl mercaptan (**3**) and 3-mercapto-3-methylbutyl formate (MMBF) (**4**)), beer (prenyl mercaptan (**5**)), vegetables (allyl isothiocyanate (**6**) (see Chapter 9), dimethyl sulfide (DMS) (**7**) and allyl disulfide (**8**)) and fruits (Grapefruit Mercaptan (**9**), 2-methyl-4-propyl-1,3-oxathiane (Tropathiane) (**10**), 1-methoxy-3-methyl-3-mercaptobutane (**11**) and 4-methyl-4-mercaptopentan-2-one (**12**)). Examples of the occurrence and uses of these materials are illustrated in Table 6.1 along with their odour thresholds. The heterocyclic sulfur compounds, *i.e.* thiazoles and thiophenes, are covered in Chapter 5.

A wider discussion on the occurrence of volatile sulfur chemicals in food stuffs and their applications in flavours has been published by Boelens and van Gemert [2].

Table 6.1 Applications and odour thresholds of some common sulfur aroma chemicals

Material	Occurrence/application	Odour threshold [1]*
Methional (**1**)	Bread/vegetables	6.3×10^{-4}
MFT (**2**)	Roast meat	1×10^{-6}
Furfuryl mercaptan (**3**)	Coffee	4×10^{-6}
MMBF (**4**)	Coffee	2×10^{-7}
Prenyl mercaptan (**5**)	Beer/coffee	2×10^{-7}+
Allyl isothiocyanate (**6**)	Mustard	3.7×10^{-2}
Dimethyl sulfide (**7**)	Vegetables (sweet-corn)	3×10^{-4}
Allyl disulfide (**8**)	Garlic	7.2×10^{-3}
Grapefruit Mercaptan (**9**)	Grapefruit	1×10^{-6}
Tropathiane (**10**)	Pineapple/tropical fruit	2×10^{-3}+
1-Methoxy-3-methyl-3-mercaptobutane (**11**)	Blackcurrants	8×10^{-8}
4-Methyl-4-mercaptopentan-2-one (**12**)	Sauvignon grape	1×10^{-7}+

* All values are given in mg/m^3 air except those marked '+' which are in mg/kg water.

Often, the levels of these highly odorous sulfur compounds can be inversely proportional to their impact. Proof of their key contribution to flavours in very low concentrations is the BBA-IFF Generessence™ programme which focused on the trace sulfur compounds in a wide variety of food stuffs using sulfur-specific GC detectors, GC-olfactometry and GC-MS to identify key aroma chemicals. This work resulted in the identification of ethyl 3-mercaptobutyrate (**13**) in mango pulp [3]. The mercaptobutyrate ester (**13**) was synthesised and, after formulation and application trials, was shown to be a key mango and tropical component in both flavours and fragrances.

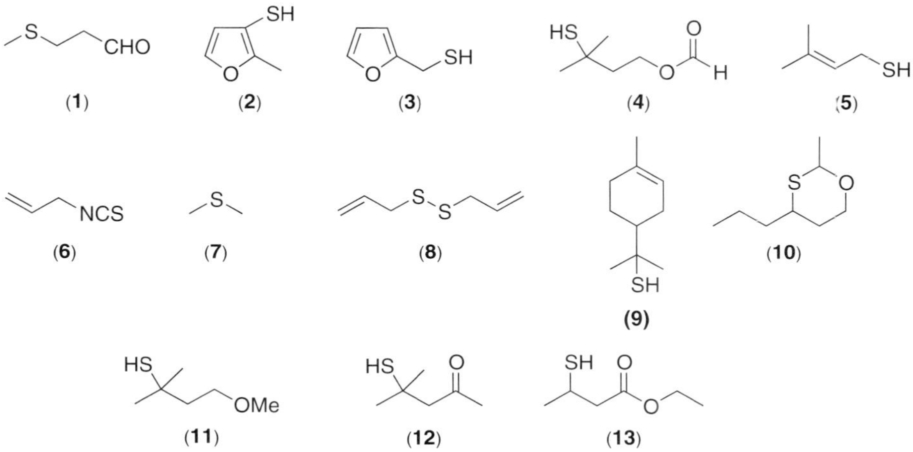

The development of new analytical methods, such as those described above, has led to an increase in the discovery of trace quantities of sulfur-containing molecules in foods over the last 30 years. Figure 6.1 illustrates how the development of new analytical techniques has allowed us to identify many more sulfur compounds in foodstuffs.

The increase in interest in these materials within the Flavour and Fragrance industry, and in particular their applications in flavours, can be demonstrated by the increasing number of sulfur compounds being registered as FEMA GRAS. Considering the eight years from 1993's GRAS 16 list to 2001s GRAS 20 list, it can be seen that sulfur compounds make up a substantial number of the new materials being registered (Figure 6.2) [4]. This is particularly true of the large GRAS 18 list. Concentrating on the families of sulfur compounds included on these lists, it is noticeable that there is a distinct lack of thiazoles and thiophenes (Figure 6.3). This is not a reflection on their importance, rather that the occurrence, chemistry and applications of these materials are well known within the F & F industry, with a large number already being listed as FEMA GRAS.

This chapter will deal with the thioesters, mercaptans (thiols), sulfides, polysulfides and the saturated heterocycles, such as dithiazines and oxathianes (Figure 6.4), the unsaturated sulfur heterocycles such as thiazoles and thiophenes being detailed elsewhere in this book.

The reactions and methodologies detailed within this chapter are those that the author believes to be the most useful to the Synthetic Organic Chemist working in the field of

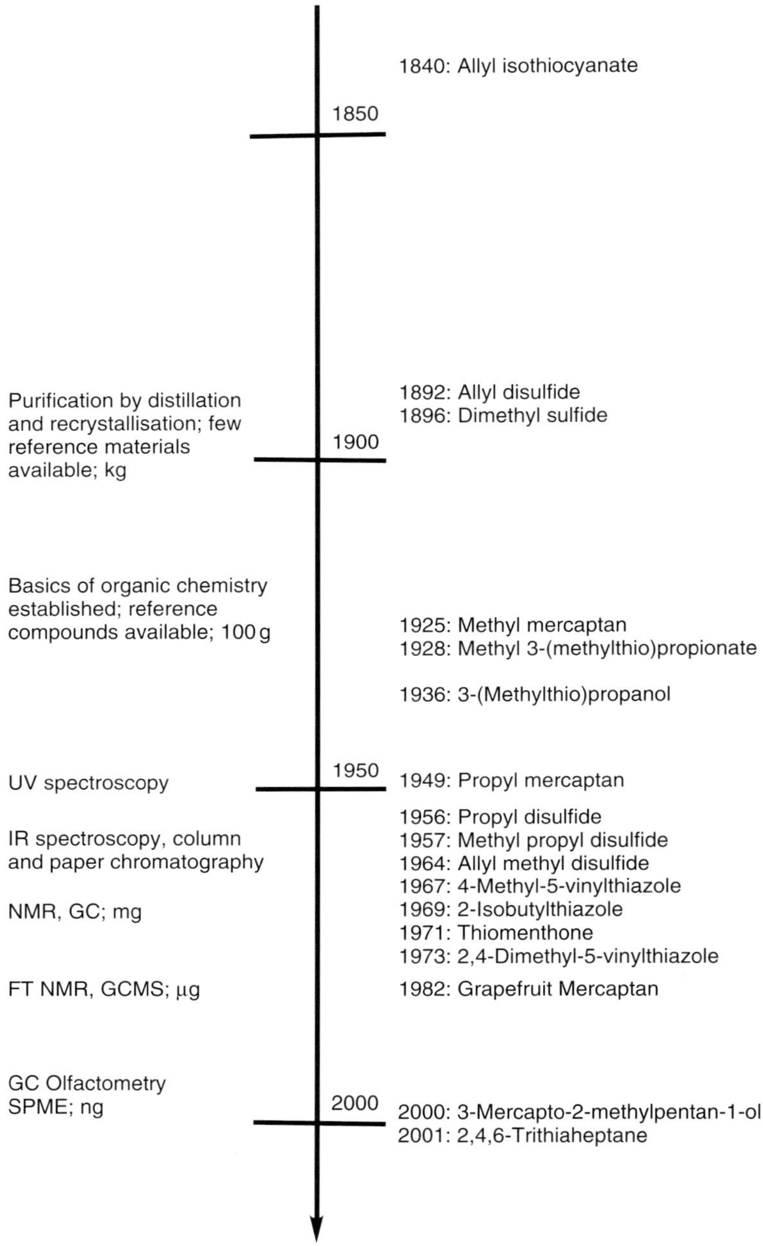

Figure 6.1 History of the discovery of sulfur aroma chemicals in foods.

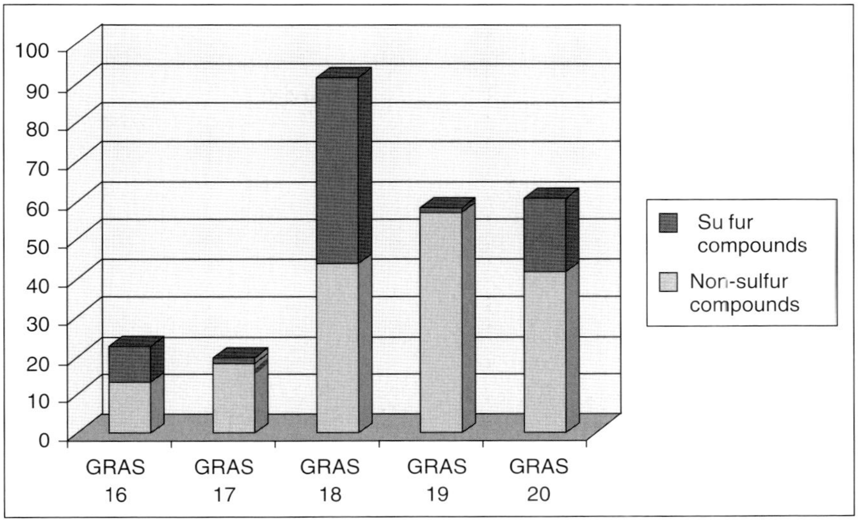

Figure 6.2 Numbers of sulfur compounds recognised on recent FEMA GRAS lists.

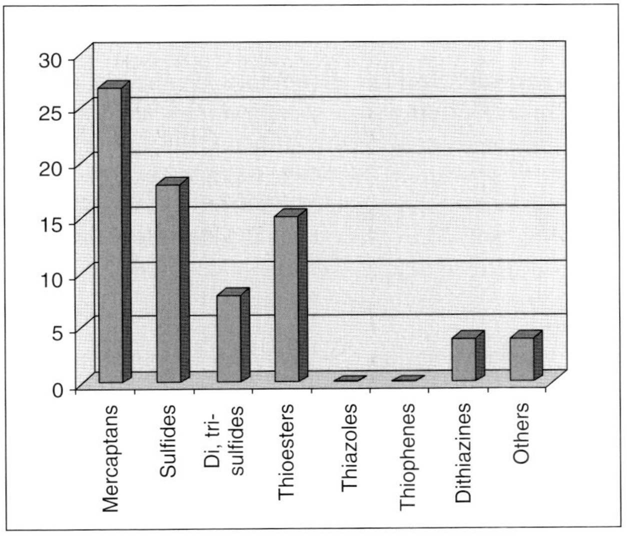

Figure 6.3 Families of sulfur compounds reported on recent FEMA GRAS lists.

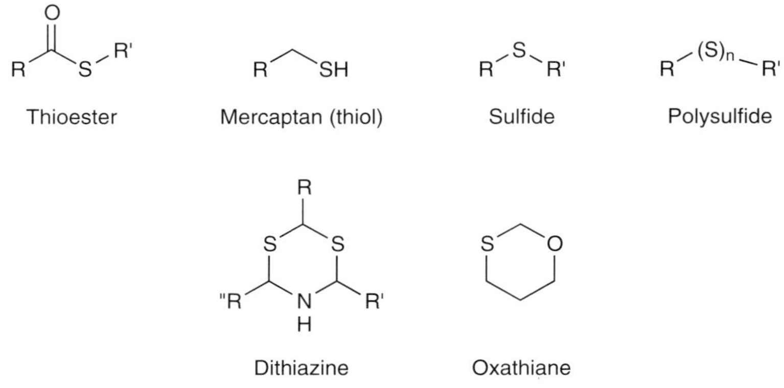

Figure 6.4

flavours and fragrances. Those wishing to read further into the subject should consult the bibliography at the end of this chapter.

6.1 Thiols and thioesters

Thiols are widely utilised in savoury flavour formulations, *e.g.* 2-methyl-3-furanthiol (MFT)(**2**) and 2-methyltetrahydrofuran-3-thiol (**14**) with their distinct aroma of roast beef and 3-methyl-2-butanethiol (**15**) with its strong onion odour. These are only used at low levels due to their low odour threshold. Furfuryl mercaptan (**3**) which possesses a rich, roast coffee odour when used at low concentrations, is also of major importance to the flavour industry where it also has applications in other flavour formulations when a 'roasted' note is required. Thiols also have applications in fruit flavours; for example, the powerful odour of 1-methoxy-3-methyl-3-mercaptobutane (**11**) is used to give the aroma of blackcurrant to soft drinks. Another example of this is 1-*p*-menthene-8-thiol (**9**) which is also widely known as Grapefruit Mercaptan. Thioesters also have their role to play with many being utilised in both tropical fruit and cheese flavourings, a prime example of the latter being methyl 2-methylthiobutyrate (**16**).

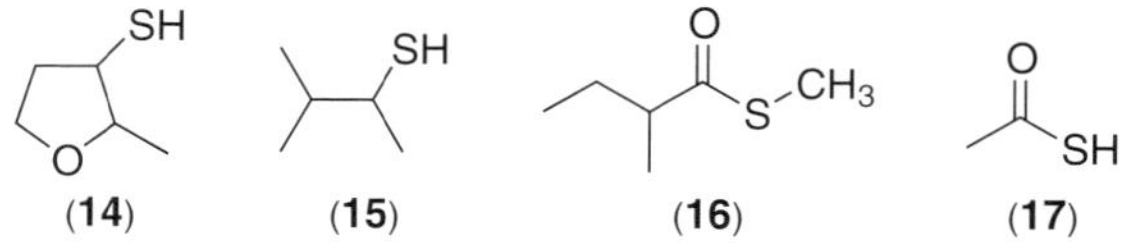

(**14**) (**15**) (**16**) (**17**)

These simple sulfur-containing compounds have been grouped together as there is a great deal of inter-conversion between the thioesters and mercaptans, a principle reagent for the introduction of sulfur into a molecule being thioacetic acid (**17**). This is a highly diverse material in that it can be employed in numerous ways to make thioesters, which can subsequently be hydrolysed or reduced to afford the desired thiol functionality.

One method of utilising thioacetic acid to make thioesters is to add it across a double bond. This can be done as a simple acid adding across a double bond or by a Michael addition to an α,β-unsaturated carbonyl (Scheme 6.1).

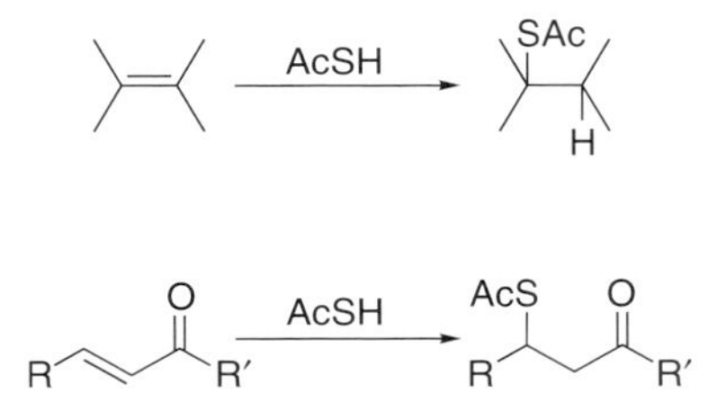

Scheme 6.1

A good example of the latter is in the synthesis of 3-mercapto-3-methylbutanol (**18**) [5], an ingredient in coffee flavourings, in which the sulfur group is introduced via addition of thioacetic acid to prenal (3-methyl-2-butenal) (**19**), the resultant thioester-aldehyde then being reduced to the alcohol which can undergo intramolecular *trans*-acylation prior to acid hydrolysis of the acetate moiety to afford the final product (Scheme 6.2). A word of caution should be added in that thioacetic acid readily forms radicals which under the right conditions can undergo anti-Markovnikov addition to give undesired by-products. In the aforementioned case, unless reaction conditions are modified to prevent radical formation (*e.g.* nitrogen atmosphere or the addition of a radical scavenger/anti-oxidant to the reaction mix), the by-product (**20**) is generated [6].

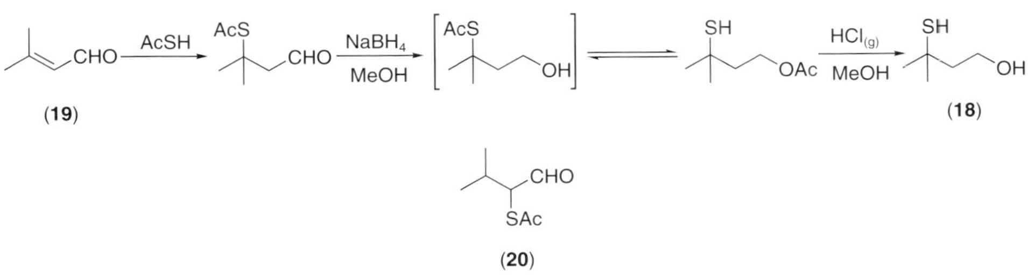

Scheme 6.2

While in some instances, this radical addition is undesirable, the propensity of thioacetic acid to form radicals may be exploited by the Synthetic Chemist. For example, 3-methyl-2-butanethiol (**15**) may be made *via* radical addition of thioacetic acid to 3-methyl-2-butene, utilising iron(II)sulfate and hydrogen peroxide to promote radical formation (Scheme 6.3) [6].

Another prime example of the use of thioacetic acid in the synthesis of a key aroma compound is in IFF's patent of the earlier mentioned MFT(2) [7]. The thioacetate moiety is again introduced *via* Michael addition to an α,β-unsaturated carbonyl. The synthesis (Scheme 6.4) starts with the chlorine-mediated or electrochemical oxidation

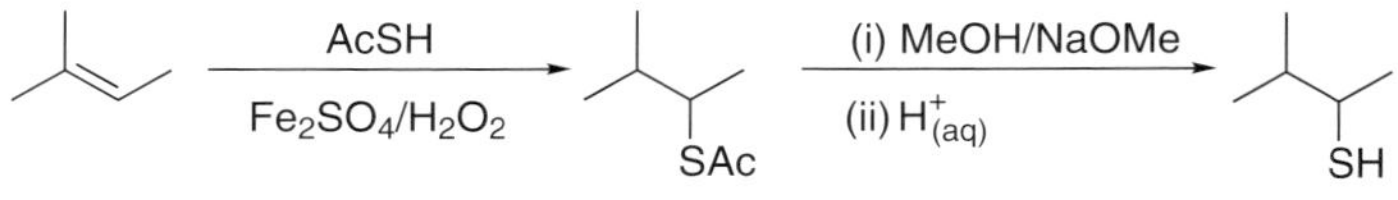

(15)

Scheme 6.3

of 2-methylfuran in methanol to give 2,5-dimethoxy-2-methyldihydrofuran. This is carefully ring-opened under acid conditions to give the keto-enal intermediate (21), which is then subjected to the Michael addition of the thioacetic acid. Although the acid can also theoretically add with the thioacetate group β to the ketone, the higher polarity of the aldehyde group dominates to give the intermediate (22). This is then subjected to acid conditions to bring about ring closure and dehydration to afford MFT acetate (23), an important aroma chemical in its own right. The desired product is then obtained by simple base hydrolysis of the thioacetate group.

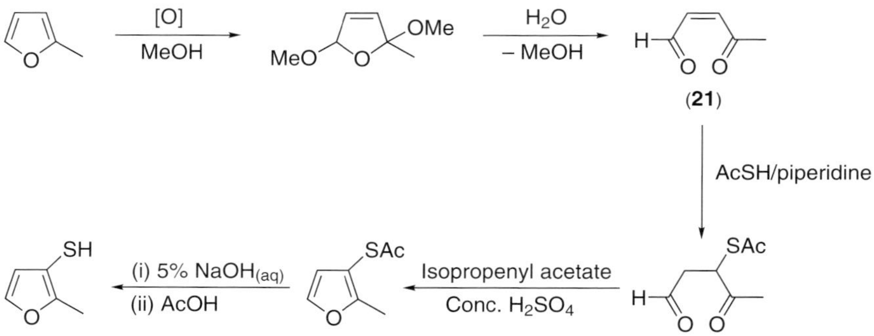

(2) (23) (22)

Scheme 6.4

This approach may also be used on pseudo-α,β-unsaturated carbonyls such as vinyl pyrazine (24) in the synthesis of pyrazine ethanethiol (25), an artificial material used in pork flavours (Scheme 6.5).

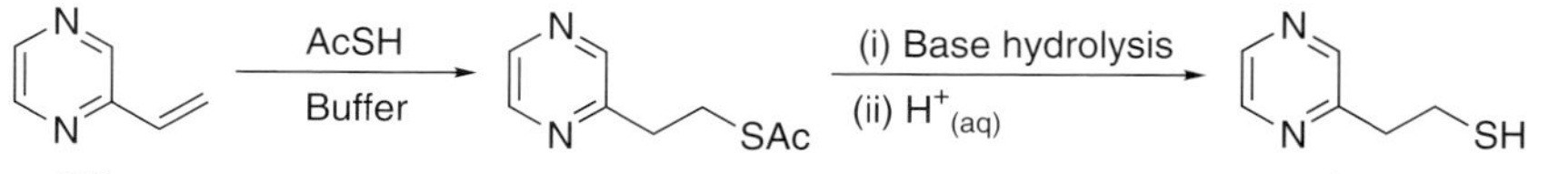

(24) (25)

Scheme 6.5

Thioacetic acid salts or thioacetic acid in the presence of an amine buffer have also been used to introduce the thioacetate group to a wide variety of molecules. A prime example is the Lever Brothers (now Quest) patent for the synthesis of

2-methyltetrahydrofuran-3-thiol (14) [8]. In this synthesis (Scheme 6.6) 2-methyl-3-chlorotetrahydrofuran (26) is reacted with potassium thioacetate in DMF to give the thioacetate (27). This is subsequently hydrolysed to the mercaptan by treatment with sodium methoxide in methanol.

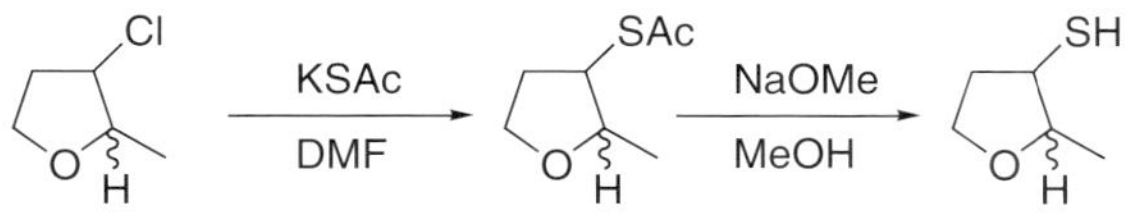

Scheme 6.6

Other leaving groups may be utilised for this transformation such as mesylate and tosylate. An example of this is the treatment of 3-methyl-2-butanol (28) with triethylamine and methanesulfonyl chloride to produce the mesylate (29), which is subsequently reacted with potassium thioacetate to give the thioester (30), which is then hydrolysed to the desired mercaptan (15) (Scheme 6.7) [6].

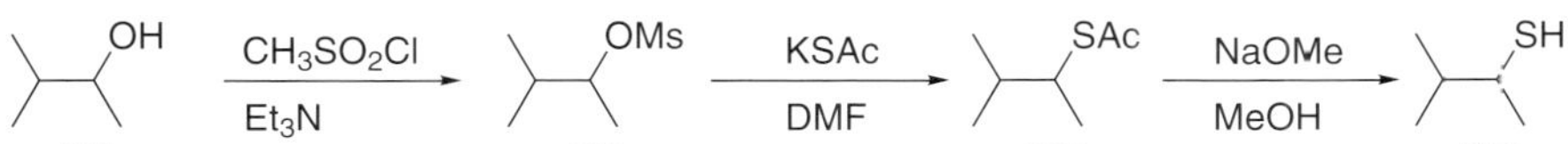

Scheme 6.7

A less well-known application that is closely related to the latter is the formation of a thioacetate from an alcohol under Mitsunobu conditions [9]. This, once again, may be applied to the synthesis of our favourite example, 3-methyl-2-butanethiol (Scheme 6.8) [6]. The key to this is the use of diethyl azodicarboxylate (DEAD) as

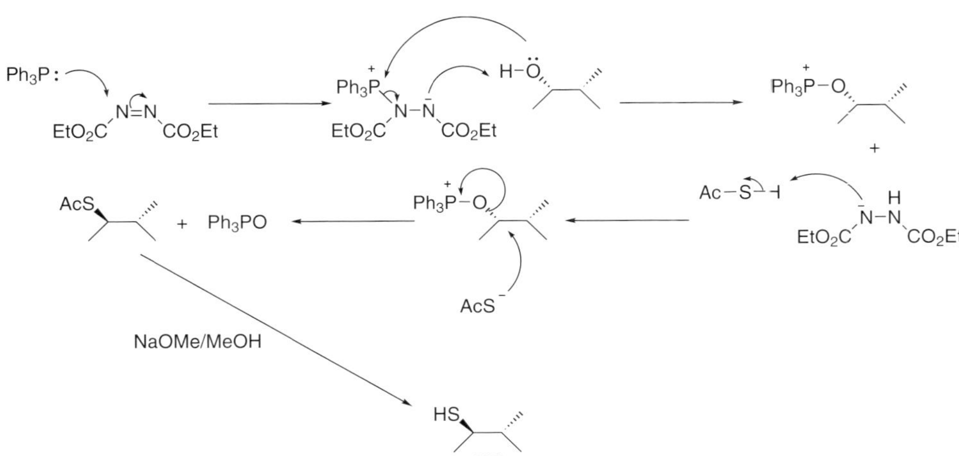

Scheme 6.8

a base to accept protons from both the acid and the alcohol. It should be noted that this reaction proceeds with complete inversion of stereochemistry about the hydroxyl-bearing carbon. The use of this approach is currently expensive; however, as the drive to pure enantiomers of a given product continues, the author expects more interest in this approach, as it allows access to chiral mercaptans from the cheaper, readily available alcohols.

The other two widely used methods for the production of thioesters involve reacting acid anhydrides or acid chlorides with either a thiol in the presence of a catalyst or with an aqueous solution of the thiolate salt, *e.g.* the sodium salt of methyl mercaptan which is readily available as a 20% solution. A good example is the synthesis of methyl thioacetate (**31**) from sodium methyl mercaptide solution and acetic anhydride (Scheme 6.9). The advantage of using an aqueous solution of the mercaptide from the Synthetic Chemist's point of view is that it also aids in the purification, allowing the product (organic) to be removed from the by-products (carboxylic acids and their salts). The anhydride is preferred as a substrate, rather than the acid chloride, due to it being easier to handle.

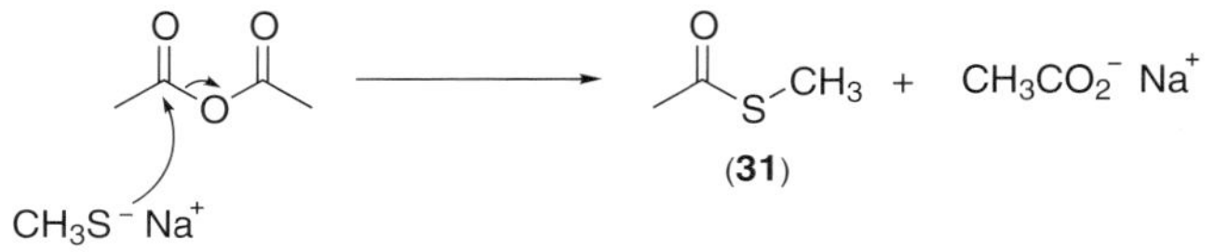

Scheme 6.9

An unusual reaction is that to make *S,S′*-ethylidene dithioacetate (**32**), a material that, although originally found in blood oranges, has applications in the more exotic fruit flavours and perfumes [10]. Sternson [11] showed that by treating thiolacetic acid with triethylamine or pyridine at ambient temperature a condensation reaction occurs between three molecules of the acid. This reaction affords the desired product in good yield with the formation of sulfur and water as by-products. To confirm the structure, Sternson also synthesised the product by treatment of acetaldehyde dimethyl acetal with thioacetic acid (Scheme 6.10).

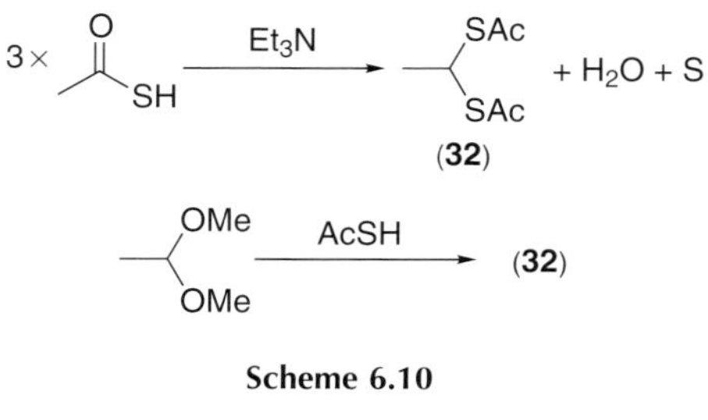

Scheme 6.10

Mercaptans can be made in a variety of ways. A popular method is to react an alcohol or alkyl halide (the former in the presence of a strong acid) with thiourea in order to obtain the isothiuronium salt. This intermediate is then hydrolysed with base to give the desired thiol functionality (Scheme 6.11).

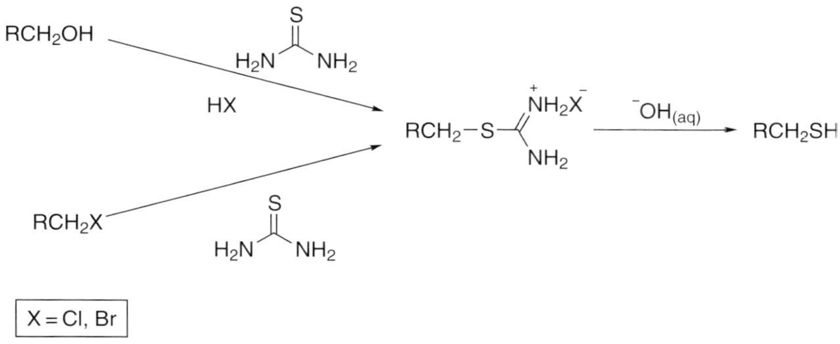

Scheme 6.11

An example of this is the synthesis of furfuryl mercaptan (3) from furfuryl alcohol [12]. In the author's experience primary alcohols react best using this method, and in the case of secondary and tertiary alcohols it is often preferable to use the alkyl halide.

Another widely applied method is the substitution of a mesylate or halide group with sodium hydrogen sulfide, *e.g.* the synthesis of allyl mercaptan (33) from allyl chloride and aqueous sodium hydrogen sulfide. This route does have its drawbacks in that the basic conditions used can, in some instances, promote oxidation to the disulfide. The formation of these undesired by-products may be overcome by reduction with zinc/acetic acid or lithium aluminium hydride reduction. Again, 2-methyl-3-butanethiol provides us with an example, the disulfide being easily reduced by zinc/acetic acid to afford the mercaptan (Scheme 6.12) [5].

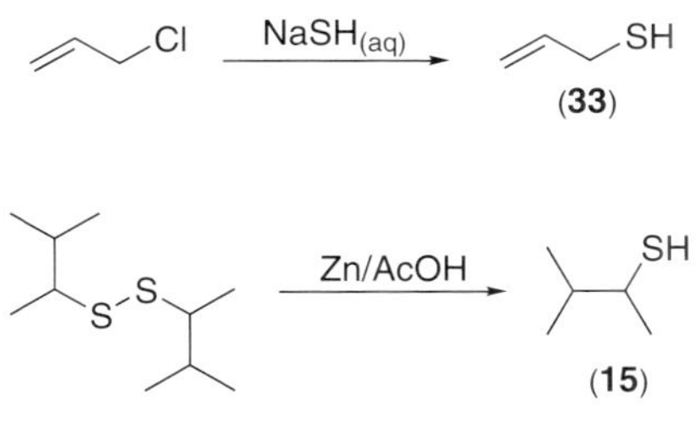

Scheme 6.12

A similar transformation may be brought about by the treatment of alcohols (particularly tertiary alcohols) with Lawesson's Reagent. Lawesson's Reagent (2,4-bis [4-methoxyphenyl]-1,3,2,4-dithiaphosphetane-2,4-disulfide) was used by Nishio in his synthesis of Grapefruit Mercaptan (9) from α-terpineol (34) (Scheme 6.13) [13].

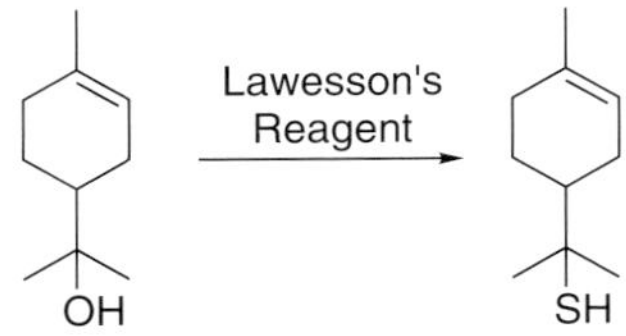

(34) (9)

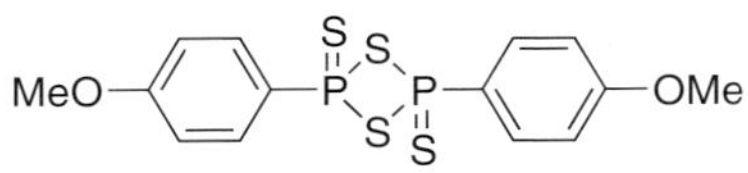

Lawesson's Reagent

Scheme 6.13

Aromatic and tertiary thiols can be made by treating the corresponding Grignard reagent with elemental sulfur then acid, a variation on two *Organic Synthesis* methods [14, 15]. A good example of this is the synthesis of thioguaiacol (**35**) [6], a constituent of coffee flavours, from 2-bromoanisole (Scheme 6.14).

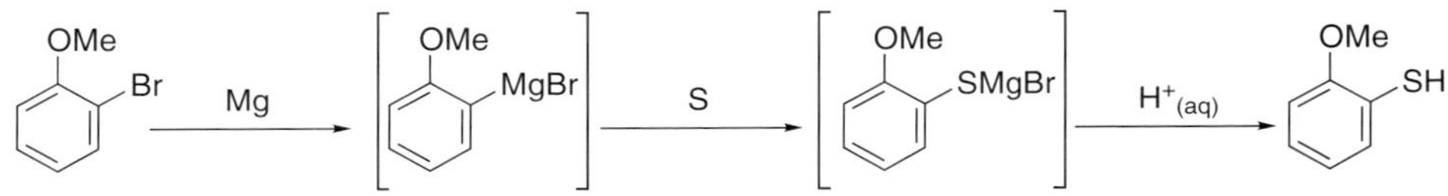

(35)

Scheme 6.14

A less common route is the ring opening of an episulfide by hydride reduction. A good example of this is the IFF synthesis of Grapefruit Mercaptan (**9**) [16]. In this synthesis, limonene (**36**), is first epoxidised preferentially at the exocyclic double bond, then treated with sodium thiocyanate to give the corresponding episulfide (**37**). This is subsequently ring-opened to the desired product by reduction with lithium aluminium hydride (Scheme 6.15).

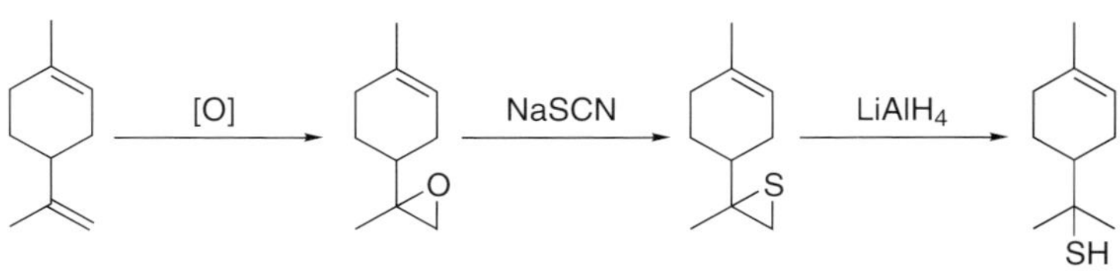

(36) (37) (9)

Scheme 6.15

A subtle variation on the above is the Firmenich route to the same compound (**9**), whereby the episulfide is formed by treatment of the epoxide with thiourea and sulfuric acid rather than by use of thiocyanate [17]. Firmenich used similar chemistry in their route to 1-methoxy-3-hexanethiol (**38**), a key odorant in clary sage oil [18]. The synthesis (Scheme 6.16) starts with the epoxidation of *cis*-1-methoxy-3-hexene using *m*-chloroperbenzoic acid (MCPBA). The resultant epoxide is then treated with thiourea and sulfuric acid to afford the desired thiirane intermediate. The final step is again ring opening with lithium aluminium hydride.

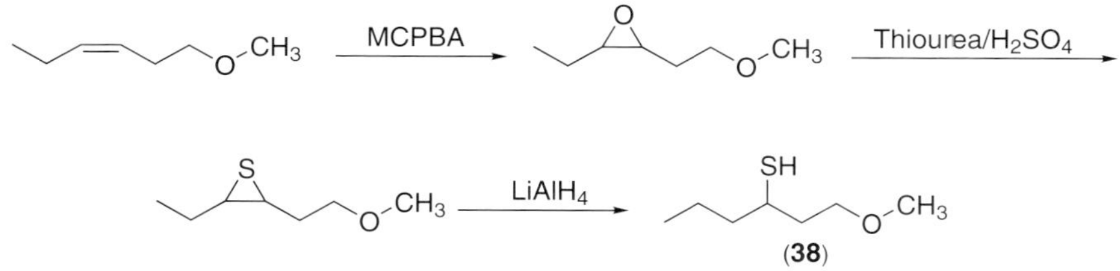

Scheme 6.16

Two useful routes exist to α-mercaptocarbonyls. First is the addition of aqueous sodium hydrogen sulfide to the halide, e.g. mercaptopropanone (**39**) from chloroacetone. This is unusual in that the product (**39**) spontaneously dimerises to (**40**) (Scheme 6.17). It is the dimer with its distinct meaty aroma that is the material of commerce.

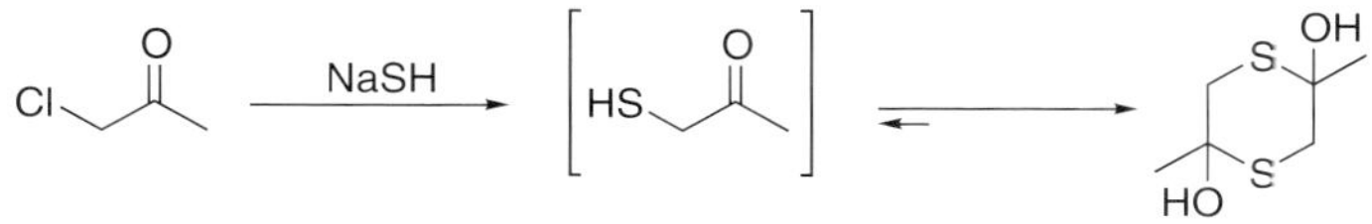

Scheme 6.17

Where the α-halocarbonyl is not available or prohibitively expensive a more complex route that may be employed is the treatment of the appropriate carbonyl compound with ammonia in the presence of elemental sulfur, affording the 3-thiazoline. This, in turn, may be decomposed to give the desired mercaptan under acid conditions [19]. An example of an important aroma chemical that can be made via this route (Scheme 6.18) is 2-mercapto-3-butanone (**41**), a material that is widely employed in savoury flavours.

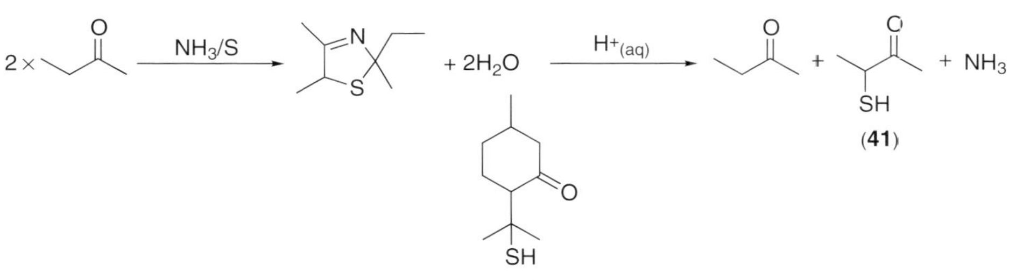

Scheme 6.18

β-Mercaptocarbonyl compounds may be made via Michael-type addition of H_2S to α,β-unsaturated carbonyls or via the thioacetate, as described earlier. An example of a material which may be made by either of these methods is thiomenthone (**42**), a material with a highly recognisable aroma of blackcurrant.

6.2 Acyclic sulfides and polysulfides

These materials can be split into two groups: those having identical hydrocarbon groups (symmetrical), *e.g.* allyl disulfide (**8**), and those having different hydrocarbon groups (unsymmetrical), *e.g.* ethyl 3-(methylthio)propionate (**43**). In terms of their synthesis, however, it is better to look at them as sulfides, disulfides and trisulfides and higher sulfides (polysulfides).

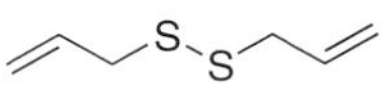

(**8**)

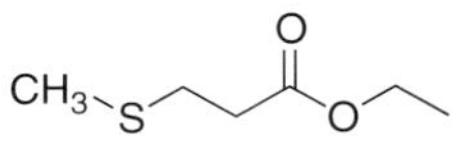

(**43**)

It has long been known [20] that simple symmetrical sulfides such as butyl sulfide (**44**), with its distinct floral odour (quite unlike other sulfides), can be made by treatment of the corresponding alkyl halide with sodium sulfide (Scheme 6.19).

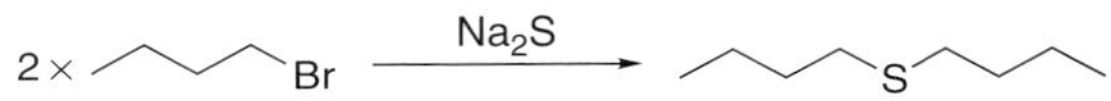

(**44**)

Scheme 6.19

There is much more scope for the synthesis of the unsymmetrical species. One method that may be used for unsymmetrical sulfides possessing a methyl group is the alkylation of a mercaptan using methyl iodide or dimethyl sulfate in the presence of base. An example of a material that may be made in this way is methyl furfuryl sulfide (**45**) (Scheme 6.20).

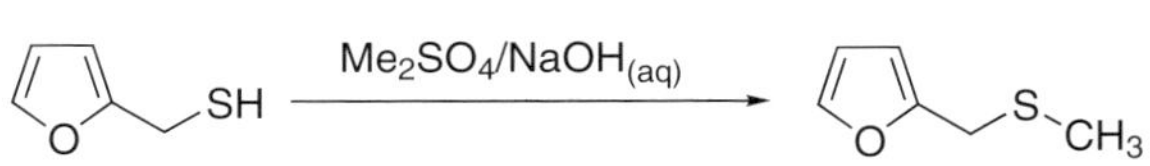

(**3**) (**45**)

Scheme 6.20

It is often preferable to employ a solution of the sodium salt of methyl mercaptan where possible. This has the advantage over using methyl mercaptan and a base of reduced odour and in being in a liquid form at room temperature. The sodium salt readily reacts with alkyl halides in a two-phase system to afford the desired unsymmetrical sulfide. A typical material that can be made by this route is methyl propyl sulfide (**46**) (Scheme 6.21), a key constituent in a wide variety of onion flavours.

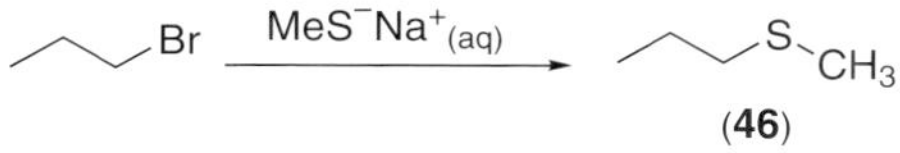

Scheme 6.21

This approach may also be utilised in the synthesis of some derivatives of heterocyclic species, *e.g.* the synthesis of 2-(methylthio)-3-methylpyrazine (**47**) (Scheme 6.22) [21].

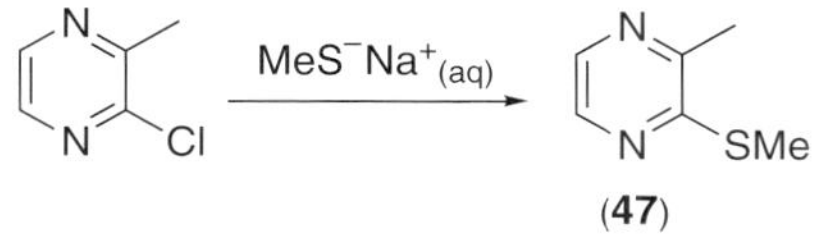

Scheme 6.22

Another chemical that may be made in this manner is *bis*-(methylthio)methane (**48**), a material that allows the distinct flavour of black truffle to be introduced to the wider population who could not normally afford this highly expensive delicacy. This use of aroma compounds to formulate more complex flavours is an area that is set to expand as the tastes of the general public become more sophisticated.

Another popular method in the making of sulfides is the Michael addition of methyl mercaptan, or more usually its sodium salt, to an α,β-unsaturated carbonyl. Materials that may be made this way include 3-(methylthio)butanal (**49**), 4-(methylthio)butanone (**50**) and 3-(methylthio)hexanol (**51**), a key component in passion fruit.

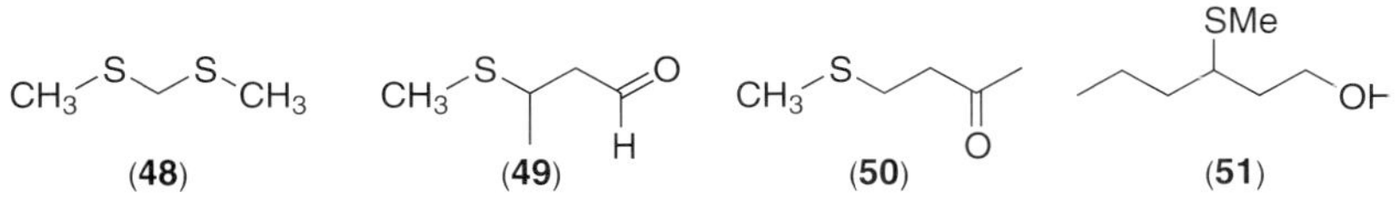

The latter may be made *via* Michael addition to *trans*-2-hexenal (**52**), itself an important aroma chemical, followed by hydride reduction to give the desired product (Scheme 6.23).

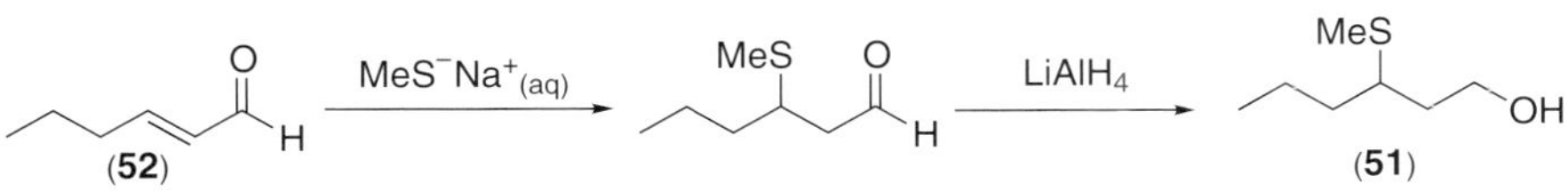

Scheme 6.23

The symmetrical disulfides can be made in a similar manner [22] to the sulfides, by first reacting the sodium sulfide with elemental sulfur, prior to reaction with the alkyl halide. It was using this method that Challenger synthesised propyl disulfide (**53**), a material that has applications in onion flavours (Scheme 6.24).

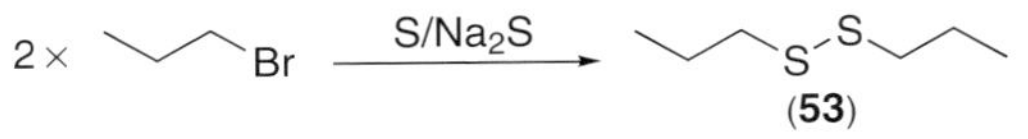

Scheme 6.24

It should be noted that the higher sulfides (trisulfides and tetrasulfides, etc.) may also be made using this method by increasing the molar ratio of the elemental sulfur employed. The result is a mix of sulfides, and it is this mix that is generally the material of commerce. The mixes of the higher sulfides can prove difficult to separate by fractional distillation due to continuous disproportionation at elevated temperatures throughout the process, even while under high vacuum or, in the case of allyl species, polymerisation. An example of this is the ready decomposition of propyl trisulfide (**54**) to the respective disulfide (**53**) and tetrasulfide (**55**) (Scheme 6.25) [6]. Where higher-purity trisulfides are required, the author recommends the Bunte salt method, described later, for the synthesis of unsymmetrical trisulfides.

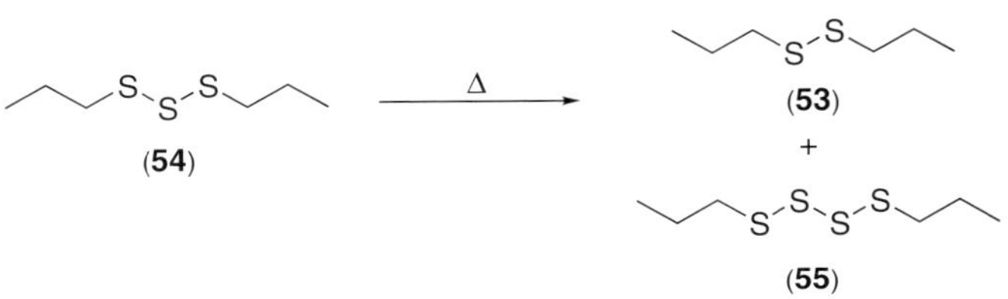

Scheme 6.25

An approach that works in some instances better than others is to simply oxidise the corresponding mercaptan to the disulfide. This can be done *via* aerial oxidation or by a chemical oxidant such as hydrogen peroxide. This reaction, an undesirable side reaction in many mercaptan syntheses, has its uses in the synthesis of the disulfides. A good example is the ease by which MFT (**2**) oxidises to its disulfide (**56**) in air (Scheme 6.26). It is thought that while the volatile mercaptan gives roast beef its aroma, it is the disulfide that gives it its flavour.

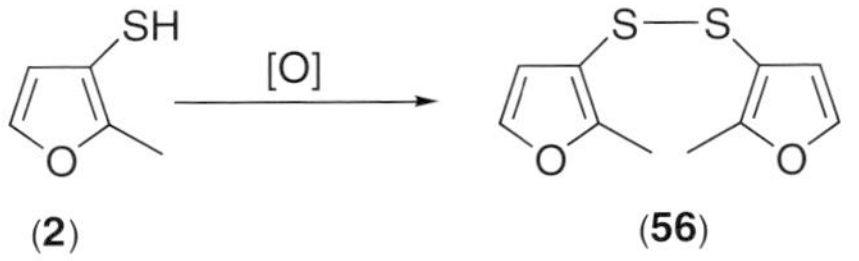

Scheme 6.26

In order to make unsymmetrical disulfides, Bunte salts have been used over the years. In his synthesis of methyl propyl disulfide (**57**) Milligan [23] reacted propyl bromide with sodium thiosulfate in order to form the desired Bunte salt intermediate. This was subsequently reacted with the sodium salt of methyl mercaptan to afford the desired product, like the afore-mentioned sulfide (**46**), an important constituent in onion flavours (Scheme 6.27).

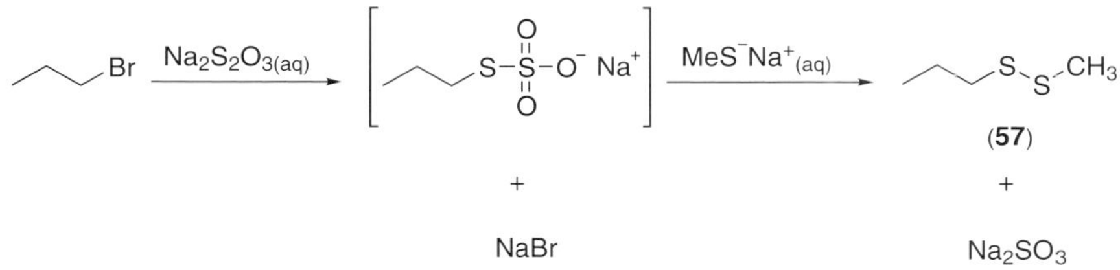

Scheme 6.27

An approach to the synthesis of symmetrical higher sulfides is that described by both Olin [24] and Westlake [25], where elemental sulfur is introduced to an existing sulfide with the aid of an amine or pyridine catalyst. A prime example of a material that may be made in this way is methyl trisulfide (**58**) (Scheme 6.28) which has applications in many savoury flavours, in particular, those of the various allium species.

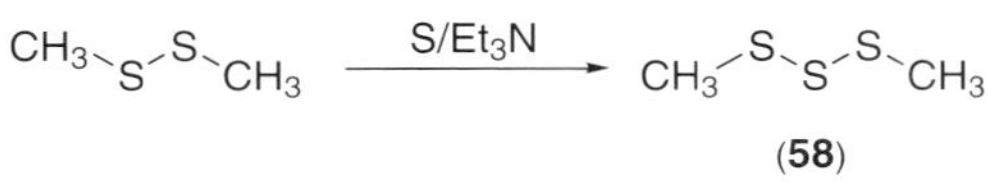

Scheme 6.28

For mixed trisulfides, a convenient method is to make both Bunte salts and then to react them with sodium hydrogen sulfide. This reaction produces a mix of sulfides, and it is this that is the material of commerce. Typical of the materials that may be made by this route is allyl propyl trisulfide (**59**), again a much-used ingredient in garlic flavourings (Scheme 6.29). When a single Bunte salt is used this reaction produces the symmetrical trisulfide in good purity.

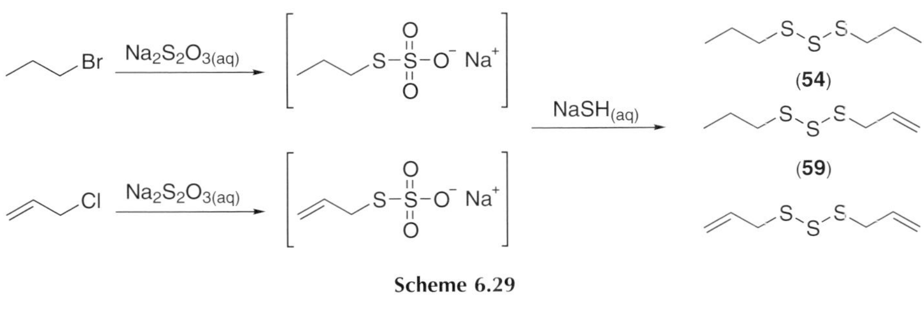

Scheme 6.29

6.3 Saturated heterocyclic sulfur compounds

These may be split into two families: those where sulfur is the only heteroatom and those where a second heteroatom, usually oxygen or nitrogen, is present. Taking the family that has only sulfur as a heteroatom first, some of the simplest materials to make are the 1,3,5-trithianes. These may be made by treating a ketone or aldehyde with hydrogen sulfide gas in the presence of either a Bronsted or Lewis acid. Two materials that may be made in this manner [26] and which are of importance in the F&F industry, are 2,2,4,4,6,6-hexamethyl-1,3,5-trithiane (trithioacetone) (**60**) and 2,4,6-trimethyl-1,3,5-trithiane (trithioacetaldehyde) (**61**), made from the carbonyl compounds suggested by their trivial names (Scheme 6.30). The former can vary in odour from minty to meaty while the latter is also used in applications where a 'meaty' aroma is required.

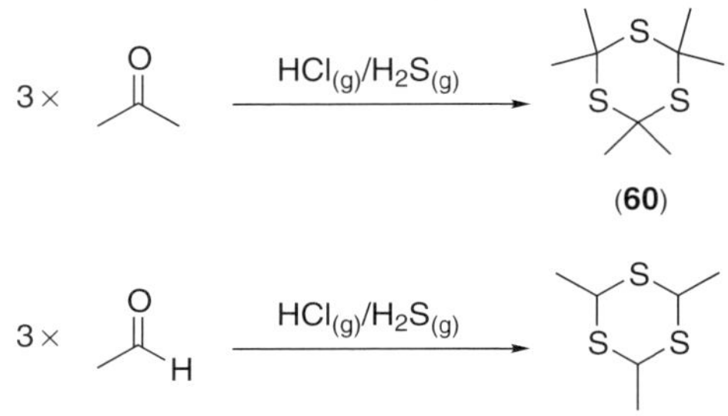

(**60**)

(**61**)

Scheme 6.30

A second group of easily accessible materials are the simple dithiolanes. These are made by treating an aldehyde with ethane-1,2-dithiol in the presence of an acid catalyst. This simple reaction has been widely exploited in synthesis as a method of protecting carbonyl species. In her book [27], Greene gives several published methods for carrying out this and similar reactions. Typical of these materials is 2-methyl-1,3-dithiolane (**62**) which has applications in savoury flavours. In principle this reaction can be used to synthesise more complex molecules by reacting propane-1,2-dithiol with an aldehyde or ketone to produce poly 1-alkylated materials should the need arise. The use of propane-1,3-dithiol instead of ethane-1,2-dithiol allows access to the series of 1,3-dithianes (Scheme 6.31).

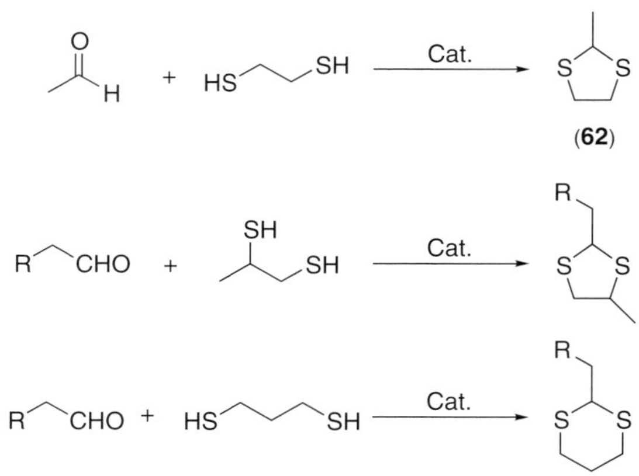

(**62**)

Scheme 6.31

More complex are the trithiolanes and trithianes. IFF has patented the use of 3,5-diisobutyl-1,2,4-trithiolane (**63**) in pork flavourings [28]. The synthesis starts with the reaction of isovaleraldehyde (**64**) with diisopropylamine to form the enamine (**65**). The enamine is then treated firstly with hydrogen sulfide gas and then sulfur, before acid work up to afford the product as a mixture of *cis* and *trans* isomers (Scheme 6.32).

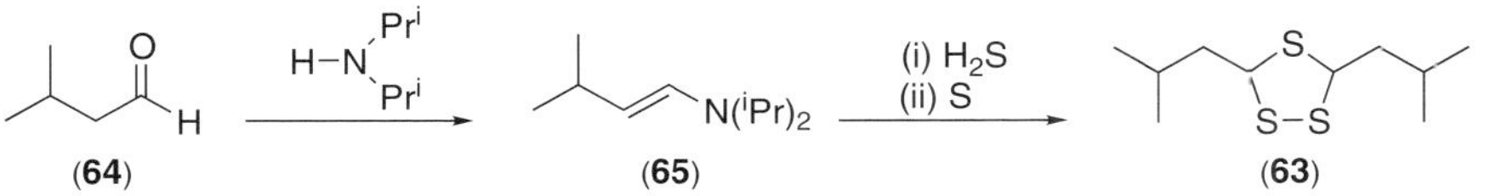

Scheme 6.32

3-Methyl-1,2,4-trithiane (**66**) has been patented as a flavouring agent by Firmenich [29] and may be used in many savoury applications. Their synthesis proceeds *via* the reaction of acetaldehyde with hydrogen sulfide in the presence of diisopropylamine and subsequent treatment of the intermediate with ethane-1,2-dithiol (Scheme 6.33).

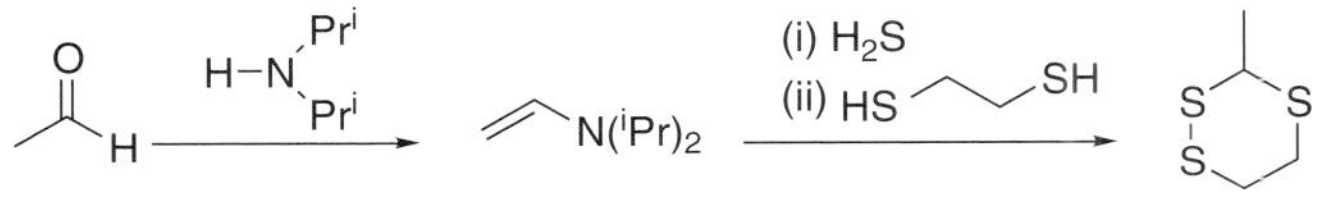

Scheme 6.33

A reaction that demonstrates the highly nucleophillic nature of sulfur is the treatment of the previously mentioned Grapefruit Mercaptan with acid. Protonation occurs in the expected Markovnikov manner at C-2 creating a positive centre at C-1. The carbonium ion is then attacked by the sulfur at C-8 to form a sulfur linkage between C-1 and C-8, thus forming thiocineole (**67**) (Scheme 6.34) [17].

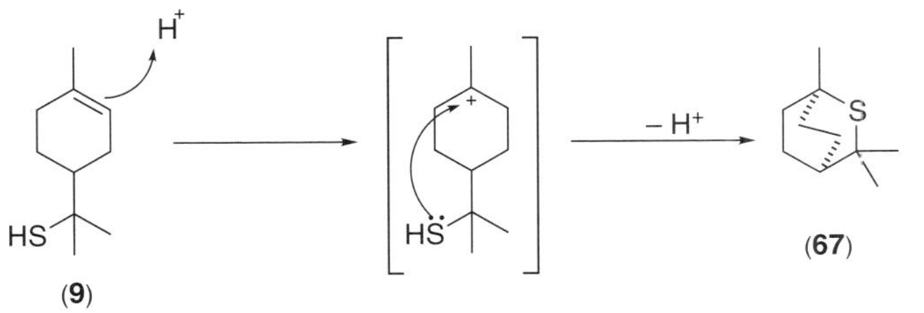

Scheme 6.34

Of the family that contains an additional heteroatom to sulfur, a material that has recently found its way to the GRAS 20 list is 2,4,6-tri-isobutyldihydro-1,3,5-dithiazine (**68**), a material with a distinct aroma of smoked bacon. The material is made by treating isovaleraldehyde (**64**) with aqueous ammonium sulfide (Scheme 6.35) [30].

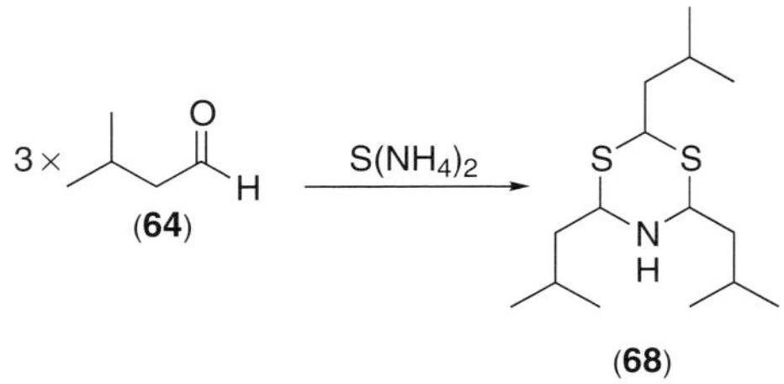

Scheme 6.35

Of perhaps the greatest importance in this family are those molecules that contain oxygen as well as sulfur and, in particular, the 1,3-oxathianes. Both IFF [31] and Firmenich [32] have patented the use of members of this family in flavourings and perfumes. The former concentrates on the 6-methyl-1,3-oxathianes, derived from the condensation of 4-mercapto-2-butanol with a variety of simple aldehydes and ketones, and their application in flavours. Both also lay claim to the synthesis of simple 1,3-oxathiolanes via reaction of an aldehyde with 2-mercaptoethanol in the presence of acid catalyst, a prime example being 2-methyl-1,3-oxathiolane (**69**) derived from acetaldehyde (Scheme 6.36). The Firmenich patent also incorporates the use of more complex members of this group of compounds in F&F applications.

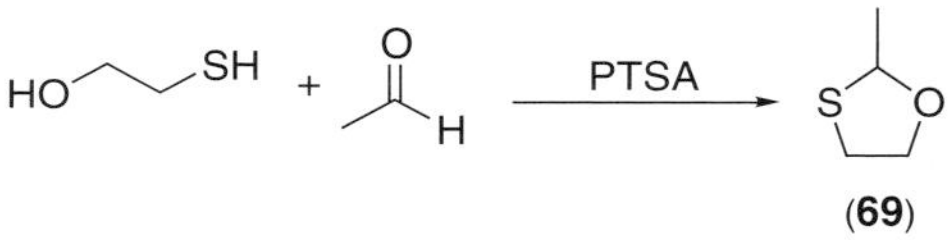

Scheme 6.36

Perhaps, the most widely used member of the 1,3-oxathiane family is 2-methyl-4-propyl-1,3-oxathiane (tropathiane) (**10**). This material is derived from the reaction of 3-mercapto-1-hexanol (**70**) with acetaldehyde in the presence of an acid catalyst (Scheme 6.37) [32]. Tropathiane is widely used in tropical fruit flavours, in particular

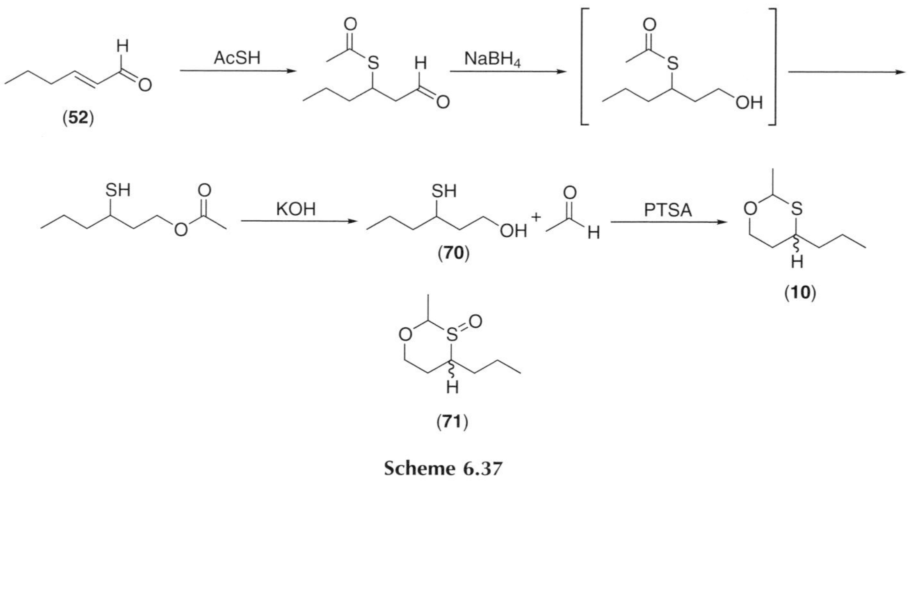

Scheme 6.37

passion fruit. It should also be noted that the Firmenich patent [32] also claims that the sulfoxide of tropathiane (71) has applications in flavourings.

6.4 Quality and stability

Taking a cue from the computer industry, the maxim of garbage in/garbage out (GIGO) also applies to chemicals. A poor-quality raw material can often have a devastating effect on the quality of the final product. This is especially true in the F&F industry where it is common practice for the raw materials as well as the product to undergo odour evaluation. The quality of the raw material is particularly crucial in the manufacture of sulfur chemicals as the presence of sulfur-based impurities, with their powerful odours, can lead to poor organoleptic properties in the product. It is the impurity and odour profiles that are crucial. A higher-grade material (*e.g.* 99.5% with 0.5% of a specific impurity) may give a worse odour profile than a lower-grade material (99%) where the level of the specific impurity is less than 0.1%.

An example of how critical these impurities are can be demonstrated with dimethyl sulfide (7). Most chemists are familiar with the foul smell, reminiscent of rotting cabbage, of this material, as supplied commercially outside the F&F industry at 99+% assay. Careful purification, however, results in a material that has the distinct odour of sweetcorn. The foul smell that we are familiar with is actually due to trace quantities of sulfur chemicals such as ethyl methyl sulfide (72) that can be identified when the material is analysed by GC-MS [6]. Another example is the sodium methyl mercaptide solution mentioned previously. In some instances this material may contain levels of dimethyl disulfide (73) that may react to form by-products that will have an impact on the odour of the final product, even after distillation.

(72) (73)

All is not lost if a problem is encountered with the odour of the final product. Treatment of the off-odour material with dilute acid or base can often remove these off-notes. Alternatively, air or an inert gas such as nitrogen may be bubbled through the material.

In terms of stability, the most common problem associated with the use of mercaptans is oxidation, usually to the corresponding disulfides, materials which typically have a higher odour threshold (and hence lower impact) due to their higher molecular weight and lower volatility. In some instances, this reaction may take place simply on contact with air. The reaction is via a free radical mechanism occurring by loss of an electron (Scheme 6.38). The reaction proceeds at an even faster rate in the case of mercaptide anions; hence the tendency to oxidise increases with pH.

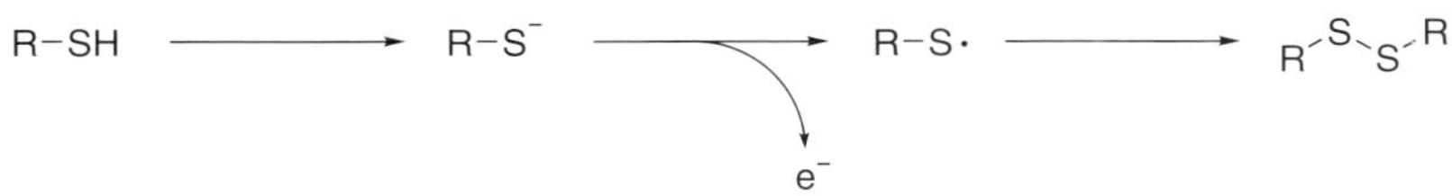

Scheme 6.38

This tendency to oxidise is variable. As the key to oxidation is electron loss, those mercaptans which are electron-rich will show a greater tendency to oxidise. This can be illustrated with methyl and dimethyl furanthiols, where the thiol function is in conjugation with the electron-rich furan ring (Schemes 6.39–6.41) [4]. Both MFT (**2**) and DMFT (**74**) oxidise rapidly in solution in the absence of any stabiliser. DMFT, whose extra methyl group contributes further to the electron density of the ring, is oxidised faster. The fact that this reactivity is not seen in tetrahydro MFT (**14**) provides support to the theory that this is due to the electron-rich nature of the system and not due to any other factors. In the latter case, there is no π-system and the reactivity is that of a secondary aliphatic mercaptan.

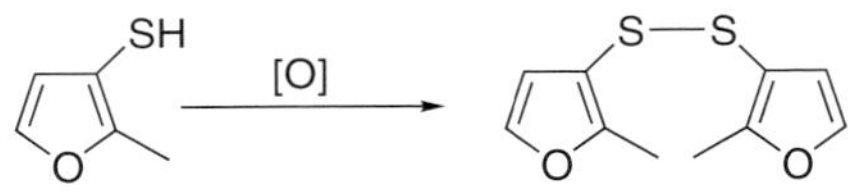

(**2**) (**56**)

In PG, ~50% oxidation, odour fades
In triacetin, slightly more stable, ~30% oxidation
Best used undiluted

Scheme 6.39

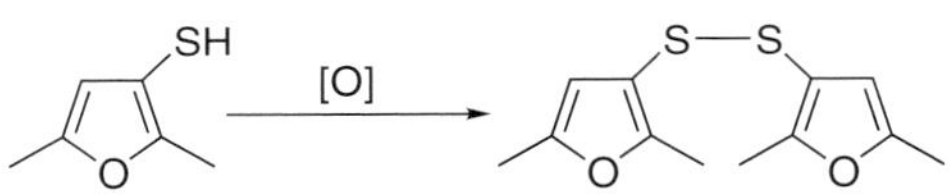

(**74**)

In PG, rapid oxidation, solution becomes dark red, precipitation of disulphide occurs
In triacetin, slightly more stable, ~60% oxidation, stays in solution
Undiluted, goes cloudy due to water formation, requires stabilising with α-tocopherol

Scheme 6.40

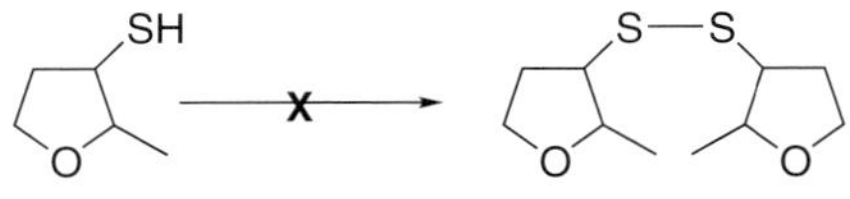

(**14**)

No oxidation, reactivity of 2° aliphatic mercaptan

Scheme 6.41

In the absence of any other factors, the tendency to oxidation is determined by steric factors, *i.e.* primary > secondary > tertiary. Tertiary mercaptans can be remarkably resistant to oxidation. For example, a sample of 1-methoxy-3-methyl-3-mercaptobutane (11) that had been stored for several years appeared to show no oxidation to the disulfide by GC. In order to establish whether this was due to the inability to detect the disulfide a sample was treated with manganese dioxide. This performed the oxidation but was less effective than expected in that only ca. 30% of the disulfide was formed. GC and GC-MS proved that oxidation had not occurred in the sample despite prolonged storage (Figure 6.5) [4]. This chemical stability means that it can be used in demanding media, especially some of those used in the fragrance industry. Stability studies showed it to give excellent results, especially under neutral and acidic conditions. Its loss of performance in alkaline conditions merely reflects the acidity of the thiol function, *i.e.* at high pH the molecule is deprotonated to give the involatile mercaptide ion.

Sulfides are generally less odorous than mercaptans, but have the advantage of greater chemical stability. The sulfide function is unaffected by acid or base and is not readily oxidised under conditions that would rapidly convert a mercaptan to a disulfide. As sulfides are much less reactive than mercaptans, this can give us a greater range of potential applications. They can still, however, be oxidised by bleach; hypochlorite can oxidise sulfides to sulfoxides and sulfones (Scheme 6.42) [33].

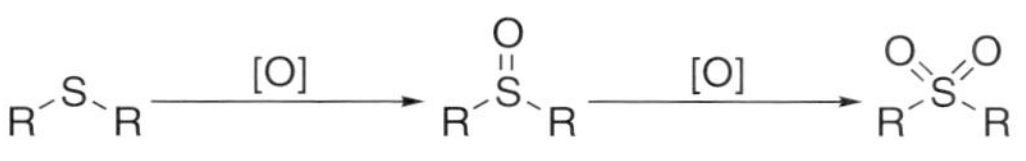

Scheme 6.42

The key stability issue associated with using thioesters is hydrolysis (Scheme 6.43). Thioesters can hydrolyse more rapidly than 'carboxylic' esters since the mercaptide ion (R–S$^-$) is a better leaving group than the alkoxide ion (R–O$^-$). This results in the formation of mercaptans. As the mercaptans are generally more odorous than the parent thioesters this can be a particular problem.

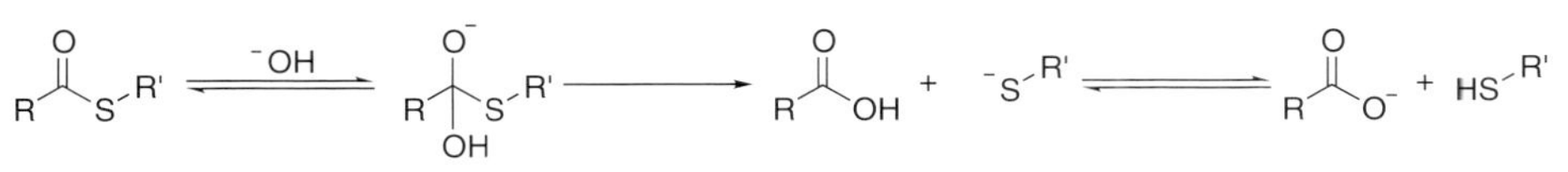

Scheme 6.43

However, this can be turned to our advantage, by using thioesters to generate mercaptans *in situ*. The tendency of MFT (2) to oxidise to the disulfide has already been noted, so a system that generates MFT may have advantages for applications where the conditions used in flavour formulations may be sufficient to release the thiol. Examination of three samples of a 10% solution of MFT acetate (23) in propylene

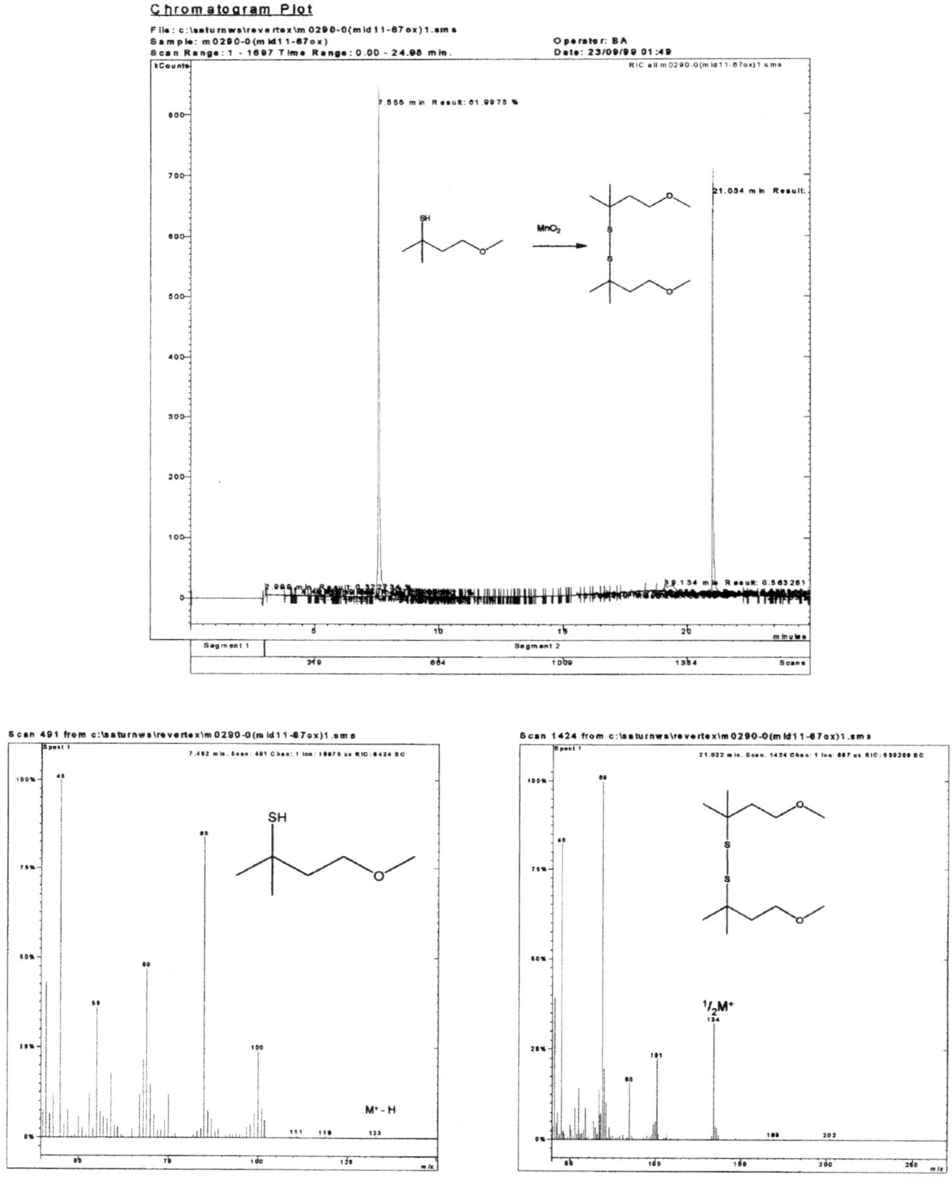

Figure 6.5 Oxidation of 1-methoxy-3-methyl-3-mercaptobutane.

glycol (PG) stored at 4°C (refrigerator), 21°C (ambient) and 40°C (accelerated ageing) revealed the formation of a wide variety of components over a period of time. GC-MS showed these to be MFT, the acetates of PG formed in the 'hydrolysis' and via transesterification, in the older samples, bis-(2-methyl-3-furyl)disulfide (**56**) formed by oxidation of the liberated MFT (Figure 6.6 a,b) [4].

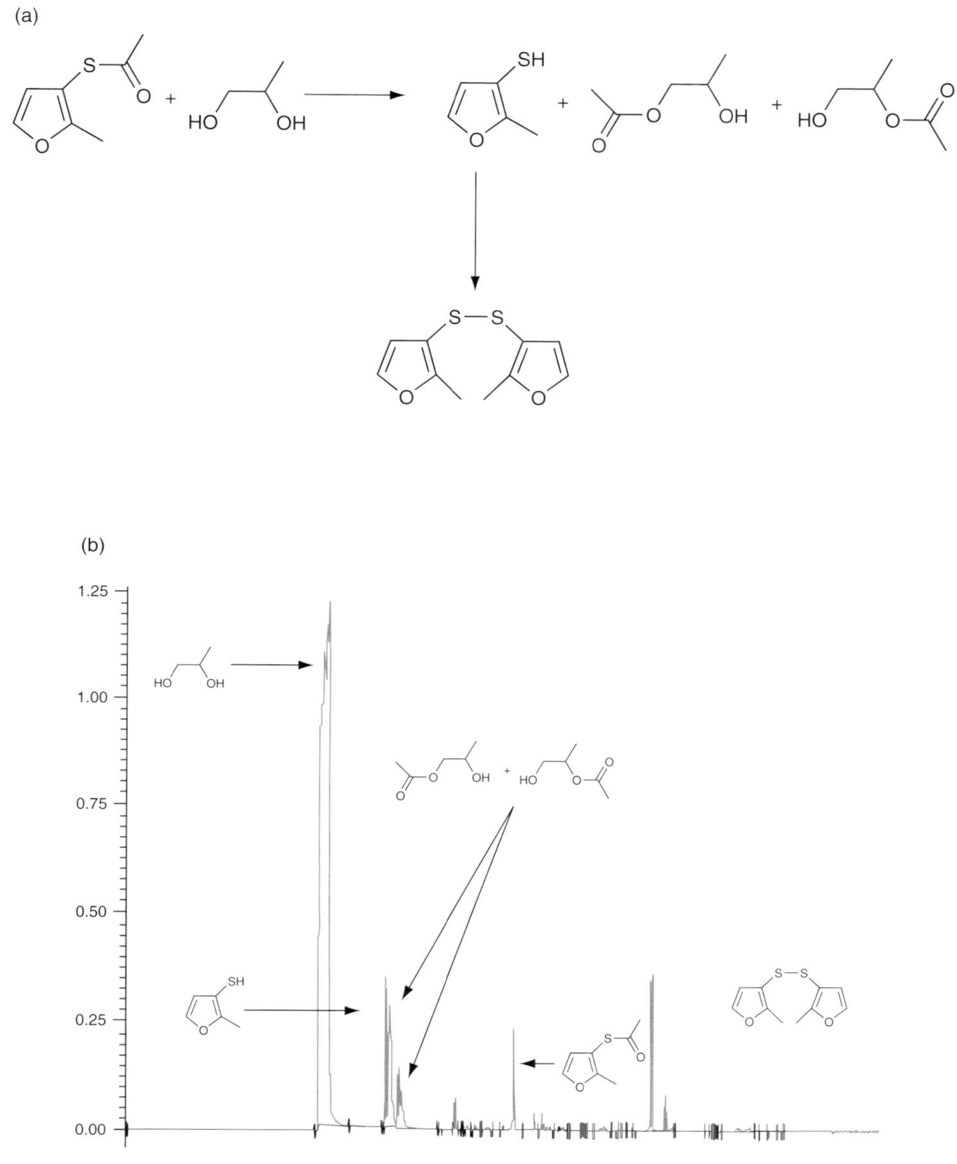

Figure 6.6 Hydrolysis of MFT acetate in PG.

Figure 6.7 demonstrates that over the typical shelf-life of a formulation of MFT acetate in PG, extensive MFT formation will take place. With care, it may be possible to develop a 'steady-state' concentration of mercaptan, where any losses due to oxidation are balanced by generation from the acetate.

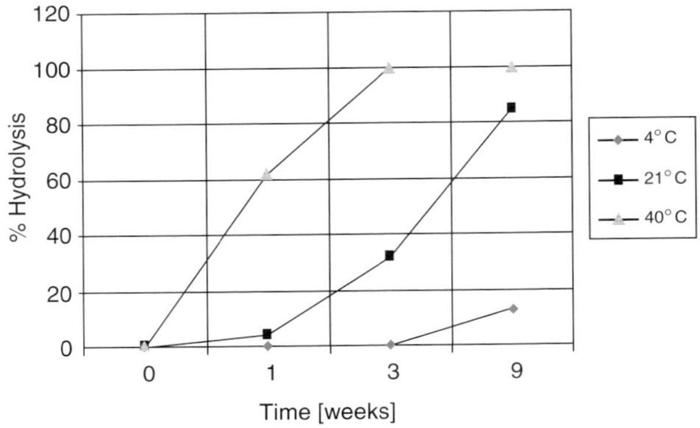

Figure 6.7 Stability of MFT acetate in PG.

Acknowledgements

The author would like to offer his thanks to all his colleagues, past and present, at Oxford Chemicals Limited for their continuing support. In particular, the author would like to thank Dr. Adrian Adamson, Robert Atkinson, Dr. Kevin Auty, Tracy Brown, Gareth Carter-Jones, Dr. Mark Dewis, Dr. Robert Gregory, Dr. John Heffernan, Lee Morgan, Dr. David Rowe and Dr. Peter Setchell for their contribution in part to the work detailed above. The author would also like to thank his associates in the Yule Catto Group at PFW for their contribution to the stability studies.

References

1. van Gemert, L.J. (1999) *Compilation of Odour Threshold Values in Air and Water*. BACIS, The Netherlands.
2. Boelens, M.H. and van Gemert, L.J. (1993) Volatile character-impact sulphur compounds and their sensory properties. *Perfumer & Flavorist*, **18**, May/June, 29–39.
3. Dewis, M.L. and Kendrick, L. (2002) Creation of flavours and the synthesis of raw materials inspired by nature. In K.A.D. Swift (ed.), *Advances in Flavours and Fragrances – From Sensation to the Synthesis*. The Royal Society of Chemistry, Cambridge, UK, pp. 147–160.
4. Rowe, D.J., Auty, K., Jameson, S.B., Setchell, P.K. and Carter-Jones, G.D. (2003) High impact aroma chemicals. Part III. "Fire and brimstone": sulfur aroma chemicals. *Perfumer & Flavorist*, 29, Jul/Aug.
5. Stoffelsma, J. and Pypker, J. (1977) Mercapto alcohols and mercaptoalkyl esters. US Pat. 4 053 656.
6. Auty, K., Jameson, S.B. and Setchell, P.K.(1996–2003) Oxford Chemicals Limited, unpublished results.
7. Evers, W.J., Heinsohn, Jr, H.H. and Mayers, B.J. (1975) Novel 3-thia furans. US Pat. 3 910 966.
8. van den Ouweland, G.A.M. and Peer, H.G. (1977) Certain lower alkyl 4,5-dihydrothiophene-3-thiols. US Pat. 4 020 170.
9. Mitsunobu, O. (1981) The use of diethyl azodicarboxylate and triphenylphosphine in synthesis and transformation of natural products. *Synthesis*, 1–28.
10. Naef, R. (1999) Use of S,S′-ethylidene dithioacetate as a perfuming and flavoring ingredient. US Pat. 5 871 721.

11. Sternson, L. (1978) 1,1-Bis(acylthio)alkanes formed by base-catalysed condensation of thiol acids. *J. Org. Chem.*, **43**(23), 4532–4533.
12. Kafod, H. (1963) 2-Furfuryl mercaptan. *Org. Syn. Coll. Vol. IV.*, 491–493.
13. Nishio, T. (1993) Direct conversion of alcohols into thiols. *J. Chem. Soc. Perkin Trans. I.* 1113–1117.
14. Foster, D.G. (1955) Selenophenol. *Org. Syn. Coll. Vol. III.*, 771–773.
15. Jones, E. and Moodie, Z.M. (1988) 2-Thiophenethiol. *Org. Syn. Coll. Vol. VI.*, 979–980.
16. Mookherjee, B.D., Chant, B.J., Evers, W.J., Wilson, R.A., Zampino, M.J. and Vock, M.H. (1985) Process for preparing mixtures containing 8,9-epithio-1-p-menthene. US Pat. 4 536 583.
17. Demole, E., Enggist, P. and Ohloff, G. (1982) 1-p-Menthene-8-thiol: a powerful flavor impact constituent of grapefruit juice (*Citrus paradisi* Macfayden). *Helv. Chim. Acta*, **65**(6), Nr.176, 1785.
18. Escher, S.D. and van der Waal, M. (2000) 1-Methoxy-3-hexanethiol and its S-acetyl derivative as perfuming and flavouring ingredients. US Pat. 6 129 941.
19. Asinger, F., Thiel, M. and Pallas, E. (1959) Process for the production of α-mercaptoketones., US Pat. 2 888 487.
20. Mann, F.G. and Purdie, D. (1935) The constitution of complex metallic salts. Part IV. The parachlors of palladium and mercury in simple and complex compounds. *J. Chem. Soc.*, 1549–1563.
21. Winter, M. (1967) Pyrazine derivatives. US Pat. 3 328 402.
22. Challenger, F. and Rawlings, A.A. (1937) The formation of organo-metalloidal and similar compounds by micro-organisms. Part V. Methylated alkyl sulphides. The fission of the disulphide link. *J. Chem. Soc.*, 868–875.
23. Milligan, B. and Swan, J.M. (1963) Unsymmetrical dialkyl disulphides from Bunte salts. *J. Chem. Soc.*, 6008–6012.
24. Olin, J.F. (1941) Sulphurization of organic polysulphides. US Pat. 2 237 627.
25. Westlake, H.E., Laquer, H.L. and Smyth, C.P. (1950) The dipole moments and the interconvertibility of diethyl disulfide and trisulfide. *J. Amer. Chem. Soc.*, 72, 436–438.
26. Wilson, R.A., Katz, I., Vock, M.H. and Shuster, E.J. (1976) Flavouring compositions and processes utilizing six membered hetero-cyclic-poly-S-compounds. US Pat. 3 958 030.
27. Greene, T.W. (1981) Protection of the carbonyl group. In *Protective Groups in Organic Synthesis.* Wiley Interscience, New York, USA.
28. Shu, C.-K., Mookherjee, B.D. and Vock, M.H. (1981) Flavoring with 3,5-di-(2-methylpropyl)-1,2,4-trithiolane. US Pat. 4 293 579.
29. Flament, I. (1981) 3-Methyl-1,2,4-trithiane as a flavouring ingredient. US Pat. 4 247 571.
30. Shu, C.-K., Mookherjee, B.D. and Vock, M.H. (1980) Preparation of 2,4,6-tri-isobutyldihydro-1,3,5-dithiazine. US Pat. 4 228 278.
31. Wilson, R.A., Pascale, J.V. and Vock, M.H. (1977) Six membered heterocyclicoxathio compounds. US Pat. 4 042 601.
32. Winter, M. (1981) Heterocyclic derivatives as flavoring agents. US Pat. 4 262 030.
33. Hudlicky, M. (1990) *Oxidations in Organic Chemistry*, ACS Monograph 186, American Chemical Society, Washington DC, USA, p. 254 and references therein.

Bibliography

Swift, K.A.D. (ed.) (1982) *Advances in Flavours and Fragrances: From the Sensation to the Synthesis*, RSC, Cambridge, UK.

Burdock, G.A. (ed.) (1995) *Fenaroli's Handbook of Flavour Ingredients. Vol II.* 3rd Edition, CRC Press, Inc., Boca Raton, USA.

Bauer, K., Garbe, D. and Surburg, H. (1997) *Common Fragrance and Flavor Materials*, 3rd edn, Wiley-VCH Verlag GmbH, Weinheim, Germany.

March, J. (1992) *Advanced Organic Chemistry: Reactions, Mechanisms and Structure*, 4th edn, Wiley Interscience, New York, USA.

Page, P. (ed.) (1995) *Organosulphur Chemistry: Synthetic Aspects*. Academic Press, London, UK.

Belen'kii, L.I. (ed.) (1990) *Chemistry of Organosulphur Compounds: General Problems*. Ellis Horwood Limited, Chichester, UK.

Hudlicky, M. (1990) *Oxidations in Organic Chemistry*. ACS Monograph 186, American Chemical Society, Washington DC, USA.

Websites

www.oxfordchemicals.com
www.perfumerflavorist.com
www.leffingwell.com
www.uspto.gov
gb.espacenet.com

Chapter 7
Aroma Chemicals IV: Musks

Philip Kraft

7.1 Introduction

> Monsieur Eme s'expliquait à présent pourquoi son choix s'était à l'époque – et si spontanément – porté sur le musc à l'exclusion de tout autre parfum, et il comprenait aussi pourquoi cette eau de toilette en particulier lui convenait si bien et servait si parfaitement sa stratégie de séduction et de conquête. La connotation sexuelle de son parfum lui apparut soudain dans toute sa clarté. L'espace d'un instant, Monsieur Eme fut lui-même un chevrotain porte-musc en rut . . .
>
> P. Kemp, *Musc* [1]

Musk is the dried secretion from an internal pouch found between the hind legs of the male musk deer *Moschus ssp.*, which resembles an ordinary deer with hare-like long ears, grey-brown body colour and two canine teeth protruding from its upper jaw. The musk deer measures approximately 60 cm, is 10–13 kg in weight, and inhabits the mountain forests of at least 13 countries in southern and eastern Asia and eastern parts of Russia. Its intensely smelling secretion serves for both territorial boundary marking and for attracting female partners over large distances, i.e. as a pheromone. To harvest the secretion, the animal was traditionally hunted. The main hunting season was in February and March, when hunger forces the animals to leave their inaccessible hiding-places in the mountains [2]. After killing the animal, the pouch was removed with the outer covering of hair and hide, and dried in the sun or on a heated stove. Most of the hair and hide was removed, and the musk pods (Figure 7.1) were soaked in water and then opened to yield fatty brownish grains. By extraction of these musk grains with alcohol, tinctures were prepared that were directly used in perfumery.

The best quality was Tonquin musk from Tibet and China, followed by Assam and Nepal musk, while Carbadine musk from some Russian and Chinese Himalayan regions was considered inferior [3, 4]. To obtain 1 kg of musk grains, between 30 and 50 animals had to be sacrificed, and thus musk tinctures were very expensive perfumery ingredients. At the beginning of the nineteenth century, the price of Tonquin musk grains was about twice their weight in gold. But despite this high price, musk tinctures were still used in perfumery till about 1979, when musk deers were protected from extinction by the Convention on International Trade in Endangered Species of Wild Fauna and Flora (CITES) and complimentary national laws. Today, natural musk, the quantity legally permitted by CITES as well as that poached and illegally traded, continues to be used only in traditional east Asian medicine, as a sedative or stimulant to the heart and nerves, and most importantly as an aphrodisiac.

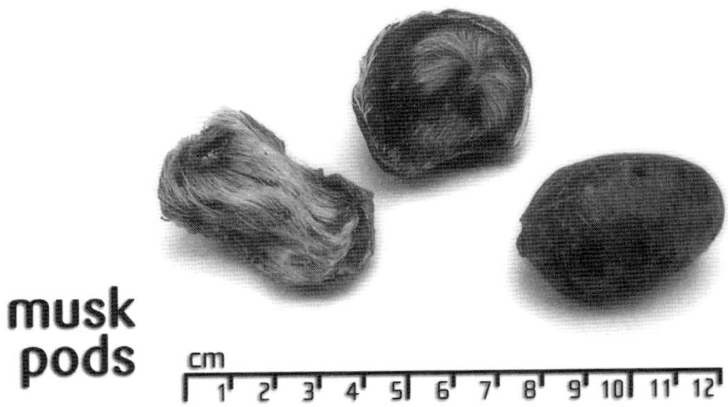

Figure 7.1 Dried musk pods in different views with hair and hide partially removed, approx. 4 cm (1.5 inches) in diameter.

The first olfactory impression when smelling musk tinctures is *animalic*, *sweet* and *ammoniacal*. But the more one studies its character, the more contrasting, vibrant and oscillating it becomes: *repulsive–attractive*, *chemical–warm*, *sweaty–balmy*, *acrid–waxy*, *earthy–powdery*, *fatty–chocolate-like*, *pungent–leathery*, *resinous–spicy*, *fig-like*, *dry*, *nutty* and *woody*, to give just some impressions [5]. To get an idea about this olfactory richness, smell for instance »Musc Ravageur« of *Maurice Roucel* (Editions de Parfums Frederic Malle, 2000) [6], which extrapolates these facets into a well-balanced perfume.

Considering this diversity of adjectives, it is understandable why biologists usually used the term '*moschata*' or '*moschatellina*' simply to indicate a strong smell, and thus several plant and animal species bear this designation: *Abelmoschus moschatus*, *Fragaria moschata*, *Adoxa moschatellina*, *Achillea moschata*, *Malva moschata*, *Mimulus moschatus*, *Fusarium moschatum*, *Aromia moschata*, *Cairina moschata*, *Hypsiprymnodon moschatus*, *Eledone moschata*, *Chrysolampus moschitus*, *Sterotherus moschatus*, *Nosotragus moschatus*, *Desmana moschata* and *Ovibus moschatus* [5]. The latter, musk ox, is even often incorrectly associated as the source of natural musk. In perfumery, however, musk is generally considered a prime olfactory descriptor. What then is '*musky*', and what are '*musks*'?

The term '*musk*' is an abstraction from the complex odour impressions of natural musk tinctures, especially from the dry-down, after the more volatile parts are evaporated. It refers to the *warm*, *sensual*, *sweet-powdery* tonality of the dry-down. This abstraction began in 1888, with the discovery of '*Musk Bauer*' (**1**) by Albert Bauer [6]. The introduction of this musk as a 10% solution in acetanilide for about $500 per kilo [7], which was half the price of musk tincture at that time, was such an immediate commercial success that several small fragrance companies were formed all over Europe. We will return to '*Musk Bauer*' later, but for the moment let us keep in mind that the term '*musk*' was coined even before the odorous principle of Tonquin musk was known.

Today, younger perfumers hardly know Tonquin musk, and chemical benchmarks define the '*musky*' odour impression, an impression that is most often associated with the odour of *baby skin*. Hence, the smooth, soft and intimate '*skin-on-skin*' connotation of musks and their indispensability in perfumery to impart sensuality. Musks refine, exalt, fix, stimulate, balance and harmonise perfumery compositions; they have an erotic effect, bring volume and diffusivity, and impart warmth and liveliness. Or in one phrase, they form the skeleton of a perfume. There is probably not a single fragrance on the market without containing any musk odorant. Actually, there is even one perfume with *just* musk: »Velviona« (Helmut Lang, 2001).

In terms of benchmarks, 15-pentadecanolide (**2**, Thibetolide®, Exaltolide®) is probably the commercial musk with the least side notes; and thus, it will serve as the reference to what we understand as a *musk odorant*. Some steroids were claimed to emanate *musky* odours; however, they generally smell *urinous* and *animalic*, aspects present in natural musk tinctures, but not musky in terms of the above definition, even in dilution. An interesting steroid in terms of odour is α-androst-16-en-3α-ol (**3**), but also this is not musky but *sandalwood-like* in dilution. That should suffice to avoid any confusion, which sometimes found its way into the literature. So let us now move on to the chemical structures of natural musk odorants.

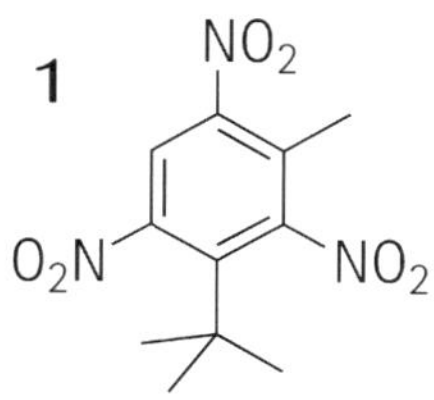

Musk Bauer

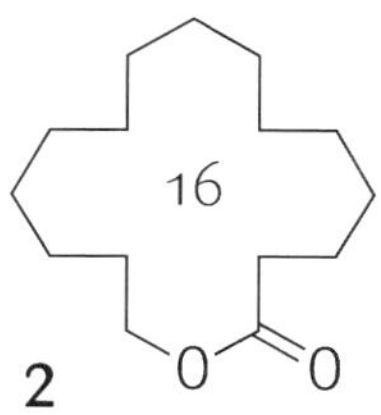

Thibetolide®
Exaltolide®

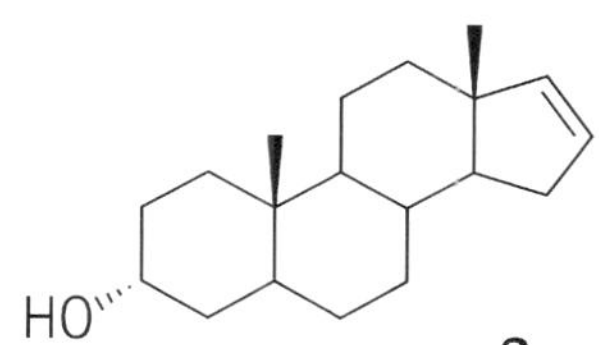

3

α-androst-16-en-3α-ol

7.2 Natural musks

Bearing the commercial success of Musk Bauer (**1**) in mind, the apparent next question was, which compounds were responsible for the musk note of the natural Tonquin musk? Were these simple aromatic compounds like **1** or vanillin? Were they easy to manufacture? This was the motivation for Heinrich Walbaum [8] of Schimmel & Co, at that time the biggest fragrance manufacturer, to investigate musk-deer glands at the beginning of the twentieth century [9]. In 1906, he isolated the principal musk odorant, established that it was a ketone of the empirical formula $C_{16}H_{30}O$ with two double-bond equivalents, and named this new ketone 'muscone' (**4**) [10]. However, he had no clue about the chemical structure of muscone (**4**), the content of which varies from 0.5 to 2% in musk grains. And thus, the structure elucidation and synthesis of **4** remained high on the wish list of the young fragrance industry; and also on the wish list of Philippe Chuit, who had founded a small fragrance company with Martin Naef in Geneva. In 1920, he contacted Leopold Ruzicka at the ETH Zurich, who was looking for financial support and who previously had been working on irone in collaboration with the German company Haarmann & Reimer (now Symrise). In 1921, Ruzicka and Naef et Cie. (the later Firmenich) closed a co-operation contract [11] and, after work on nerolidol and farnesol, Ruzicka started the structure elucidation of muscone (**4**) at the end of 1922. Vigorous chromic acid oxidation of **4** yielded a series of aliphatic dicarboxylic acids that finally led to the conclusion that muscone was 3-methylcyclopentadecan-1-one (**4**) [12]. According to Baeyer's theory of strain in multimembered rings [13], such macrocyclic compounds were thought to be too unstable to exist; thus, Ruzicka's finding opened a new chapter of organic chemistry, which, amongst other achievements, earned him a Nobel prize in 1939.

Ruzicka also investigated and elucidated the structure of civetone (**5**) [14], the musk odorous principle of the glandular secretion of the civet cat (*Viverra civetta*), which had been isolated in 1915 [15]. In comparison with **4**, civetone (**5**) possesses a far more animalic musk character, perhaps due to a steroid-type folding that **5** could adopt. Even though such a steroid-type folding had already been proposed by Vladimir Prelog and Leopold Ruzicka [16] in 1943, and found its way into many organic chemistry textbooks, we shall not forget about the severe transannular strain it implies. It is virtually impossible for an odorant receptor to compensate for the resulting steric hindrance of the transannular hydrogens. The concept that due to attractive van der Waals interactions two opposite sides of a macrocycle would align just like aliphatic lipid chains, is simply wrong. To avoid transannular interactions, the *gauche* atoms in a macrocyclic ring tend to be separated as far as possible from each other. They form the corners of a triangle, a square or a pentagon, the sides of which consist of *trans*-configured aliphatic chains [17].

But let us not forget that Chuit was aiming at an industrial synthesis of muscone (**4**), no matter how interesting these macrocycles were from an academic point of view. Therefore, Ruzicka attempted the synthesis of macrocyclic ketones by pyrolysis of heavy metal salts of dicarboxylic acids, a method previously applied by Zelinsky [18] and Willstätter [19] to the synthesis of smaller ring systems. However, this cyclisation method [20] employing thorium salts gave only low yields of macrocyclic ketones in the order of 5–6%, and did not tolerate the β-methyl group of muscone (**4**). Therefore, in 1926 the *nor*-muscone, cyclopentadecanone (**6**, Exaltone®) was introduced to the market instead of muscone (**4**), at the exorbitant price of 50 000 CHF/kg [11] due to the low yields and the difficult purification. Cyclopentadecanone (**6**) possesses a pleasant musk note, and was later even identified in nature; it occurs in the scent glands of the Louisiana muskrat [21].

In-depth modern analyses of the odoriferous principles of Tonquin musk were carried out by Mookherjee and co-workers [22, 23] of IFF in 1970 and 1981. In the course of these studies, several new macrocycles were identified, the most interesting one in terms of odour being (5*Z*)-14-methylcyclopentadec-5-en-1-one (**7**). Other structurally interesting constituents are muscoxide (**8**) and muscopyridine (**9**), the latter of which was found by Ruzicka and co-workers [24] in 1946, and had been synthesised by Büchi and co-workers [25] in 1957.

The musk odorants of the animal kingdom are exclusively macrocyclic ketones. Today, *rac*-muscone (*rac*-**4**), civetone (**5**) and Exaltone® (**6**) are all important commercial products, and muscone continues to be a popular scientific target structure, especially for enantioselective synthesis of its naturally occurring (*R*)-(–)-isomer **4**. With an odour threshold of 4.5 ng/l air [26] **4** is not even an extremely potent musk, and in water the threshold of the enantiomers differ only by a factor of 4, i.e. 61 ppb for **4** and 233 ppb for *ent*-**4**, while being similar in odour character [27]. However, the fascination for the genuine natural odorant motivated chemists to investigate several stereoselective syntheses, of which we present three recent ones.

Alexakis and co-workers [28] investigated the copper-catalysed enantioselective Michael addition of dimethyl zinc to cyclopentadec-2-en-1-one (**10**). Out of various 2-arylcyclohexal-substituted phosphites the ligand **11** turned out to be particularly

4 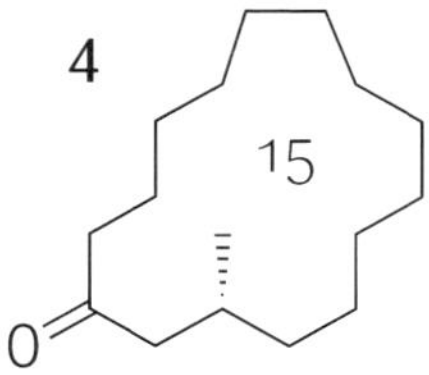 **5** 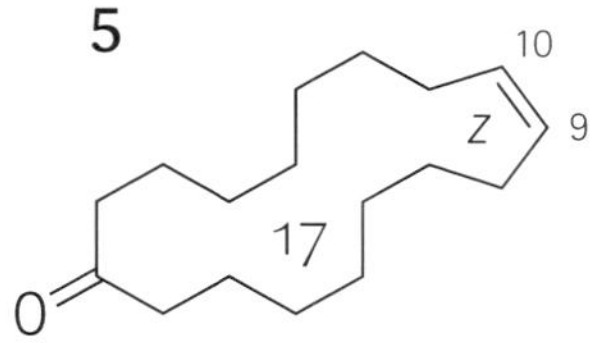 **6** 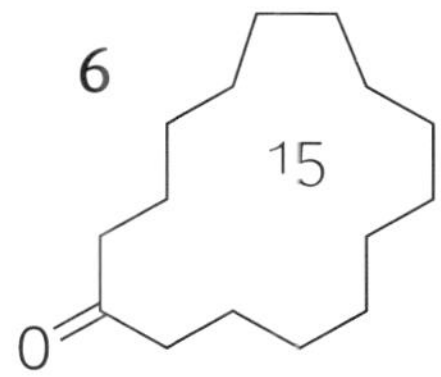

(*R*)-(-)-muscone civetone Exaltone®

7 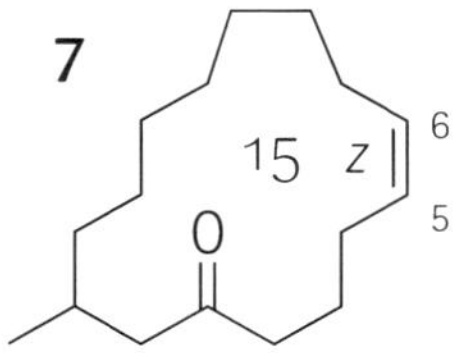 **8** 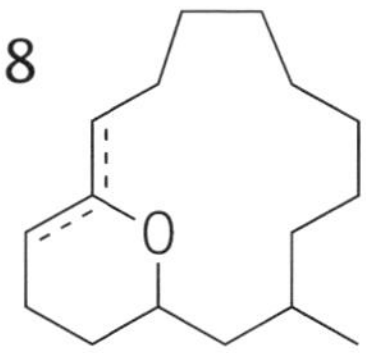 **9** 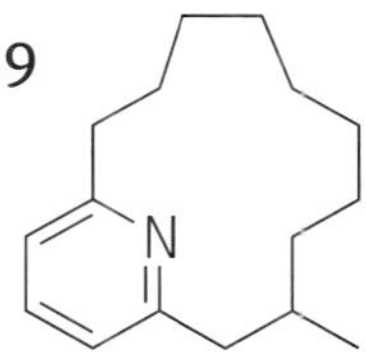

muscoxide muscopyridine

efficient for the 1,4-addition of organozinc reagents and furnished **4** with an enantiomeric excess of 79%. A very elegant synthesis of **4** was reported by Grubbs and co-workers [29]. The stereochemical information stems from (*R*)-(+)-citronellal, which is commercially available in high enantiomeric purity. A simple Grignard reaction with 1-bromodec-9-ene provides **12** that was subjected to ring-closing metathesis (RCM) in the presence of the second-generation ruthenium alkylidene **13**. Following RCM, sodium hydroxide and 3-pentanone were added to initiate Ru-catalysed transfer hydrogenation to provide the macrocyclic ketone. Addition of gaseous hydrogen then resulted in the Ru-catalysed hydrogenation of the formed double bond and afforded **4** in overall 56% yield. Another approach was recently developed at Takasago [30]. Intramolecular aldol condensation of 2,15-hexadecadienone (**14**), prepared from 1,10-dibromodecane by double acetoacetic ester synthesis, furnished an (*E*/*Z*)-mixture of 3-methyl-2-cyclopentadecen-1-one. Silica-gel column chromatography (CC) afforded the pure (*E*)-isomer, which was required for the enantioselective reduction using the $Ru_2Cl_4[(S)$-*p*-tolyl-BINAP$]_2NEt_3$ catalyst system. Employing this catalyst, hydrogenation at room temperature at 70 atmospheres provided (*R*)-(−)-muscone (**4**) with >98% *ee*. If some day one of these three strategies will be applied on an industrial scale is an open question.

In the plant kingdom, musk odorants are exclusively macrolides, macrocyclic lactones. The most important one we have already met, 15-pentadecanolide (**2**, Thibetolide®, Exaltolide®) was discovered in 1927 in Angelica root oil (*Archangelica officinalis* Hoffm. syn. *Angelica archangelica* L.) by Max Kerschbaum [31] of Haarmann & Reimer. At the same time, the competitor Firmenich introduced it into perfumery as Exaltolide®, at the exorbitant price of 100 000 CHF/kg [11]. They had subjected their Exaltone® (**6**) to subsequent Baeyer–Villiger oxidation. The 15-pentadecanolide (**2**), thus obtained, has a more distinct and delicate musk note and is less animalic than **6**, and like most macrolides, it smells slightly floral. The odour threshold of **2** was measured as 2.1 ng/l air [10].

In a recent in-depth analysis [32], 15-pentadecanolide (**2**) was shown to account for about 42% of the macrolide fraction of Angelica root oil. Besides some traces of

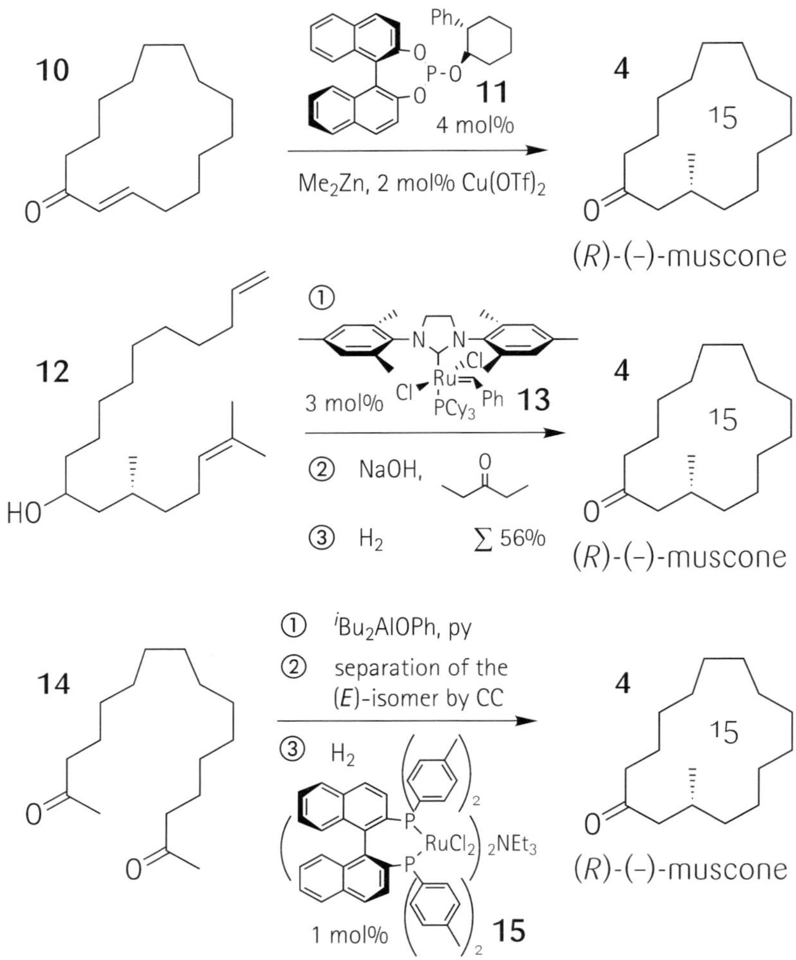

16-hexadecanolide, several other macrolides were detected. The main component is 13-tridecanolide (**16**), which makes about half of the macrolide fraction, accompanied by its 12-methyl derivative **18/19** occurring in around 4% of the macrolide fraction. This is followed by 3.5% of 17-heptadecanolide (**17**) and 0.5% of 14-tetradecanolide (**20**), so that the main components possess an odd number of ring-carbon atoms. However, excepting **18/19**, no further ($\omega - 1$)-methyl macrolides occur, not even of the second-most abundant component 15-pentadecanolide (**2**). Though not occurring in nature, the corresponding 14-methyl-15-pentadecanolide is a very interesting musk odorant [33], and in an obvious analogy it has been termed '*muscolide*' by Günther Ohloff [34]. Nevertheless, 12-methyl-13-tridecanolide (**18/19**), discovered in Angelica root oil by Taskinen [35], is an interesting musk odorant too, especially since it was the first example for which the importance of methyl substituents on the character of macrocyclic musks was demonstrated [36, 37]. The (*R*)-isomer **18** possesses a *clean musk note with strong sandalwood accents and a slight fruity undertone of pear*, while the enantiomer **19** differs unambiguously by a *pronounced animalic musk note accompanied by camphoraceous aspects*. The *nor*-derivative **16** is much more *camphoraceous and rather cedarwood-like than musky*. Both enantiomers **18** and **19** occur in Angelica root oil [32] in the ratio of 72:28, and the major enantiomer **18** recently has become an attractive target for synthetic organic chemists [38]. These were the most important aspects on Angelica root oil, and if you are interested in an olfactory impression, smell »Angeliques sous la pluie« (Editions de Parfums Frederic Malle, 2000) [6], in which *Jean-Claude Ellena* interprets this theme.

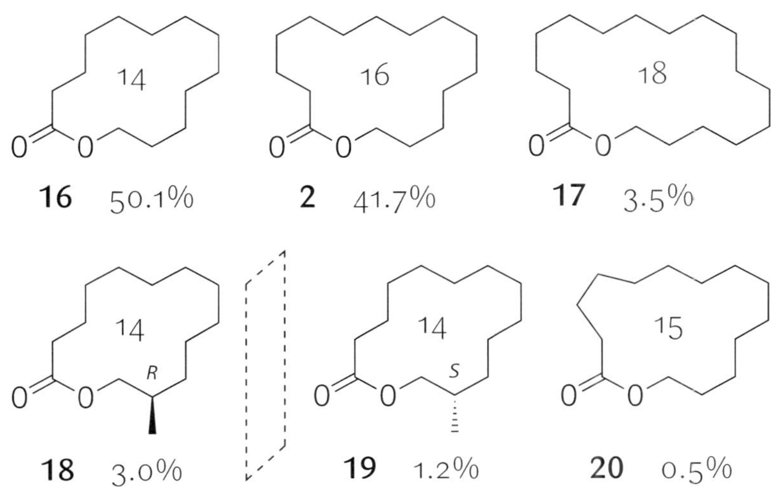

16 50.1% **2** 41.7% **17** 3.5%

18 3.0% **19** 1.2% **20** 0.5%

Next to 15-pentadecanolide (**2**), the commercially second-most important natural musk is ambrettolide (**21**), or, more strictly, **21** rather than **22**, which is sold as Ambrettolide®, i.e. a double-bond isomer. The reason for this is that it is not synthesised from ambrettolic acid, but from aleuritic acid, the main acidic constituent of shellac [39]. So do not get confused, what was isolated from Ambrette seed oil (*Hibiscus abelmoschus* L. syn. *Abelmoschus moschatus* M.) by Kerschbaum [31] in 1927 has the structural formula **21**. However, both compounds are relatively similar in smell: *Powerful musky, warm-aromatic with fruity-floral nuances and some reminiscence to genuine Ambrette seed oil.*

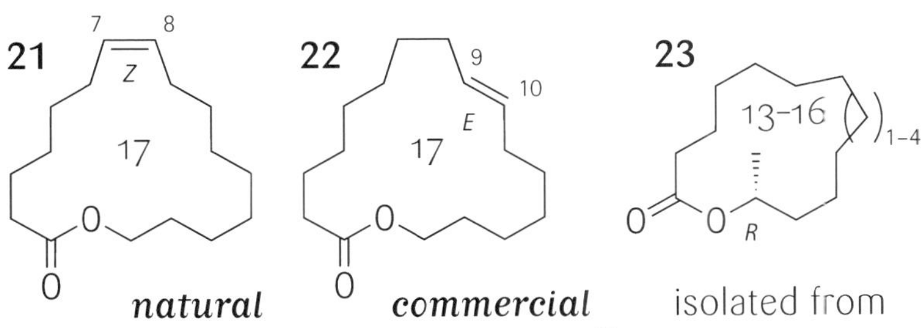

21 ***natural*** ambrettolide

22 ***commercial*** Ambrettolide®

23 isolated from ***galbanum oil***

The four macrolides of the general formula **23**, which were discovered by Roman Kaiser and Dietmar Lamparsky [40] of Givaudan in Galbanum oil (*Ferula galbaniflua* Boiss. et Buhse and *Ferula rubicaulis* Boiss. syn. *Peucedanum galbanifluum* Baill. and *Peucedanum rubicaule* Baill.), do not at all possess the character of their parent essential oil. Galbanum oil has a very *characteristic sharp spicy-green, leaf-like character with balsamic nuances*. The odour characteristics of the isolated ω-methyl macrolides **23** range from *fruity-woody* (12-tridecanolide) via *woody-cedar, balsamic* (13-tetradecanolide) to *distinct musky* (14-pentadecanolide and 15-hexadecanolide), and again the compounds with an odd number of ring-carbon atoms are the more abundant in the oil. It is interesting to note that, in comparison with **18** and **19**, both enantiomers of 13-tetradecanolide smell very similar and both possess complex cedarwood-type odours [37], so an ω-methyl group probably sterically hinders the interaction with the musk receptor. We clearly see that methyl groups do matter.

This is stressed, because Max Stoll [41] of Firmenich summarised in 1936 his data on macrocycles derived from the natural leads in the following empirical rules (slightly abbreviated and adapted):

(1) To smell musky the macrocyclic ring must contain 14–18 members.
(2) One carbonyl or imino group must be present, but a second one will completely destroy the musk odour.
(3) Lactones and oxa ketones are stronger than ketones, yet further oxa substituents diminish the odour intensity.
(4) Methyl groups have almost no influence on the odour.

As a rule of thumb, these statements are quite practical, but one must be very careful as there are important exceptions to each of them. Above, we saw the importance of methyl groups. These exceptions provide interesting insights into the design of new powerful macrocycles. However, for the moment we leave it here, since for a long time macrocyclic musks had economically no chance against the easier-to-produce and consequently much cheaper nitro and polycyclic musks.

7.3 Nitro musks

So back to the nitro musks. Musk Bauer (**1**) had simply been discovered by serendipity [10], Albert Bauer had actually been searching for new explosives and had alkylated trinitrotoluene (TNT) by Friedel–Crafts reaction with *tert*-butyl halides [7]. Together with the *Société des produits chimiques de Thann et de Mulhouse* he patented and produced Musk Bauer (**1**), and motivated by its immediate success, systematically studied further derivatives. In short succession he discovered Musk xylene (**24**), Musk ketone (**25**) [42] and Musk ambrette (**26**) [43], all before the dawn of the twentieth century. These three musks **24–26** soon replaced Musk Bauer (**1**), and became the epitome of musks for the next 50 years. Many perfumes of that period, like for instance »Chanel N°5« by *Ernest Beaux* (Chanel, 1921), had over 10% of nitro musks incorporated, mainly **25**. Musk ketone (**25**), with an excellent odour threshold of 0.1 ng/l air [10], is considered to resemble the natural Tonquin musk character quite closely, with a *sweet, powdery, mild animalic and warm character*. Musk xylene (**24**) is somewhat harsher, and due to its cheaper price found more application in the soap and detergent segment. Musk ambrette (**26**) was special in adding a *heavy floral side note* to the *persistent sweet musk tonality* of the family. Consequently, it was often employed in floral perfumes, like for instance the floral salicylate prototype »L'Air du Temps« by *Francis Fabron* (Nina Ricci, 1948). Especially in these perfumes, Musk ambrette (**26**) was difficult to replace, when it was severely restricted [3] in use in 1981, and subsequently withdrawn from the market. Reasons for the restrictions of the nitro arenes were mainly a certain toxicity (neurotoxicity) and their phototoxicity, but besides they also caused ecological concerns due to their poor biodegradability.

Extensive research on nitro arenes – in the course of which also the wrong structures assigned to Musk ketone (**25**) and Musk ambrette (**26**) by Bauer were corrected – led to the introduction of three further nitro musks by Givaudan: Musk tibetene (**27**) [44], Musk *alpha* (**28**) and Moskene (**29**) [45]. Musk tibetene (**27**) discovered by M. S. Carpenter [44] in 1937 became the commercially most important one of these, being closer in odour to **25** than to **24**. Musk *alpha* (**28**) is interesting for its unique blend of marine and musky tonalities, while Moskene was the first musk with an indanic skeleton [46].

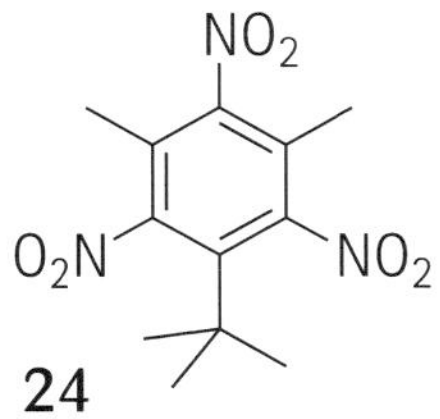

24
Musk xylene

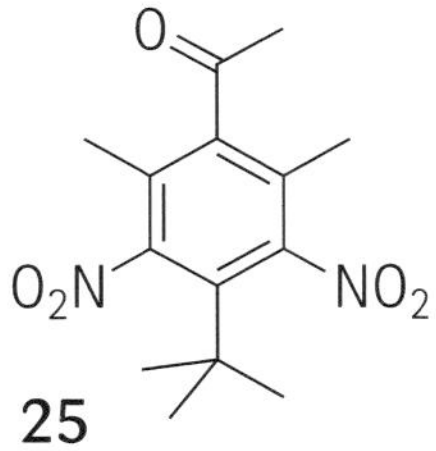

25
Musk ketone

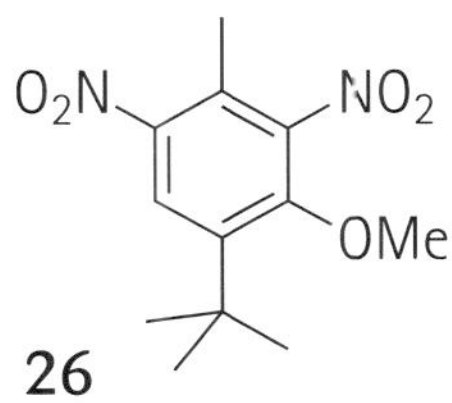

26
Musk ambrette

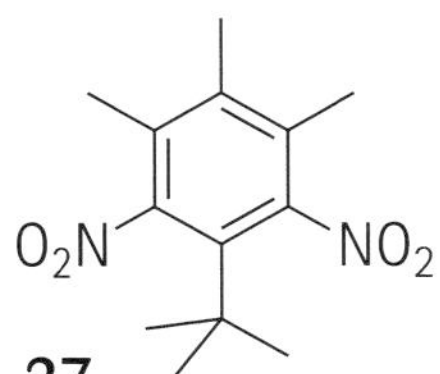

27
Musk tibetene

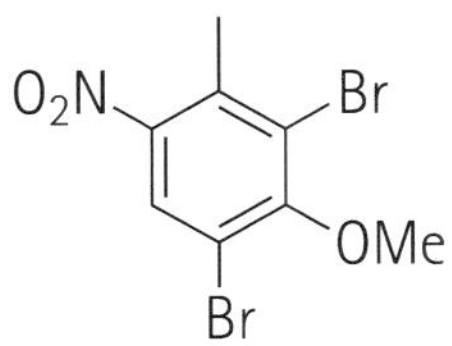

28
Musk *alpha*

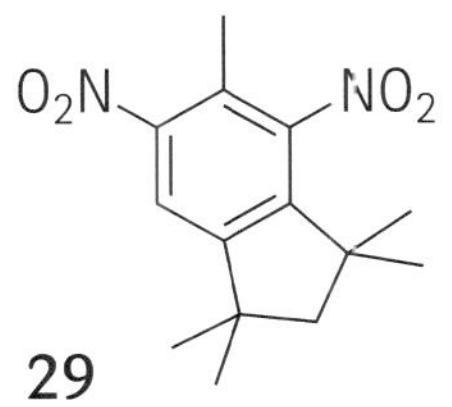

29
Moskene

In 1951, Carpenter [47] of Givaudan summarised his data on nitro musks in the following empirical rules (slightly modified):

(1) To smell musky, nitro arenes must possess a *tert*-alkyl group (*tert*-butyl or *tert*-amyl) and a molecular weight below 300 u.
(2) Either two nitro groups need to be present, or a nitro group and an alkoxy group (MeO, EtO or ^{i}PrO).
(3) In alkoxy nitro arenes, the *tert*-alkyl group must be *ortho* to the alkoxy group in order to smell musky.

Of course, there are also exceptions to these simple rules: 2-Bromo-1-*tert*-butyl-4-methoxy-3,5-dinitrobenzene smells musky, while 2-bromo-5-*tert*-butyl-4-methoxy-1,3-dinitrobenzene with the *tert*-butyl group *ortho* to the methoxy group does not, and Musk *alpha* (**28**) is devoid of any *tert*-alkyl moiety.

7.4 PCM – Polycyclic aromatic musks

In the beginning of 1948, Carpenter [47] himself had found the most striking exception to his postulated rules: 2,4-Di-*tert*-butyl-5-methoxybenzaldehyde (**30**), a fine persistent musk odorant without a nitro group. Though it was never introduced to the market, it became known as Ambral within Givaudan, and inspired chemists to synthesise many related nitro-free aromatic musks. Due to their photochemical reactivity and their instability in alkaline media, nitro musks decomposed and caused discolouration problems in functional products. If it were possible to synthesise aromatic musks without nitro groups, it would be possible to derive stable musks for functional perfumery, for cosmetics, laundry and detergents, for a huge market with enormous potential.

The first nitro-free aromatic musk to get introduced to perfumery was discovered by Kurt Fuchs in 1951, who assigned his invention to Polak's Frutal Works. It was introduced to the market one year later, even without the knowledge of its correct structure, which was reflected in its name 'Phantolide®' (AHDI, **31**). When four years later [48] its structure was elucidated as acetyl indane, polycyclic arenes became the new lead structure for musk odorants. With an odour threshold of 6.7 ng/l air, Phantolide® (AHDI, **31**) was not superior in terms of strengths; yet, due to its stability and hydrophobicity, its performance in washing powders and detergents was outstanding. Now, besides hedonics (*the pleasance*) and odour strength, the evolution in musk odorants was mainly directed towards stability in harsh media and hydrophobicity, a key factor in the deposition of odorants on the fabric during the washing process. This evolution gave rise to numerous polycyclic musks (PCM): Traseolide® (ATII, **32**), Celestolide® (ADBI, Crysolide®, **33**), Fixolide® (AHTN, Tonalide®, *rac*-**34**), Vulcanolide® (FHTN, *rac*-**35**), Versalide® (ATTN, **36**) with the isochromane musk Galaxolide® (HHCB, Pearlide®, *rac*-**37**) as the culmination of this evolution to stability and hydrophobicity.

Galaxolide® (*rac*-**37**) was the result of systematic studies of M. G. J. Beets of IFF on the osmophoric group of PCMs, the most polar functional group of odorants, the one that is proposed to orient the odorant on the receptor site. So far, the known PCMs had all carbonyl functions, but Beets and Heeringa wanted to replace the carbonyl group by other functional groups in order to make them even more stable and hydrophobic. The recipe was an ether oxygen, placed in a rigid tetrahydropyran ring at the same position they assumed for the carbonyl oxygen of the other known PCMs. Galaxolide® (*rac*-**37**) was first synthesised in 1965, and already in the late 1960s, used in dosages up to 40% in fabric softeners such as ›Comfort‹ and ›Softlan‹, and in detergents like ›Coral‹ at 27%. But high doses were also incorporated in fine fragrances, for instance »Trésor« by *Sophia Grojsman* (Lancôme, 1990) with its 21.4% of Galaxolide® (*rac*-**37**).

The odour threshold of the commercial isomeric mixture of Galaxolide® (*rac*-**37**) is 0.9 ng/l air. However, the different stereoisomers differ distinctly in their thresholds [49], and this provides some interesting insight into the molecular features of musks.

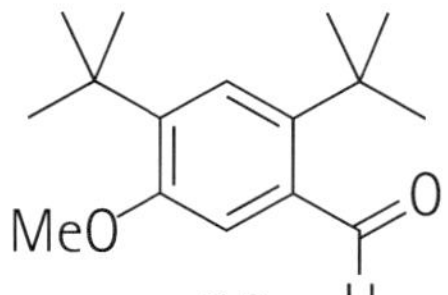

30 ambral

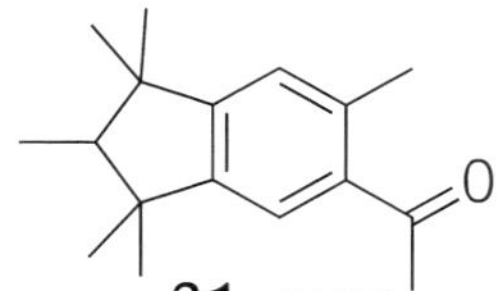

31 AHDI Phantolide®

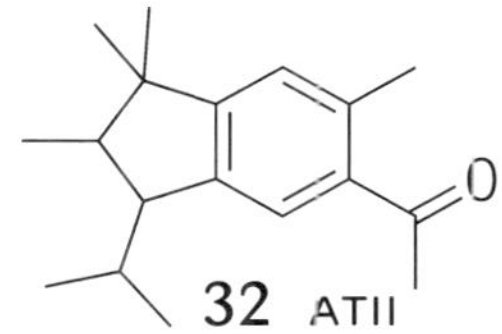

32 ATII Traseolide®

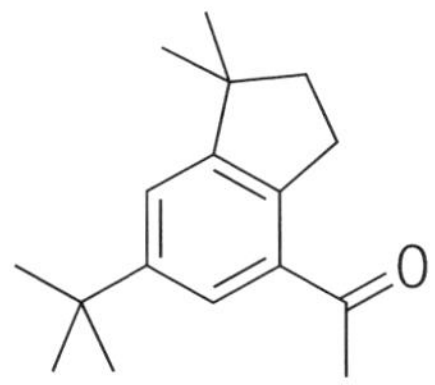

33 ADBI Celestolide®

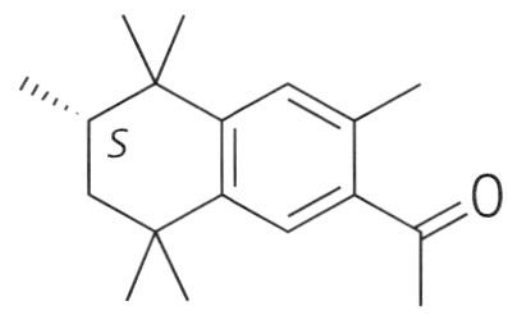

34 AHTN (*S*)-(-)-Fixolide®

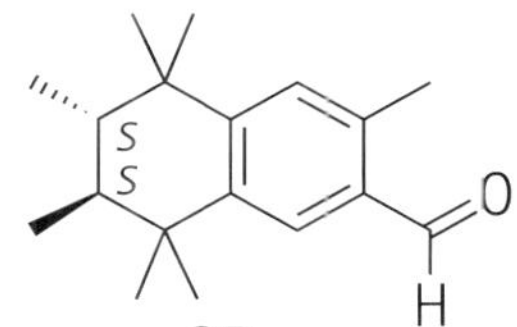

35 FHTN (*S*,*S*)-(-)-Vulcanolide®

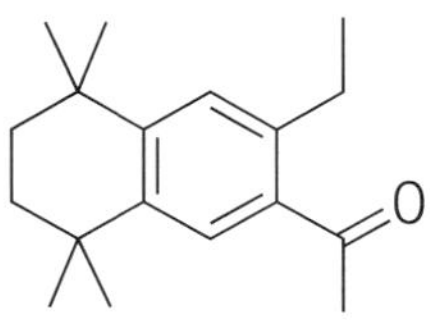

36 ATTN Versalide®

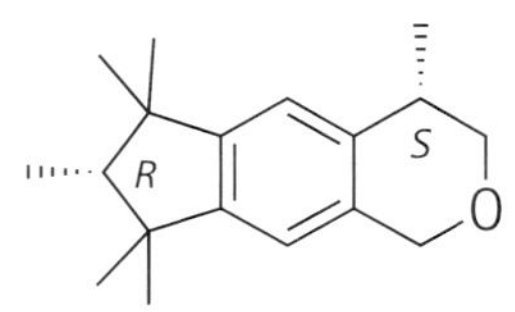

37 HHCB (4*S*,7*R*)-(-)-Galaxolide®

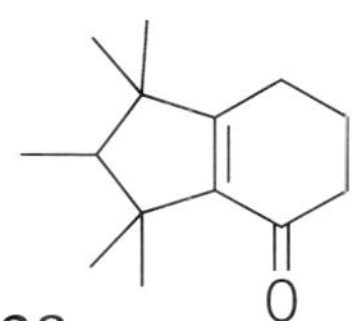

38 DPMI Cashmeran®

All four stereoisomers were synthesised by Friedel–Crafts alkylation of 1,1,2,3,3-pentamethylindane with (*S*)- and (*R*)-methyloxirane, acid-catalysed reaction with paraformaldehyde and separation of the diastereomeric pairs via their tricarbonyl chromium complexes. Interestingly, the configuration at C-4 was much more important than that at C-7, reflected in the odour threshold values of the four isomers: 0.63 ng/l air for **37**, 130 ng/l air for *ent*-**37**, 1.0 ng/l air for the (4*S*,7*S*)-(−)- and 440 ng/l air for the (4*R*,7*R*)-(+)-isomer [49], respectively. So the odour of commercial Galaxolide® (*rac*-**37**) is determined by the (4*S*)-(−)-isomers, while the 7-Me group seems to be situated in a hydrophobic bulk region of the receptor site, which is not very shape-dependent [26]. Transferred into the old (1894) but vivid 'lock-and-key' model (Figure 7.2) of Emil Fischer, the osmophoric ether group *inserts* the molecule into the receptor, the profile-determining 4-Me group *codes* the odour information and the bulky substituents provide a *firm grip*. Together, these molecular parameters make up the olfactophore, the *key* for the musk receptor.

But is there really only one musk receptor? In the series, Musk ketone (**25**), Phantolide® (**31**) and Galaxolide® (*rac*-**37**), there is some apparent molecular similarity, but do these benzenoid musks use different receptors than, for example, the macrocyclic musks? This is a very difficult question to answer. To gain some insight, the

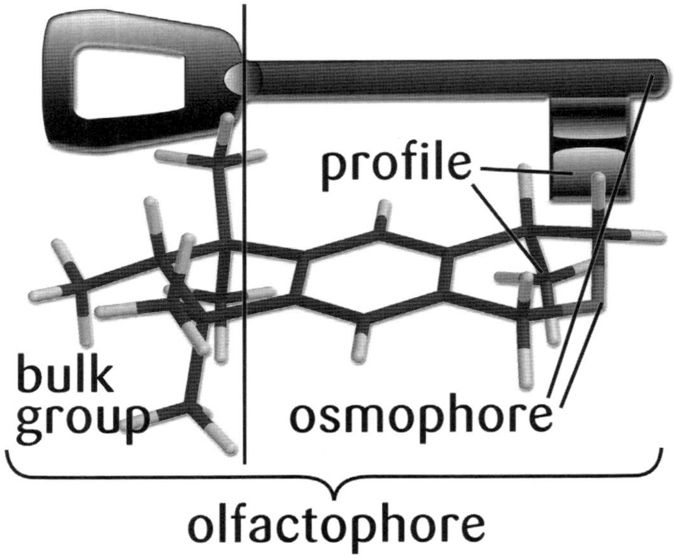

Figure 7.2 The molecular parameters of (4*S*,7*R*)-(–)-galaxolide (**37**) transferred into the 'lock-and-key' model.

macrolide **18** was conformationally constrained into a Galaxolide-like shape by introduction of methylene bridges between C-3 and C-8 or C-9. It was possible to obtain X-ray data of these rigid macrobicycles, shown here for (1*S*,6*R*,10*R*)-(+)-6-methyl-4-oxabicyclo[8.4.1]pentadecan-3-one (**39**). This macrobicycle was the most powerful odorant, yet the least musky of the series, smelling *woody, cedarwood-like, ambery and slightly animalic*. The most musky macrobicycle investigated [50] was **40**, and in the superposition analysis on the crystal structure of **37**, the methyl substituent of **40** superimposes very well with the (4*S*)-Me group of **37**. So, similar structural features may be key for a musk odour in both macrocycles and benzenoid musks. However, **40** is weaker than **18**, and the conformation that **18** would need to adapt to mimic **40** or **37** is energetically quite unfavourable. So severe doubts remain, but it might be possible that there is just one musk receptor.

As for Galaxolide (**37**), the structures of the most powerful stereoisomers of Fixolide (**34**) [26] and of Vulcanolide (**35**) [51] are shown; however, no exact threshold data were reported in the literature for these compounds. Also given are the abbreviations that ecotoxicologists introduced for **31–38**. As a consequence of their massive production volumes, their excellent chemical stability, their non-biodegradability and their high octanol/water partition coefficient, PCMs have bioaccumulated in fish and other marine organisms, human fat and human milk [52]. Although no apparent toxicity has been reported [10], possible long-term effects [53] are difficult to foresee, and more and more PCM-free formulations appear on the market. DPMI (Cashmeran®, **38**) is often considered a PCM by ecotoxicologists as well. Like HHCB (Galaxolide®, *rac*-**37**) its industrial synthesis starts from pentamethylindane; yet, DPMI (**38**) is neither aromatic nor a typical musk odorant. It possesses a *woody-ambery and leathery odour with some*

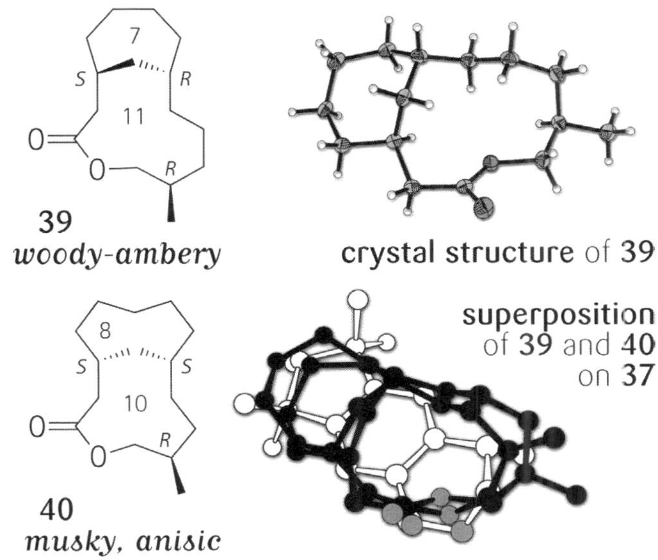

powdery, vanilla-type, velvet nuance that has only a very faint musk-like tonality. That is why, from a perfumer's point of view, **38** is not considered a PCM.

Today, Galaxolide® (HHCB, *rac*-**37**) is still the most widely used musk with some 7000–8000 t/a production volume and a market price of around 15 CHF [10]. But macrocycles are gaining substantial market share. Whilst in 1998, 75% of the musk odorants used were polycyclic, it is expected that in 2008 macrocycles will make up 60–65% of the global musk market [10]. But to make this possible, far better synthetic processes for industrial macrocyclisation have had to be discovered. So what has happened in the meantime with the industrial macrocyclisation processes?

7.5 Evolution of the industrial synthesis of macrocycles

The first improvements in macrocyclisation were the Thorpe–Ziegler reaction for macrocyclic ketones, and acid-catalysed cyclisations of hydroxycarboxylic acids according to the Ruggli–Ziegler dilution principle. The next breakthrough was the Stoll–Hansley–Prelog process, an intramolecular version of the acyloin condensation, which does not require high-dilution conditions, as it is a heterogeneous reaction. As a result, the price of Exaltone® (**6**) dropped from 50 000 CHF/kg in the 1930s to about 500 CHF/kg in 1947 [11].

At that time, Exaltolide® (**2**) was still manufactured by Baeyer–Villiger oxidation of **6**. This changed when E. W. Spanagel and W. H. Carothers at Du Pont [54] found a way to circumvent the necessity of high dilution also in the synthesis of macrocyclic lactones. 15-Hydroxypentadecanoic acid was polycondensed in the presence of Lewis acids at 180–250°C, and afterwards the monomeric 15-pentadecanolide (**2**) was distilled off *in vacuo* at 270°C/1 Torr. Consequently, the price of **2** decreased by a factor of ten, and its production rose from kilogram to ton scale. The real revolution in the synthesis of macrocycles started however in the 1960s, when cyclododecanone (**41**) became an inexpensive industrial building block as the result of the cyclooligomerisation of butadiene, a process pioneered by G. Wilke of the MPI für Kohlenforschung [55]. Thus, ring expansion reactions based on cyclododecanone (**41**), which today reaches production volumes of 100 000 t/a, became the method of choice for the industrial synthesis of macrocyclic lactones and ketones [56].

Radical addition of allyl alcohol to **41** with subsequent acid-catalysed cyclisation provides the bicyclic enol ether **42**, which is transformed to **43** by addition of hydrogen peroxide. Fragmentation of this hydroperoxide **43** in the presence of cupric acetate and iron sulfate furnishes an isomeric mixture of 11/12-pentadecen-15-olides (**44**), which already possesses an elegant musk note with metallic nuances and is commercially known as Habanolide® and Globalide® (**44**). Catalytic hydrogenation of **44** with Raney nickel completes the current industrial process [34] for 15-pentadecanolide (**2**), which is sold as Cyclopentadecanolide® (Symrise), Exaltolide® (Firmenich), Pentalide® (Soda Aromatics) or Thibetolide® (Givaudan) at around 60 CHF/kg.

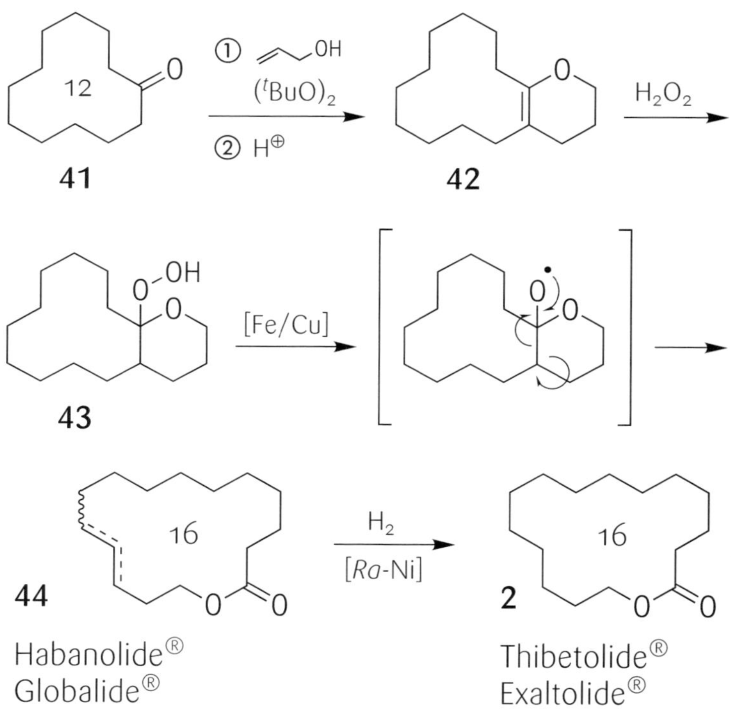

Habanolide® and Globalide® (**44**) are only slightly less expensive, but became trendy because of their *metallic character reminiscent of hot-ironed linen*, usually considered an undesirable side aspect by perfumers. However, this daring '*urban high-tech*' facet can provide some special radiant freshness for which the term of '*white musks*' was coined. The prototype *white musk* accord was created by Alberto Morillas in answer of the brief for »Emporio Armani White for Her« (Armani, 2001). His own code name for the winning submission was »cotton and linen«, and bergamot as well as mandarin oil, cassis, ginger and woody notes together with mint and fig leaves join in the *white musk* theme. The traditional hesperidic (*citrus-fruit like*) yet modern green »Cologne« (Thierry Mugler, 2002) and the intense white floral »Glow by J. Lo« (Lancaster, 2002) also feature *white musk* accords, which of course have nothing to do with the rare Arctic white musk-oxen, as some may have believed.

The Story process [57] developed in 1968 constitutes an industrial alternative to the presented ring-enlargement route to 15-pentadecanolide (**2**). In terms of production costs this process is very attractive; however, the handling of the trimeric peroxide **46** is quite tricky, and the industrial Story synthesis of **2** was discontinued after a severe explosion. The trimeric peroxide **46** was synthesised via the crystalline 1,1′-dihydroperoxydicyclohexyl peroxide, which is formed by treatment of cyclohexanone (**45**) and 90% hydrogen peroxide in the presence of perchloric acid at room temperature. Reacting this intermediate peroxide with cyclopentanone at 0–4°C in the presence of anhydrous cupric sulfate yielded the trisperoxide **46**, which was thermally decomposed in boiling decane, presumably via decarboxylation of the intermediate diacyl peroxide **47**. 15-Pentadecanolide (**2**) was obtained in about 25%, accompanied by about the same quantity of cyclotetradecane and smaller amounts of isomeric dilactones. The hazardous nature of the peroxides involved is obviously a severe restriction on this elegant and economic approach.

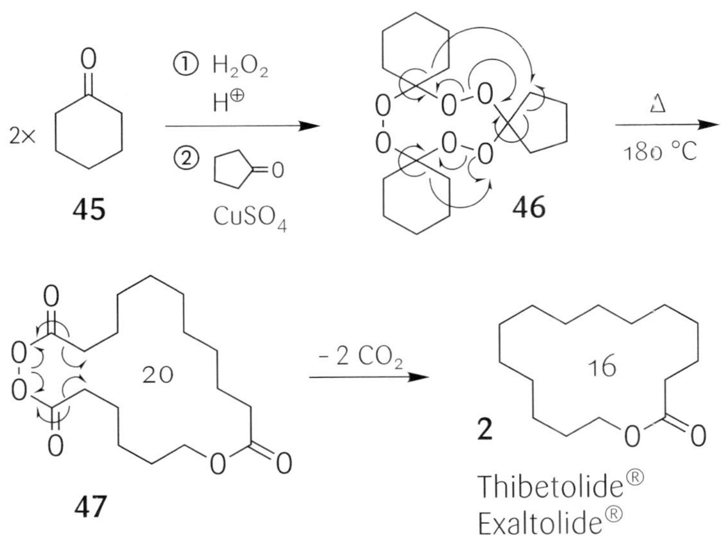

In the academic world, the RCM constitutes today the preferred method for the synthesis of macrocyclic compounds. For industrial syntheses of macrocyclic musks, Grubbs' catalysts are however still too expensive to compete with the established processes. Yet, one industrially important process involves ring-opening RCM of cyclooctene (**48**), employing not a Grubbs' catalyst but dirhenium heptoxide on alumina in flow reactors. Monohydroboration of the resulting cyclohexadeca-1,9-diene (**49**) is followed by oxidation of the cycloalkenylborane to the corresponding alcohol and then further to the cylcoalkenone **50** [56], which possesses a *slightly aldehydic and waxy musk odour* and has been introduced to perfumery as Globanone®.

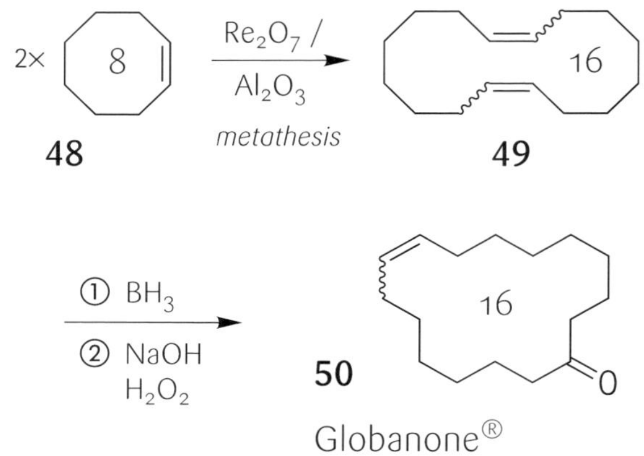

Shifting the double bond from C-8 to C-5 intensifies the musk odour of cyclohexadecenone, and renders the odour note more towards a *powdery nitro-musk* character. This product is known in perfumery as Ambretone (Takasago), Musk TM II (Soda Aromatics) and Velvione (Givaudan), and it is produced by oxy-Cope ring enlargement of cyclododecanone (**41**) [39]. Chlorination of **41** in the presence of copper(I)chloride–DMF affords 2-chlorocyclododecanone (**51**), which is reacted with two equivalents of vinyl magnesium chloride to provide the divinyl alcohol **52**. Thermal oxy-Cope rearrangement of **52** leads to the final product **53**. Velvione (**53**) was extensively used in the fine fragrance »Velviona« (Helmut Lang, 2001), but it is also economic and substantive enough to provide powdery volume and musky softness in laundry-care products.

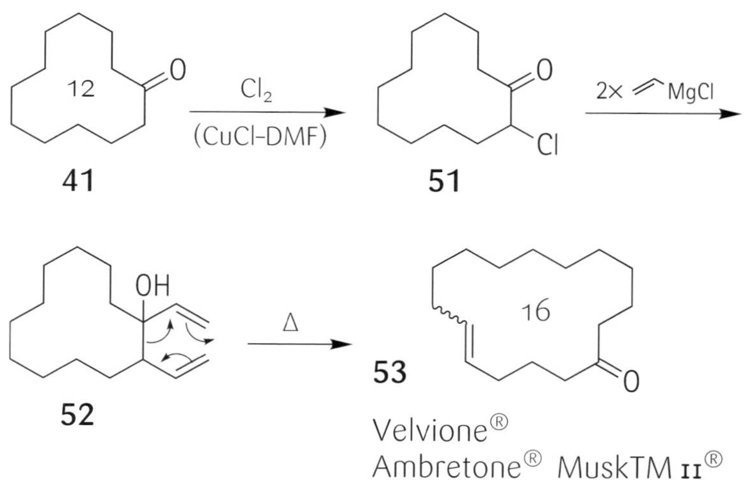

7.6 Modern macrocyclic musks

To replace or even outperform PCMs and nitro musks in diverse applications, one can either lower the production price of macrocycles or increase their odour intensity, which means lowering their odour threshold. The latter option allows for more complex synthetic approaches and accordingly higher production costs. Muscenone (57), which had its debut in perfumery in 1993 with the floriental feminine fine fragrance »Jean-Paul Gaultier« (Shiseido, 1993) created by Jacques Cavallier, was the first example for such high-performance modern musks.

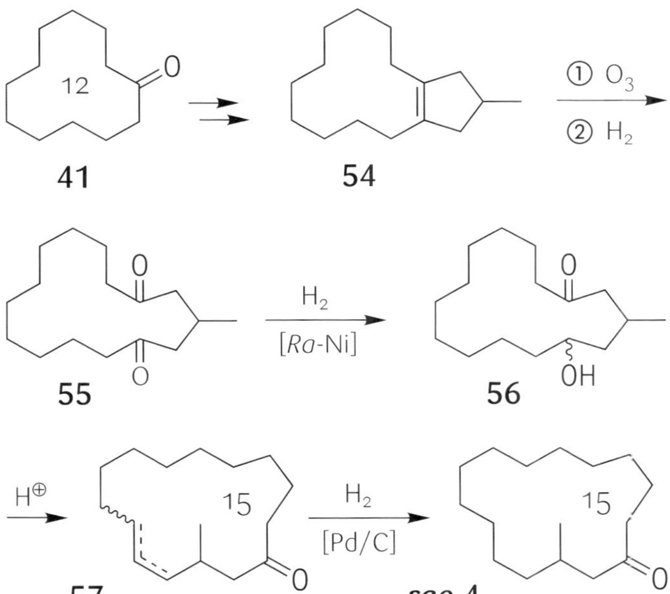

Muscenone (**57**) was first reported as intermediate in a muscone synthesis by A. Eschenmoser, G. Ohloff and co-workers [58] in 1971, but its value in perfumery was discovered only much later. Muscenone (**57**) is still an intermediate in the current industrial synthesis of muscone (*rac*-**4**), which commences from 14-methylbicyclo-[10.3.0]pentadec-1(12)-ene (**54**), accessible by annulation of cyclododecanone (**41**). Ozonolysis of **54**, followed by reductive work-up and further hydrogenation of one carbonyl group of **55** with *Raney* nickel, provides **56**. Acid-catalysed elimination of the hydroxy function of **56** completes the synthesis of muscenone (**57**), which can then optionally be hydrogenated to muscone (*rac*-**4**). Muscenone (**57**) possesses a *very elegant and diffusive musk odour reminiscent of Musk ketone* (**25**), and with an olfactory threshold of 0.9 ng/l air, is as powerful as the benchmark musk Galaxolide® (*rac*-**37**).

Besides the ring size and methyl substituents, double bonds are an important feature to influence both character and strength of macrocyclic musks. Is the function of double bonds restricted to conformational effects, or is it also of electronic importance? Can more powerful macrocyclic musks be designed by introducing additional polar groups?

To answer these important questions, we developed a simple sequence [59] and synthesised a series of thiamacrolides **58**. As demonstrated by the olfactory similarity of *p*-xylene and 2,5-dimethylthiophene as well as by several (3*Z*)-hexenyl derivatives, a sulfur atom is able to mimic a double bond. For the thiamacrolides **58**, we found that a 1,7-distance of the sulfur atom and the carbonyl group in even-membered rings, and a 1,6-position of the sulfur atom and the carbonyl group in odd-membered rings significantly lowered the odour thresholds. An olfactophore model, featuring two hydrogen-bond acceptors and three hydrophobic-binding pockets, was generated with the Catalyst software [60], which predicts the experimental odour thresholds of these thiamacrolides **58**, quite accurately. The threshold of a given molecule is calculated on the basis of its degree of mapping with the hypothetical features of the receptor model [38]. The depicted olfactophore model clearly suggests that a second electronegative feature, besides a carbonyl or oxycarbonyl function, would be advantageous for the musk odour, and it also suggests a hydrophobic group right in the middle of these two hydrogen-bond acceptors.

58

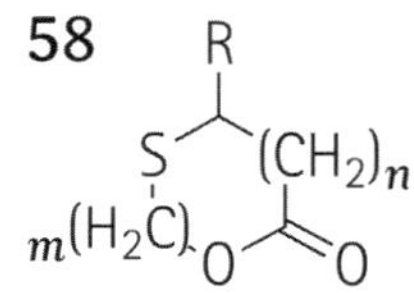

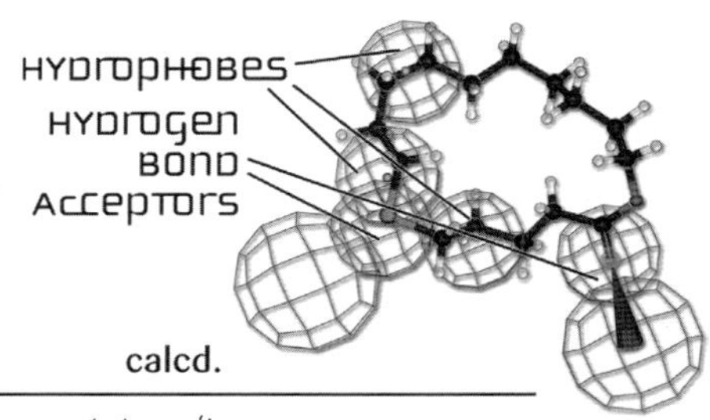

m-n-R	exper.	calcd.	
4-7-H	2 ng/L	1.1 ng/L	*musky, woody, green*
4-8-H	0.2 ng/L	0.36 ng/L	*musky, waxy-allylic*
4-9-H	0.4 ng/L	0.34 ng/L	*musky, animalic, sulfurous*
7-4-H	1.0 ng/L	0.81 ng/L	*musky, medicinal*
8-3-H	0.3 ng/L	0.45 ng/L	*musky, sweet, green*
8-4-H	0.2 ng/L	0.32 ng/L	*musky, green-mossy*
9-2-H	2.5 ng/L	3.4 ng/L	*musky, animalic, powdery*
9-2-Me	23 ng/L	23 ng/L	*musky, sweet, powdery, green*
9-3-H	0.4 ng/L	0.33 ng/L	*musky, green*
9-4-H	0.5 ng/L	0.48 ng/L	*musky, waxy-allylic*
10-2-H	0.3 ng/L	0.32 ng/L	*musky, powdery, fruity*
10-2-Me	1.8 ng/L	0.8 ng/L	*musky, metallic, sweet*
10-3-H	0.1 ng/L	0.072 ng/L	*musky, powdery, dry*

This was the inspiration behind the synthesis of 4-substituted 1,7-dioxacycloalkan-8-ones [61], which indeed possessed very low odour thresholds in the range of 0.1–0.2 ng/l air. Of particular interest was 12-methyl-9-oxa-14-tetradecanolide; firstly because just the (*R*)-enantiomer **59** is responsible for the odour of the racemate, while the enantiomer is odourless on GC-olfactometry, and secondly because its odour is surprisingly reminiscent of the very characteristic Musk ambrette (**26**). One can indeed superimpose both molecules so that the *tert*-butyl group of **26** and the 12-methyl group of **59**, the nitro group of **26** and the oxycarbonyl group of **59**, and also the ether oxygens of both molecules overlie very well [38].

In Nirvanolide® (**60**), the latest captive macrocyclic musk of Givaudan, the second electronegative feature is a double bond. Again, the odour is due to the (*R*)-enantiomer [26]; however, even the racemate **60** has an odour threshold of 0.1 ng/l, just like Musk ketone (**25**). And with its *clean and sweet, powdery and persistent, and slightly animalic musk odour*, **60** is also odourwise quite close to Musk ketone (**25**), just like muscenone (**57**), but more powerful. Nirvanolide® (**60**) had its debut at 6.7% in the perfume »Forever Elizabeth« (Elizabeth Taylor, 2002) created by David Apel.

There are some other creative ways of making use of double bonds for modern macrocyclic musks. Williamson reaction of 1,10-decandiol with allyl bromide, followed by RCM, furnished **61**, possessing a pleasant powdery musk note without a carbonyl function being present. This work of Takasago [38] also demonstrates the importance of two hydrogen-bond acceptors in a certain distance. Methylene macrolides have been described by W. Tochtermann and co-workers [62]. Especially interesting is the 9-methylene macrolide **62** with an *intense sweet musk note of animalic tonality*. This group also reported [63] the first double-unsaturated macrocyclic musk **63**, which possesses a strong and erogenous natural musk note. Two stereochemically defined double bonds in a certain distance are an efficient way of confining the conformational freedom of macrocycles, and thus designing powerful musks. However, the industrial-scale synthesis of such dienolides is quite costly. C. Fehr *et al.* [64] recently developed an iterative fragmentation of tricyclic systems, by which they synthesised the cyclopentadecadienenone **64**. Yet, in comparison with the pleasant musk note of **63**, the greasy, waxy, relatively weak musk odour of **64** is rather disappointing.

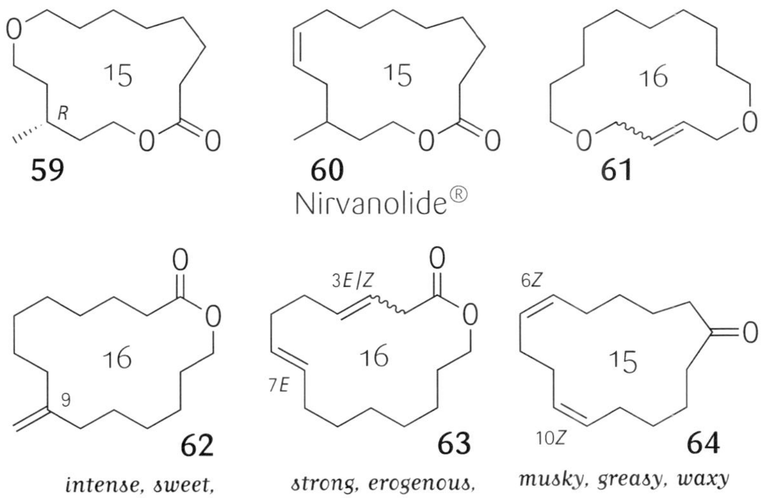

59 **60** Nirvanolide® **61**

62 *intense, sweet, musky, animalic*

63 *strong, erogenous, natural musk*

64 *musky, greasy, waxy*

Today, Muscenone (**57**) and Nirvanolide® (**60**) are the quintessence of modern musk accords, in terms of power, diffusivity and character. Muscenone (**57**) and Nirvanolide® (**60**) are the perfumery materials of choice when it comes to replacing PCMs and nitro musks in older formulations.

7.7 New musk structures

In 1979, M. Klaus of F. Hoffmann-La Roche discovered by serendipity a new intense musk odorant with some damascone-like and woody nuances. In the course of synthetic work on aromatic retinoids he had synthesised the hexahydroindene **65**, which because of its powerful odour soon became internally known as klausenone (**65**) in Hoffmann-La Roche and also at Givaudan, at that time the flavour and fragrance division of Roche. As **65** is structurally not an aromatic polycycle, the musk odour of **65** was rather surprising. Attempts to develop an industrially feasible route to klausenone (**65**) turned out to be more difficult than anticipated, and thus derivatives were also investigated [38]. The corresponding alcohol, obtained by LAH reduction of **65**, possessed as well a pronounced musk note, indicating the carbonyl function was no prerequiste. Surprisingly, the epoxide intermediate **66**, *en route* to a methyl klausenone by acid-catalysed rearrangement, turned out to be the best musk odorant of the whole series and it was introduced into perfumery as Moxalone® (**66**). Moxalone® (**66**) had its debut in »Eden« (Cacharel, 1994) by Jean Guichard and was even at 0.62% an important cornerstone in the unisex fragrance »CK be« (C. Klein, 1996) created by René Morgenthaler.

In the late 1960s, J. Kula was doing his master's degree in the lab of Professor J. Kulescza on derivatives of carotol. Ozonolysis with reductive work-up and intramolecular aldol condensation of the intermediate oxo aldehyde gave a dihydroxy ketone, which was dehydrated to provide a mixture with a pleasant musky smell. Investigating this mixture, they believed the musk note to be due to its main component **67** which they called mageritone [65]. However, more than 30 years later J. Kula reinvestigated this chemistry, and found mageritone (**67**) only weak and uncharacteristic in smell. Instead the isomeric dienone **68**, present at only 5%, was found responsible for the musky odour of the mixture [66]. The tetrahydroindene musk **68** possesses a *dry musky odour* with an odour threshold of around 1 ng/l.

Caution is advisable, when the descriptor '*musky*' appears in the chemical literature, especially if perfumers were not involved in the evaluation of the compounds. For instance, K. H. Ney [67] reported an intense musk odour for ethyl 2-hydroxyimino-

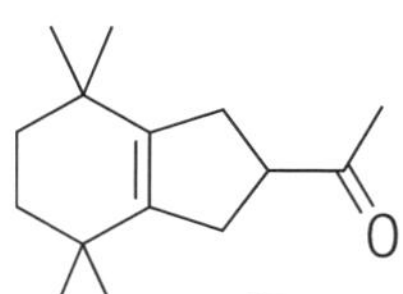

65
klausenone

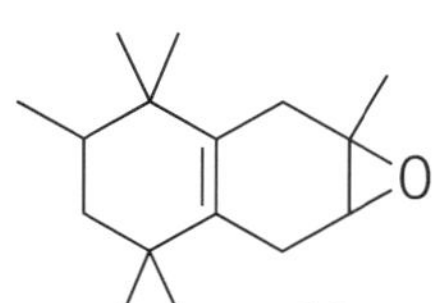

66
Moxalone®

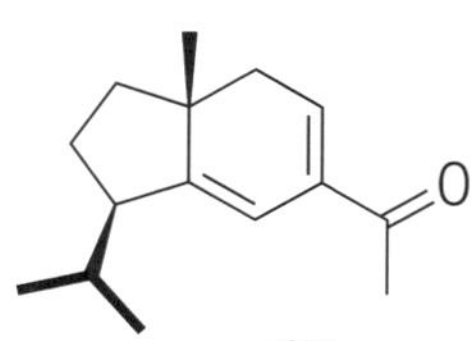

67
mageritone *very weak*

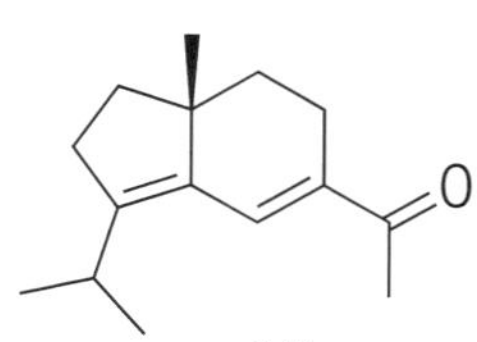

68
pronounced musky

4-methyl-pentanoate (**69**), but when we resynthesised this compound, our perfumers described it *as relatively weak, green and only reminiscent of olives, but not of musk*.

A musky smell was also reported for the secretion of the alligators *A. sclerops* and *A. latirostris*, which live in the Paraná river region of Argentina [68]. The intense smelling secretion serves as mating pheromone, and had also been used as perfume by the natives. A '*musk-like*' odour principle was even isolated and named *yacarol* (**70**). However, investigations by G. Fester *et al.* [69] proved the identity of *yacarol* with (*R*)-(+)-citronellol (**70**), which of course is no musk odorant but possesses a *sweet, typically rosy odour*. Probably due to animalic-smelling impurities, this rosy odour had been mistaken for musk.

However, the ethyl oxalate ester of (*R*)-(+)-citronellol **71** possesses some musky facets, though they are barely noticeable. If the product **71** is completely free from traces of **70**, it smells *powdery, fatty and slightly musky* with an odour threshold of over 14 ng/l air. These musky facets have even been rationalised by molecular modelling calculations [70], but we should not consider **71** a musk odorant.

Yet, structurally not too far away, there is a class of compounds – neither macro- nor polycyclic or benzenoid – that indeed possesses a pronounced musk odour. Already in 1975, W. Hoffmann and K. von Fraunberg of BASF discovered the cyclopentenyl ester **72**, which emanates a *warm, powdery musk odour with fruity, strawberry-like nuances* [38]. This compound **72** was introduced to perfumery as Cyclomusk®, even though it is rather linear than cyclic in structure. But though its industrial synthesis, which commenced with the thermal cyclisation of dedydrolinalool, was not too expensive, **72** had no chance against

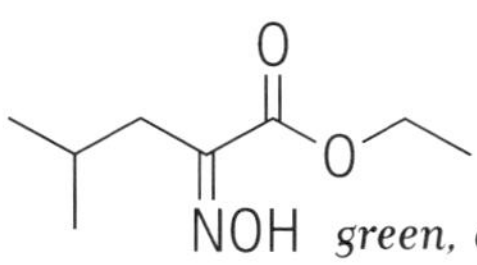

69 *green, olives*

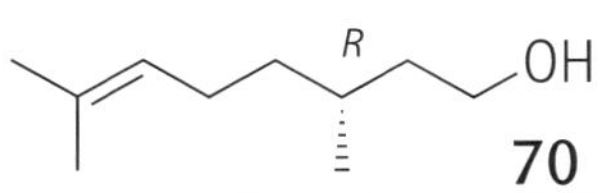

70
yacarol = citronellol

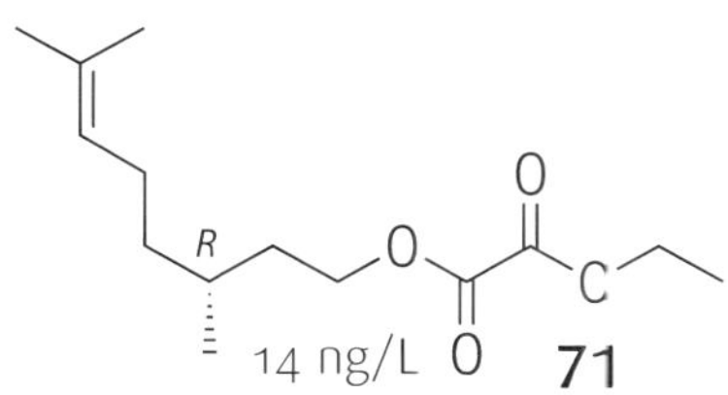

71
powdery, fatty, slightly musky

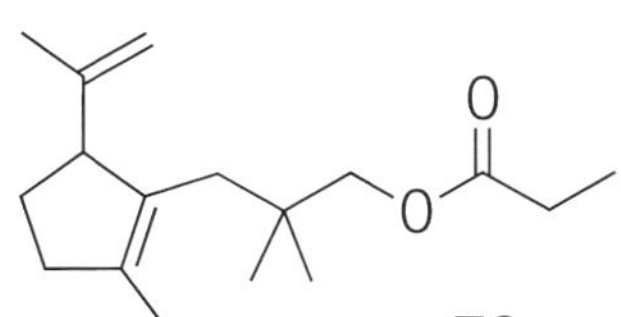

72
Cyclomusk®

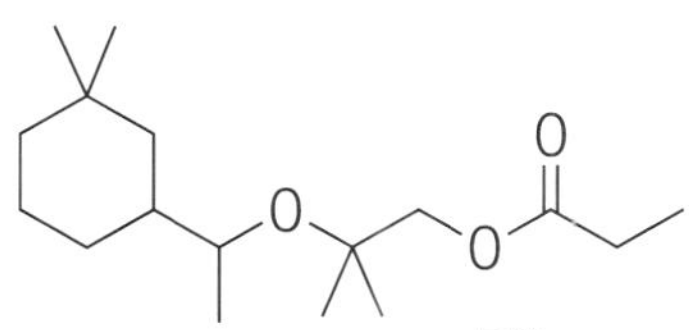

73
Helvetolide®

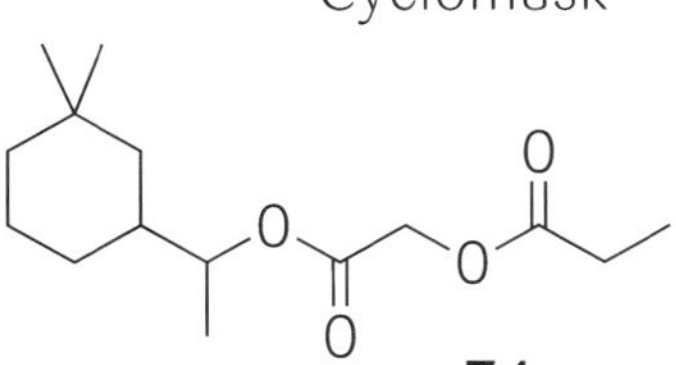

74

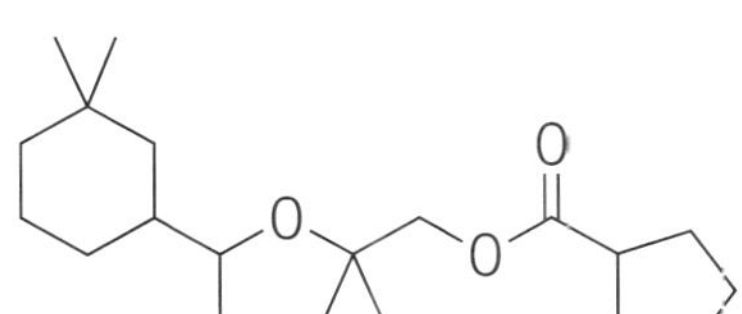

75

Galaxolide® (*rac*-37) in the late 1970s and was subsequently withdrawn from the market. Some 15 years later, however, Firmenich introduced another representative of this class to perfumery – Helvetolide® (73), which turned out to be far more successful. It was, for instance, used by Alberto Morillas at 8.8% in »Emporio Armani White for Her« (Armani, 2001), at 6.1% in »Miracle« (Lancôme, 2000) and at 3.8% in »Flower« (Kenzo, 2000). Interestingly, one *gem*-dimethyl group of **73** can be replaced by a carbonyl group without losing the musk odour. The fruity, blackberry-type nuance of **73** then turns into the more camphoraceous side note of **74**. This 'oxo-Helvetolide' **74**, which is known as Romandolide in Firmenich, was for instance used by Jacques Cavallier at 5.0% in »Absolu« (Rochas, 2002) and at 1.0% in »Murmure« (Van Cleef & Arples, 2002). Surprisingly, we have found that the corresponding cyclopentanoate of Helvetolide (**75**) still smells musky with an odour threshold of 1.4 ng/l air, and with a molecular weight of 324.27 u(!) for $C_{20}H_{36}O_3$, it holds the world record for the heaviest musk odorant, if not the heaviest odorant of all. Previously, one would not have believed it possible to construct musk odorants over 300 – so surprises are still possible in fragrance chemistry!

Let us conclude this chapter with an overview of the most important milestones in musk chemistry (Figure 7.3). Even before the odorous principle (*R*)-(−)-muscone (**4**) of

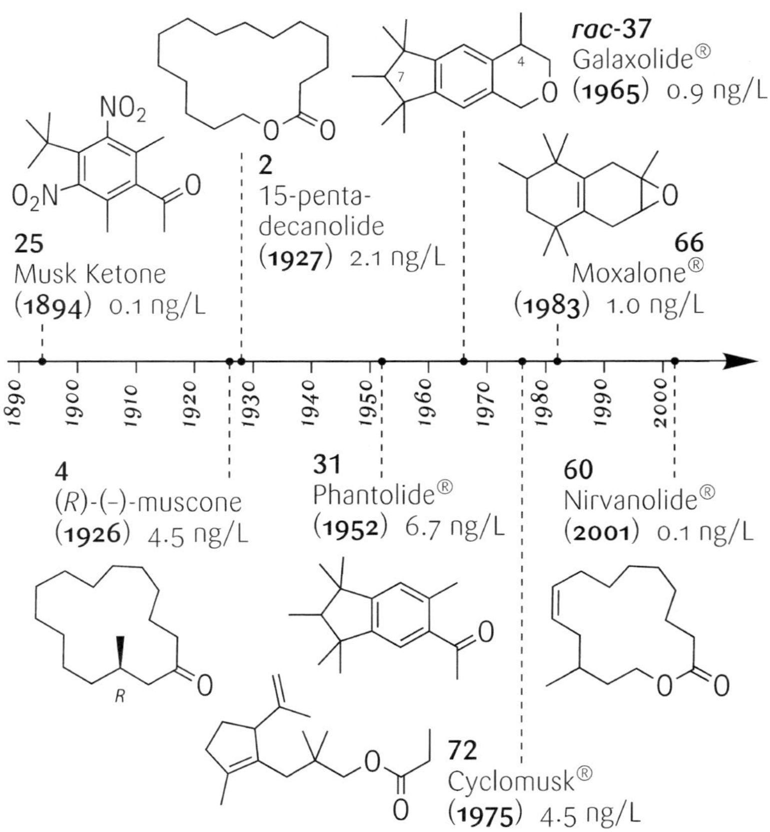

Figure 7.3 The timeline of musk odorants.

the Tonquin musk was known, nitro musks like Musk ketone (**25**) became the epitome of musks. And with a threshold of 0.1 ng/l air for **25**, they became a very hard-to-beat benchmark, compared to the 4.5 ng/l air for **4** and 2.1 ng/l air for 15-pentadecanolide (**2**). The latter natural macrolide from the vegetal kingdom can be considered the most typical musk, but was, for a long time, too expensive to allow its use in consumer products. But discolouration problems in consumer products caused the decline of the nitro musks and PCMs like Phantolide® (**31**), as their first representative became the new standard. The evolution towards increased stability culminated in Galaxolide® (*rac*-**37**) with a threshold of 0.9 ng/l air. Yet, the lack of biodegradability of *rac*-**37** ignited the search for new musk structures, for which Cyclomusk (**72**) and Moxalone (**66**) were the first commercial examples. But it also led to modern macrocyclic musks, with improved thresholds in the range of nitro musks, for which Nirvanolide (**60**) with 0.1 ng/l air is the prominent example.

Acknowledgements

Without the enormous help of Dr Roman Kaiser, Françoise Gay and Jonathan Leighton it would have been impossible to take an original photo of musk pods for this chapter. Thanks are also due to Professor Dr Jozef Kula for supplying a sample of compound **68**, to Heinz Koch, Adriana Flückiger and Katarina Grman for threshold measurements, to Beat Studinger and Kevin Dastis for data on perfumery formulae, and to Dr Markus Gautschi and Sofia Gallo for careful proof-reading of the manuscript. And finally, very special cheers to Axelle Beernaert.

References

1 Kemp, P. *Musc*, Roman; Éditions Albin Michel: Paris, 2000; pp 51–52. German Ed.: Kemp, P. *Musk*, Roman; Argon Verlag: Berlin, 2002; pp 45–46.
2 Rimmel, E. *Das Buch des Parfums*, Sachbuch; Verlag Ullstein: Frankfurt M. / Berlin, 1988; pp 282–287.
3 Poucher, W. A. *The Raw Materials of Perfumery*, 9th ed., rev. by Jouhar, A. J.; Chapman & Hall: London, 1991; pp 236–237.
4 Groom, N. *The New Perfume Handbook*, 2nd ed.; Blackie Academic & Professional: London, 1997; p 219.
5 Kastner, D. Moschus – immer wieder Moschus. *Fragrances–SÖFW-Journal* **2001**, *126*, 71–82.
6 Editions de Parfums Frederic Malle; 37, Rue de Grenelle; F-75007 Paris; France. http://www.editionsdeparfums.com/.
7 Bauer, A. Studien über den künstlichen Moschus. *Ber. Dtsch. Chem. Ges.* **1891**, *24*, 2832–2843.
8 Wood, T. F. Synthetic Nonbenzenoid Musks. In *Fragrance Chemistry – The Science of the Sense of Smell*; Theimer, E. T., Ed.; Academic Press: Orlando, 1982; pp 495–507.
9 Walbaum, H. Das natürliche Moschusaroma, *J. Prakt. Chem.* **1906**, *73*, 488–493.
10 Gautschi, M.; Bajgrowicz, J. A.; Kraft, P. Fragrance Chemistry – Milestones and Perspectives, *Chimia* **2001**, *55*, 379–387.
11 Ohloff, G. Der Forscher Leopold Ruzicka als Erfinder, *Chimia* **1987**, *41*, 181–187.
12 Ruzicka, L. Zur Kenntnis des Kohlenstoffringes VII. Über die Konstitution des Muscons, *Helv. Chim. Acta* **1926**, *9*, 715–729.

13 Baeyer, A. Über Polyacetylenverbindungen, *Ber. Dtsch. Chem. Ges.* **1885**, *18*, 2269–2281.
14 Ruzicka, L. Zur Kenntnis des Kohlenstoffringes I. Über die Konstitution des Zibetons, *Helv. Chim. Acta* **1926**, *9*, 230–248.
15 Sack, E. Zur Kenntnis des Zibeths, *Chem. Ztg.* **1915**, *39*, 538.
16 Prelog, V.; Ruzicka, L. Über zwei moschusartig riechende Steroide aus Schweinetestes-Extrakten, *Helv. Chim. Acta* **1944**, *27*, 61–66.
17 Dale, J. Exploratory Calculations of Medium and Large Rings, *Acta Chem. Scand.* **1973**, *27*, 1115–1129.
18 Zelinsky, N. Über Cyclononanon und Cyclononan. *Ber. Dtsch. Chem. Ges.* **1907**, *40*, 3277–3279.
19 Willstätter, R.; Kametaka, T. Notiz über Cyclononanon. *Ber. Dtsch. Chem. Ges.* **1907**, *40*, 3876.
20 Ruzicka, L.; Stoll, M.; Schinz, H. Zur Kenntnis des Kohlenstoffringes II. Synthese der carbocyclischen Ketone vom Zehner- bis zum Achtzehnerring. *Helv. Chim. Acta* **1926**, *9*, 249–264.
21 Stevens, P. G.; Erickson, J. L. E. American Musk. I. The Chemical Constitution of the Musk of the Louisiana Muskrat, *J. Am. Chem. Soc.* **1942**, *64*, 144–147.
22 Mookherjee, B. D.; Trenkle, R. W.; Wilson, R. A. New insights in the three most important natural fragrance products: Wood, Amber and Musk. *Indian Perfumer* **1992**, *36*, 313–328.
23 Mookherjee, B. D.; Wilson, R. A. The Chemistry and Fragrance of Natural Musk Compounds. In *Fragrance Chemistry – The Science of the Sense of Smell*; Theimer, E. T., Ed.; Academic Press: Orlando, 1982; pp 433–494.
24 Schinz, H.; Ruzicka, L.; Geyer, U.; Prelog, V. Muscopyridin, eine Base $C_{16}H_{25}N$ aus natürlichem Moschus. *Helv. Chim. Acta* **1946**, *29*, 1524–1528.
25 Biemann, K.; Büchi, G.; Walker, B. H. The Structure and Synthesis of Muscopyridine, *J. Am. Chem. Soc.* **1957**, *79*, 5558–5564.
26 Kraft, P.; Fráter, G. Enantioselectivity of the Musk Odor Sensation, *Chirality* **2001**, *13*, 388–394.
27 Pickenhagen, W. Enantioselectivity in odor perception. In *Flavour Chemistry: Trends and Developments*; Teranishi, R.; Buttery, R. G.; Shahidi, F., Eds.; ACS Symposium Series No. 388. American Chemical Society: Washington, DC, 1989; pp 151–157.
28 Alexakis, A.; Benhaim, C.; Fournious, X.; Van der Heuvel, A.; Levêque, J.-M.; March, S.; Rosset, S. Catalytic Asymmetric Conjugate Addition on Macrocyclic and Acyclic Enones. Synthesis of R-(−)-Muscone, *Synlett* **1999**, 1811–1813.
29 Louie, J.; Bielawski, C. W.; Grubbs, R. H. Tandem Catalysis: The Sequential Mediation of Olefin Metathesis, Hydrogenation, and Hydrogen Transfer with Single-Component Ru Complexes, *J. Am. Chem. Soc.* **2001**, *113*, 11312–11313.
30 Yamamoto, T.; Ogura, M.; Kanisawa, T. A novel synthetic method for (*R*)- and (*S*)-muscones by enantioselective hydrogenation of (*E*)- and (*Z*)-3-methyl-2-cyclopentadecen-1-ones catalyzed by *p-tolyl*-BINAP-Ru(II) complexes, *Tetrahedron* **2002**, *58*, 9209–9212.
31 Kerschbaum, M. Über Lactone mit großen Ringen – die Träger des vegetabilischen Moschus-Duftes, *Ber. Dtsch. Chem. Ges.* **1927**, *60*, 902–909.
32 Schultz, K.; Kraft, P. Characterization of the Macrolide Fraction of Angelica Root Oil and Enantiomeric Composition of 12-Methyl-13-tridecanolide, *J. Essent. Oil Res.* **1997**, *9*, 509–514.
33 Bollbuck, B.; Kraft, P.; Tochtermann, W. Nature-like Odorants by Stereoselective Ring Enlargement of Cyclohexanone and Cyclododecanone, *Tetrahedron* **1996**, *52*, 4581–4592.
34 Becker, J.; Ohloff, G. Eine neuartige Fragmentierung bicyclischer Enolether. Verfahren zur Darstellung macrocyclischer Lactone, *Helv. Chim. Acta* **1971**, *54*, 2889–2895.
35 Taskinen, J. 12-Methyl-ω-tridecanolide, A New Macrocylic Lactone from Angelica Root Oil, *Acta Chem. Scand., Ser. B* **1975**, *29*, 637–638.
36 Kraft, P.; Tochtermann, W. Ring Enlargement of Cyclodecanone by a Chiral Building Block: Synthesis and Olfactory Properties of (12*R*)-(+)-12-Methyl-13-tridecanolide, *Liebigs Ann. Chem.* **1994**, 1161–1164.

37 Kraft, P.; Tochtermann, W. Stereoselective Synthesis of Both Enantiomers of 13-Tetradecanolide by Ring Enlargement with Different Building Blocks and Olfactory Comparison with (12*R*)-(+)- and (12*S*)-(–)-12-Methyl-13-tridecanolide, *Liebigs Ann.* **1995**, 1409–1414.

38 Kraft, P.; Bajgrowicz, J. A.; Denis, C.; Fráter, G. Odds and Trends: Recent Developments in the Chemistry of Odorants, *Angew. Chem. Int. Ed.* **2000**, *39*, 2980–3010. German Ed.: Kraft, P.; Bajgrowicz, J. A.; Denis, C.; Fráter, G. Allerlei Trends: die neuesten Entwicklungen in der Riechstoffchemie, *Angew. Chem.* **2000**, *112*, 3106–3138.

39 Fráter, G; Bajgrowicz, J. A.; Kraft, P. Fragrance Chemistry, *Tetrahedron* **1998**, *54*, 7633–7703.

40 Kaiser, R.; Lamparsky, D. Neue Makrolide und einige Sesquiterpen-Derivate aus dem Galbanum-Harz, *Helv. Chim. Acta* **1978**, *61*, 2671–2680.

41 Stoll, M. Many membered rings and musk odor, *The Drug and Cosmetic Industry* **1936**, *38*, 334–337.

42 Bauer-Thurgau, A. Ueber Keton Moschus, *Ber. Dtsch. Chem. Ges.* **1898**, *31*, 1344–1349.

43 Bauer, A. Über die Phenole des Butyltoluols und dessen Aether, *Ber. Dtsch. Chem. Ges.* **1894**, *27*, 1614–1619.

44 Fuson, R. C.; Mills, J.; Klose, T. G.; Carpenter, M. S. The structures of Musk ketone and Musk tibetene, *J. Org. Chem.* **1947**, *12*, 587–595.

45 Barbier, H. Les tert.-butyl-cymènes et leurs produits de nitration, *Helv. Chim. Acta* **1932**, *15*, 592–596.

46 Teisseire, P. J. *Chemistry of Fragrant Substances*; VCH Publishers: New York, 1994; pp 333–358. Original Ed. in French: Teisseire, P. J. *Chemie des substances odorantes*; Technique et Documentation – Lavoisier: Paris, 1991; pp 359–386.

47 Carpenter, M. S.; Easter, W. M.; Wood, T. F. Nitro Musks. I. Isomers, Homologs, and Analogs of Musk Ambrette, *J. Org. Chem.* **1951**, *16*, 586–617.

48 Weber, S. H.; Spoelstra, D. B.; Polak, E. H. The structure of tert. amylated p-cymene. A new synthesis of polyalkylindans. I. *Recl. Trav. Chim. Pays-Bas* **1955**, *74*, 1179–1196.

49 Fráter, G.; Müller, U.; Kraft, P. Preparation and Olfactory Characterization of the Enantiomerically Pure Isomers of the Perfumery Synthetic Galaxolide®, *Helv. Chim. Acta* **1999**, *82*, 1656–1665.

50 Kraft, P.; Cadalbert, R. Constructing Conformationally Constrained Macrobicyclic Musks, *Chem. Eur. J.* **2001**, *7*, 3254–3262.

51 Fehr, C.; Chaptal-Gradoz, N.; Galindo, J. Synthesis of (–)-Vulcanolide by Enantioselective Protonation, *Chem. Eur. J.* **2002**, *8*, 853–858.

52 Rimkus, G.; Brunn, H. Synthetische Moschusduftstoffe – Anwendung, Anreicherung in der Umwelt und Toxikologie; Teil 1: Herstellung, Anwendung, Vorkommen in Lebensmitteln, Aufnahme durch den Menschen, *Ernährungs-Umschau* **1996**, *43*, 442–449.

53 Rimkus, G.; Brunn, H. Synthetische Moschusduftstoffe – Anwendung, Anreicherung in der Umwelt und Toxikologie; Teil 2: Toxikologie der synthetischen Moschusduftstoffe und Schlußfolgerungen, *Ernährungs-Umschau* **1997**, *44*, 4–9.

54 Spanagel, E. W.; Carothers, W. H. Preparation of Macrocyclic Lactones by Depolymerization, *J. Am. Chem. Soc.* **1936**, *58*, 654–656.

55 Wilke, G.; Bogdanovic, B.; Borner, P.; Breil, H.; Hardt, P.; Heimbach, P.; Herrmann, G.; Kaminsky, H.-J.; Keim, W.; Kröner, M.; Müller, H.; Müller, E. W.; Oberkirch, W.; Schneider, J.; Stefeder, J.; Tanaka, K.; Weyer, K. Cyclooligomerisation von Butadien und Übergangsmetall-π-Komplexe, *Angew. Chem.* **1963**, *75*, 10–20; *Angew. Chem. Int. Ed. Engl.* **1963**, *2*, 105–115.

56 Williams, A. S. The Synthesis of Macrocyclic Musks, *Synthesis* **1999**, 1707–1723.

57 Story, P. R.; Busch, P. Modern Methods for the Synthesis of Macrocyclic Compounds, *Adv. Org. Chem.* **1972**, *8*, 67–95.

58 Felix, D.; Schreiber, J.; Ohloff, G.; Eschenmoser, A. α,β-Keton→Alkinon-Fragmentierung I: Synthese von Exalton und *rac*-Muscon aus Cyclododecanon, *Helv. Chim. Acta* **1971**, *54*, 2896–2912.

59 Kraft, P.; Cadalbert, R. The Thia-Analog of Ambrettolide. Synthesis and Odor of 1,8-Oxathiacyclohexadecan-2-one, *Synlett* **1997**, 600–602.
60 Accelrys Catalyst software; http://www.accelrys.com/catalyst/.
61 Kraft, P.; Cadalbert, R. 4-Substituted 1,7-Dioxacycloalkan-8-ones: Preparation and Olfactory Properties, *Synthesis* **1998**, 1662–1669.
62 Rodefeld, L.; Heinemann, I.; Tochtermann, W. Synthesis and Olfactory Properties of New Macrolides from Unsaturated Fatty Acids and 1,ω-Diols, *Tetrahedron* **1998**, 5265–5286.
63 Tochtermann, W.; Kraft, P. Our Tactics in Ring Enlargement – Construction of Medium and Large Ring Compounds, *Synlett* **1996**, 1029–1035.
64 Fehr, C.; Galindo, J.; Etter, O.; Thommen, W. Access to C-15 Macrocyclic Ketones by Iterative Fragmentation of A Tricyclic System, *Angew. Chem.* **2002**, *114*, 4705–4708; *Angew. Chem. Int. Ed.* **2002**, *41*, 4523–4526.
65 Kulesza, J.; Kula, J.; Kwiatkowski, W. Anwendung des Carotols bei der Synthese der Riechstoffe, *Riechstoffe, Aromen, Koerperpflegemittel* **1973**, *23*, 34, 36 and 37.
66 Kula, J.; Bonikowski, R.; Staniszewska, M.; Krakowiak, A.; Wieczorek, M. W.; Majzner, W. R.; Bujacz, G. D. Transformation of Carotol into the Hydrindane-Derived Musk Odorant, *Eur. J. Org. Chem.* **2002**, 1826–1829.
67 Ney, K. H. α-Hydroxyiminoisocapronsäureäthylester – ein Synthesezwischenprodukt mit intensivem Moschusgeruch, *Z. Naturforsch.* **1970**, *25b*, 555.
68 Fester, G.; Bertuzzi, F. Drüsen-Sekret der Alligatoren (Yacarol), *Ber. Dtsch. Chem. Ges.* **1934**, *67B*, 365–370.
69 Fester, G.; Bertuzzi, F.; Pucci, D. Drüsen-Sekret der Alligatoren (Yacarol) (II. Mitteil.), *Ber. Dtsch. Chem. Ges.* **1937**, *70B*, 37–41.
70 Yoshii, F.; Hirono, S.; Moriguchi, I. Relation between the Odor of (R)-Ethyl Citronellyl Oxalate and its Stable Conformations, *Quant. Struct.-Act. Relat.* **1994**, *13*, 144–147.

Chapter 8
Aroma Chemicals V: Natural Aroma Chemicals

John Margetts

8.1 Introduction

The demand for natural flavourings, and hence natural chemicals and extracts, continues to increase worldwide. This trend is being fuelled by a variety of factors including legislation, marketing, the move towards more environmentally friendly processes and the somewhat illogical belief that natural is always healthier and safer than the synthetic equivalent. As this trend continues, so the industry responds through research into natural products and the processes by which nature produces the tastes and aromas we all know so well. The majority of flavour chemicals available to the flavourist and perfumer today are produced by chemical synthesis. An alternative approach for natural flavourings and fragrances is to extract the same chemicals from plant or animal sources, but nature's resources are finite and climatic conditions are variable, affecting economics, availability and yield. Therefore the range of natural materials available from these sources is limited. There is an increasing demand for biotechnology to provide suitable natural alternatives for the production of natural flavourings and fragrances.

This chapter aims to give a brief overview of the history of natural ingredients, the present state of the industry and the types of natural materials now available to the flavourist and perfumer for the creation of natural flavourings and fragrances. The main focus is on the production of natural chemicals. These may be extracted from natural sources such as plants, manufactured through the use of biotechnology or manufactured using so-called soft chemistry. Current research and commercial process are discussed. My apologies to all purist chemists and biochemists for the nomenclature (or lack of) throughout this chapter. The flavour and fragrance industry is particularly lacking in its use of a nomenclature, and a mixture of common names, trade names and more formal nomenclature will be found throughout. As a first step in understanding natural aroma chemicals it is important to understand the natural concept. What is the meaning of the term 'natural' from a food-flavouring perspective?

8.2 The natural concept

One might expect that being 'natural' would be a very simple and clear concept. This is not the case, and much work has been done in the US and Europe to legally define the term natural for food flavourings. In the US a distinction is made between artificial (synthetic) and natural flavours. Natural flavours or flavourings are governed by the Code of Federal Regulations (CFR) 21, Part 101.22 (a) (3). This states that 'the term natural flavor or natural flavoring means the essential oil, oleoresin, essence or extractive, protein hydrolysate, distillate, or any product of roasting, heating or enzymolysis, which contains the flavoring constituents derived from a spice, fruit or fruit juice, vegetable or vegetable juice, edible yeast, herb, bark, bud, root, leaf or similar plant material, meat, seafood, poultry, eggs, dairy products, or fermentation products thereof, whose significant function in food is flavoring rather than nutritional'. By definition, this regulation covers the ingredients used in flavourings. In addition, flavour materials must be 'generally recognised as safe' by being on the US positive list of flavour materials known as the FEMA GRAS list. In Europe, flavour chemicals are generally categorised into three classes: artificial chemicals that have yet to be found in nature, nature identical chemicals that occur in nature but are produced synthetically and natural chemicals. The natural chemicals and other natural ingredients used in flavourings were only defined as recently as 1988 by Community Directive 88/388/EEC as amended by 91/71/EEC. Broadly, natural ingredients used in flavourings must be obtained 'by appropriate physical processes (including distillation and solvent extraction) or enzymatic or microbiological processes from material of vegetable or animal origin either in the raw state or after processing for human consumption by traditional food-preparation processes (including drying, torrefaction and fermentation)'. Europe is currently moving towards a positive list system, and all chemicals used in flavourings must now appear on the European Inventory.

Both the above definitions for natural are rather broad and leave room for interpretation. Industry and regulators continue to debate the meaning of many of the words and phrases used within these definitions such as 'edible', 'appropriate' and 'traditional food-preparation processes'. There is also significant debate on the actual definition of what constitutes a food. Each year we are introduced to new foods, not necessarily through advanced food technology, but more likely the introduction of a new tropical fruit found in the Amazon rain forest or the latest herbal remedy that will extend our lives and make us look younger (I hope!). Couple this with the technological changes that are taking place within agriculture and the food industry, and you have a continually moving definition of natural and hence of natural aroma chemicals.

It is generally accepted that a natural flavour material must meet the following criteria [1]:

(1) Source materials must occur in nature.
(2) Water and atmospheric air can be used during production.
(3) When considering a process, the resultant material must be found in nature or as an artefact in traditional foods.
(4) In biotechnological processes, the substrate must be natural, but the nutrients can be used regardless of source.
(5) Processing conditions are allowed so long as they meet the criteria of traditional or in-home preparation techniques.

8.3 Chirality

Many natural aroma chemicals occur in nature as a specific chiral isomer, and the aroma of that specific enantiomer can be very distinctive and characteristic. Chemical synthesis tends to result in racemic mixtures and this can be extremely detrimental to the sensory characteristics of the aroma chemical. It may be impossible to develop a stereo-specific synthetic route, the separation of the resultant mixture may not be possible or the subsequent separation process may be very expensive. Some examples of the differences in aroma between stereoisomers are shown in Table 8.1. In addition, the aroma of stereoisomers may be perceived at very different levels of concentration. For example (4*S*)-(+)-carvone (D-carvone) has an aroma threshold of 600 ppb and is reminiscent of caraway, whilst (4*R*)-(−)-carvone (L-carvone) has an aroma threshold of just 43 ppb and has the aroma of spearmint [2]. Natural chemicals, isolated from plant sources or produced through a bioprocess, show a high degree of entantiomeric selectivity, and consequently, natural processes for obtaining chiral aroma chemicals can provide an efficient and cost-effective source. Biotechnology may also be considered for the purification of racemic mixtures where the substrate for the biotransformation is stereo-specific. One example is the production of L-menthol which is the major constituent of peppermint oil. Microbial lipases have been found that predominantly hydrolyse the L-menthyl esters (from the D, L-racemic mixture) into L-menthol, leaving the D-menthyl esters intact [3]. The L-menthol may then be readily separated from the D-menthyl ester. This type of stereo-specific process finds application in synthetic chemistry as well as for natural aroma chemicals.

8.4 Isolation from natural sources such as essential oils

As early as 3000 BC, the Egyptians were using papyrus manuscripts to record the use of fragrant herbs, oils, perfumes and temple incense, and told of healing salves made of

Table 8.1 Examples of chiral aroma chemicals

Chemical	Aroma
(*R*)-(+)-2-methyl butanol	Fermented, fatty
(*S*)-(−)-2-methyl butanol	Ethereal, fresh
(1*S*,3*S*,4*R*)-(+)-menthol	Minty, musty, bitter
(1*R*,3*R*,4*S*)-(−)-menthol	Cooling, minty, fresh
(*R*)-(+)-limonene	Citrus, orange
(*S*)-(−)-limonene	Citrus, harsh, turpentine
(4*S*)-(+)-carvone	Caraway, herbaceous
(4*R*)-(−)-carvone	Spearmint
(1*S*,4*R*)-*cis*-*p*-menthan-8-thiol-3-one	Blackcurrant leaf, tropical fruit, fruity
(1*R*,4*S*)-*cis*-*p*-menthan-8-thiol-3-one	Rubber, burnt, sulphurous
(+)-(*E*,*S*)-5-methyl-2-hepten-4-one	Hazelnut, fatty, metallic, strong
(−)-(*E*,*R*)-5-methyl-2-hepten-4-one	Hazelnut, soft, butter, chocolate, weak

fragrant resins [4]. References to the use of herbs, spice, oils and resins for fragrances and foods can be found throughout Chinese, Indian, Greek and Roman histories. Spices from across the Roman Empire were used to flavour foods. This may have been for their inherently pleasant flavour, to help preserve foods for storage or to mask off-tastes as meat or fish decayed. Throughout history, herbs and spices have played an important role in the development of commerce and trade across the world. The isolation of essential oils has a similar and long history. However, it was not until the early nineteenth century that developments in distillation and chemistry led to significant refinements of essential oil processing and the separation and identification of the natural components responsible for the taste and aroma.

The flavour and fragrance industry uses an extensive range of sources for its natural ingredients. Traditional materials used, which are extracted from plants, include essential oils, extracts, oleoresin, tinctures, distillates and juice concentrates. Definitions for these terms can be found widely in the literature and, in particular, in an excellent industry standard reference published by Steffan Arctander in 1960 [5] which has been reprinted several times. Parts of the plant used to form these extracts vary from the bud, flower, fruit, peel, seed, kernel, stem, leaf, bark, exudate to root. Methods of extraction include expression, distillation, concentration and extraction with organic solvents (including super-critical and sub-critical fluids). Distillation yields the volatile constituents whereas expression and solvent extraction (based on the polarity of the solvent/s used) yield both volatile and non-volatile constituents. Depending on the plant species and variety, its geographic origin and the growing conditions, the part of the plant used and the method of extraction, a different range of natural aroma chemicals will be produced with a distinctive organoleptic profile and a unique range of organic chemical constituents. A few examples of plants and the parts used are given in Table 8.2.

Essential oils contain the majority of the volatile, low molecular mass aroma constituents derived from a plant (as opposed to the 'fixed' or 'vegetable' oil that is generally lacking in flavour). They are typically extracted from plant material using a variety of distillation techniques, many of which are traditional processes having been used for many years. The most common technique is steam distillation, where steam is generated in a separate boiler and then blown through the botanical material which has been placed on racks in a still [6]. Water, and water and steam distillation methods are also commonplace. The steam and the volatile essential oil are condensed, and the oil is then separated from the water. The essential oil may be recovered as a single fraction or as different fractions collected over the duration of the distillation, as is the case with Ylang ylang oil [7]. Essential oils produced using these methods differ from the original oil in the botanical material as a result of changes that take place during the distillation. These changes are due to a number of factors such as the volatility of the components in the steam, hydrolysis due to the mildly acidic distillation conditions, further chemical rearrangements, interaction between the components and losses of components due to their solubility in the distillation water.

An alternative method often used for citrus oils is expression. This requires abrasion and pressing of the peel to yield a so-called cold-pressed oil. Expressed oils more closely resemble the original oil in the botanical material since they are not exposed to the heat processing involved in distillation. Additionally, they will contain both highly volatile components and non-volatile components not found in distilled oils. A second type

Table 8.2 Plant material used in flavourings

Common name	Botanical name	Parts used
Allspice (Pimento)	*Pimento dioica*	Berry, leaf
Almond, bitter	*Prunus amygdalus* var. *amara*	Nut
Balsum tolu	*Toluifera balsamum*	Gum exudate
Blackcurrant	*Ribes nigrum*	Bud, leaf
Cinnamon	*Cinnamomum zeylanicum*	Bark, leaf
Davana	*Artemisia pallens*	Herb
Eucalyptus	*Eucalyptus globulus*	Leaf
Fennel, sweet	*Foeniculum vulgare* var. *dulce*	Seed, herb
Garlic	*Allium sativum*	Bulb
Ginger	*Zingiber officinale*	Rhizome
Hops	*Humulus lupulus*	Female flower
Jasmine	*Jasminum officinale*	Flower
Lemon	*Citrus limon*	Fruit, Peel
Lemongrass	*Cymbopogon citratus, C. flexuosus*	Grass
Locust	*Ceratonia siliqua*	Bean
Mint	*Mentha* spp.	Flowering top and leaf
Peach	*Prunus persica*	Leaf, fruit, kernel
Pine, scotch	*Pinus sylvestris*	Needle, twig
Rose	*Rosa alba*	Flower, hip, leaf
Vanilla	*Vanilla planifolia*	Fruit/pod

of citrus oil can be derived from citrus fruit juice processing. For the production of citrus juice concentrates the juice is pressed from the fruit and may then be further processed in a number of ways. The final stage is the concentration of the fruit juice by the evaporation of water. During evaporation, fractions of the condensate may be taken and the volatile fruit or 'essence' oil recovered. Essence oils contain high levels of volatile 'fruity, juicy' components such as acetaldehyde and ethyl butyrate, 'green' components such as *cis*-3-hexenol and lack the non-volatile components found in cold-pressed peel oils. The natural aroma volatiles present in other fruits and vegetables may be extracted using similar techniques. Depending on the level of concentration these yield aqueous solutions and/or essential oils.

Extraction of botanical material using organic solvents yields a broader range of components, specifically including non-volatile components, many of which are responsible for taste rather than aroma. The range of compounds extracted is highly dependent on the polarity of the solvent(s) used. These vary from non-polar solvents such as hexane, through to highly polar solvents such as ethanol. As with expression, the crude extract produced will more closely resemble the original botanical material, particularly if little or no heating is used during processing. These advantages may be lost during concentration of the extract to remove the solvent/s. In this respect, sub-critical and super-critical carbon dioxide extractions have a definite advantage over traditional extraction techniques. Processing is carried out at a relatively low temperature and the carbon dioxide is easily removed by pressure reduction following the extraction process.

The disadvantage of this method is that the process equipment required is expensive which limits the possible applications.

Further processing of the above essential oils and extracts may be undertaken. Simple redistillation of essential oils may take place to clean an oil of impurities. Alternatively, some form of fractionation may be required. The simplest form of fractionation is known as rectification, where the initial (top) distillate and the residue (tail) in the still are removed. When 'topped and tailed' the oil has a cleaner, less-harsh aroma. Further fractionation of essential oils and extracts may take place to remove unstable or undesirable components, to improve solubility, to increase the concentration of the principal aroma components or to derive particular aroma profiles. In addition to a wide variety of distillation techniques, further concentration and fractionation may be achieved using, for example, reverse osmosis, solvent extraction and preparative chromatography. The principal by-products of the concentration of essential oils are terpenes. These may be used as a feedstock for biotransformation, as described later in this chapter.

The above techniques yield an extensive array of natural ingredients that may be used in both synthetic and natural flavourings. In many cases, this level of refinement is sufficient for use in a wide range of formulations. This is particularly true where an essential oil is used in the named flavouring, e.g. orange oil in an orange flavouring, or when the essential oil is almost exclusively composed of one chemical such as bitter almond oil which is in the order of 99% pure benzaldehyde. In other cases, there may be desirable chemicals present in an essential oil, but the purity is not sufficiently high so as to provide the desired aroma. One example would be *Litsea cubeba* oil which contains approximately 75% citral (a mixture of geranial and neral), the characterising chemical found in lemons. In its natural form, the other components in *Litsea cubeba* oil give it a floral, green character that is not characteristic of lemon. Therefore, the oil is fractionated to yield citral of high purity and this is used extensively in lemon flavourings. Examples of essential oils and their principal components are given in Table 8.3.

Globally, the market demand for natural flavourings continues to increase. As a consequence, natural sources such as essential oils, extracts, distillates and tinctures are being looked at to provide sources for the isolation of natural chemicals. Additionally, materials once considered as waste streams or by-products are now finding value as sources of natural chemical isolates. One example would be the 'tops and tails' from peppermint oil now used as a source of 'green notes', such as *cis*-3-hexenol. Aldehydes such as citral may be extracted from a range of essential oils including *Litsea cubeba* and lemongrass by treating the oils with bisulfite to form the addition compounds and then regenerating the aldehydes after extraction. Unfortunately, this is not a natural process and the resulting product cannot be described as natural. All techniques used for the isolation of natural chemicals from botanical sources must be physical processes so that the natural components are not chemically modified (within the limits described earlier for the isolation of essential oils).

Methods used for the separation of the natural isolates include virtually every imaginable physical extraction or separation technology known to man. All forms of distillation (steam, water, fractional, molecular, pervaporation (a combination of distillation and membrane filtration), etc.), solvent extraction (including sub-critical and super-critical solvents), membrane technologies (ultra-filtration and reverse osmosis), preparative chromatography, adsorption and exclusion columns, etc. Table 8.4 shows

Table 8.3 Essential oils and their principal components

Essential oil	Principal components
Bitter almond	Benzaldehyde (hydrogen cyanide is chemically removed)
Bergamot	Linalyl acetate, limonene, linalool, β-pinene, γ-terpinene, geranial, neral, geraniol, neryl acetate, geranyl acetate, bergaptene
Buchu leaf	Diosphenol, menthone, mentha-8-thiol-3-one
Caraway	D-Carvone, limonene, myrcene, α-phellandrene, α-pinene, β-pinene
Cassia	Cinnamaldehyde, *ortho*-methoxy cinnamaldehyde, cinnamyl acetate, benzaldehyde, ethyl cinnamate, salicaldehyde
Cinnamon leaf	Eugenol, caryophyllene, cinnamaldehyde, iso eugenol, linalool, cinnamyl acetate
Cinnamon bark	Cinnamaldehyde, cinnamyl acetate, eugenol, caryophyllene, linalool, α-terpineol
Clove	Eugenol, caryophyllene, eugenyl acetate, α-humulene
Coriander seed	Linalool, γ-terpinene, camphor, α-pinene, *para*-cymene, limonene, geranyl acetate
Cornmint (*Mentha arvensis*)	L-Menthol, L-menthone, iso menthone, limonene, L-menthyl acetate, piperitone, octan-3-ol
Dill (weed)	D-Carvone, α-phellandrene, limonene
Eucalyptus	1,8-Cineole, α-pinene, *para*-cymene, limonene
Fusel	Iso amyl alcohol (3-methyl butanol and 2-methyl butanol), butanol, ethanol, iso butanol, propanol, pyrazines
Garlic	Diallyl disulfide, diallyl trisulfide, diallyl sulfide
Ginger	Zingiberene, AR-curcumene, β-sesquiphellandrene, bisabolene, camphene, β-phellandrene, 1,8-cineole
Grapefruit	Limonene, myrcene, octanal, decanal, linalool, nootkatone, citronellal, neral, geranial, β-sinensal
Ho wood	Linalool
Lemon	Limonene, β-pinene, γ-terpinene, geranial, neral, neryl acetate, geranyl acetate, citronellal, linalool, nonanal
Lime (distilled)	Limonene, γ-terpinene, α-terpineol, terpinolene, *para*-cymene, 1,4-cineole, 1,8-cineole, β-pinene
Litsea cubeba	Geranial, neral, limonene, 6-methyl hept-5-en-2-one, myrcene, linalyl acetate, linalool, geraniol, nerol, caryophyllene, citronellal
Mandarin petitgrain	Limonene, dimethyl anthranilate, geraniol, citral, citronellal
Orange peel (sweet)	Limonene, myrcene, decanal, octanal, linalool, citronellal, neral, geranial, valencene, β-sinensal, α-sinensal
Peppermint (*Mentha piperita*)	L-Menthol, L-menthone, L-menthyl acetate, 1,8-cineole, menthofuran, limonene, iso menthone, octan-3-ol, oct-1-en-3-ol, mint lactone
Rose oil	Citronellol, geraniol, nonadecane, nerol, methyl eugenol, geranyl acetate, eugenol, 2-phenyl ethanol, rose oxide
Spearmint oil	L-Carvone, limonene, myrcene, 1,8-cineole, carvyl acetate, dihydrocarvyl acetate, dihydrocarveol, octan-3-ol, *cis*-jasmone
Star anise	*trans*-Anethole, limonene, anisaldehyde, linalool, methyl chavicol
Ylang ylang	Germacrene D, linalool, α-farnesene, benzyl acetate, α-caryophyllene, linalool, *para*-cresyl methyl ether, benzyl benzoate, geranyl acetate, methyl benzoate, benzyl salycilate, eugenyl acetate

Table 8.4 Examples of natural chemical isolates

Natural isolate	Source
Acetaldehyde	Orange
Iso amyl alcohol	Fusel
Anethole	Fennel
Benzaldehyde	Bitter almond, apricot, peach
α-Bisabolol	Lemon
Butanol	Fusel
Iso butanol	Fusel
L-Carvone	Spearmint
D-Carvone	Caraway
Cinnamaldehyde	Cassia, cinnamon bark
Cinnamyl cinnamate	Styrax
Citral	*Litsea cubeba*, lemongrass, lemon
Citronellal	*Eucalyptus citriadora*, citronella
Citronellol	Citronella
Decanal	Orange
Dimethyl anthranilate	Mandarin petitgrain
Dimethyl sulfide	Mint
Ethyl butyrate	Orange
Eugenol	Clove, bay, cinnamon leaf
cis-3-Hexenol	Mint
Geraniol	Citranella
Geranyl acetate	*Litsea cubeba*, lemongrass
Hydroxycitronellal	*Eucalyptus citriadora*, citronella
α-Irone	Orris
Limonene	Citrus
Linalool	Ho wood, rosewood, coriander
Linalyl acetate	*Mentha citrata*, petitgrain
L-Menthol	Mint
Methyl cinnamate	Sweet basil, *Ocimum canum*
Myrcene	Orange
Nerolidol	*Melaleuca viridiflora*, cabreuva
Nonanal	Orange
Nootkatone	Grapefruit
Phenyl ethyl alcohol	Rose, geranium rose
Pyrazines	Fusel
Sinensal	Orange
γ-Terpinene	Lemon
α-Terpineol	Lime
Valencene	Orange
Vanillin	Vanilla

examples of natural chemicals isolate from botanical sources. It should be noted that not all of these sources are commercially viable. For example, irones separated from Orris root cost tens of thousand of dollars per kilogram to produce, phenylethyl alcohol similarly and vanillin from vanilla beans is several thousand dollars per kilogram.

The range and quality of materials available by isolation is limited. Additionally nature's resources are finite and the costs of recovering natural isolates may limit their use even further. Therefore the flavour and fragrance industry has to look to other sources, such as biotechnology for its ever increasing requirement for natural ingredients.

8.5 Biotechnology

The history of biotechnology can be traced back almost to the beginnings of mankind. The Sumerians and Babylonians practised beer brewing before 6000 BC; references to wine making can be found in the Book of Genesis, and the Egyptians used yeast for baking bread [8]. Microbial transformation of ethanol into acetic acid, the production of vinegar, dates back to ca. 2000 BC. For many centuries, these traditional processes continued to be used with little or no understanding of the underlying biological mechanisms. The use of whole cell biotransformation and fermentation continued virtually unchanged until the nineteenth century when an understanding of the underlying processes began to develop. The isolation and the use of enzyme and our understanding of their functionality follow a similar path, and the manufacture of cheese, for example, was little understood, with improvements to processes being made on an empirical basis. During the past 150 years we have moved from virtually no knowledge to the development of penicillin, the discovery of the structure of DNA to recombinant DNA technology and genetic engineering. Research and development has primarily been focused on pharmaceuticals, bulk chemicals and food products.

Food products are the major output of biotechnology from both the value and the volume perspectives. Traditional processes, such as alcohol fermentation, vinegar brewing, food acid production, dairy product manufacture, etc., still dominate the biotechnology industry. For the flavour and fragrance industry the focus on natural raw materials produced by biotechnology has only become truly important during the last 20 years or so. During that time we have seen significant improvements in analytical techniques such as gas chromatography/mass spectrometry, and our understanding of the production of natural flavour compounds in foods has developed strongly. Our knowledge of the microorganisms and enzymes used by the food industry and the types of chemical reactions they catalyse has been the foundation for the research and development of biotechnology based raw materials used in or as flavourings. Biotechnology is blossoming under the pressure being exerted upon it. Areas of development include fermentation using whole organisms, the use of enzymes as natural biocatalysts to perform biotransformations and transesterification, plant cell tissue culture and genetic engineering. The majority of compounds developed to date are produced by either intact microbial cells, crude enzyme mixtures or isolated enzymes. The major enzyme classes and their functions are listed in Table 8.5. The main aims of these developments can be subdivided into two general groups: the enhancement of total flavour and the formation of individual flavour chemicals.

Table 8.5 Enzyme classification

Class	Function	Example
Oxidoreductases	Oxidation–reduction involving transfer of electrons (hydride ion or hydrogen atoms). Cofactor may be required	Catalase, glucose oxidase, lipoxy genases
Transferases	Transfer of functional groups such as acyl, glycosyl (sugars), phosphate, methyl, sulfur groups. Cofactor may be required	Glycosyl transferases, methyl transferase
Hydolases	Hydrolysis reactions (addition of water) with substrates such as amides, esters, epoxides, glycosides, peptides. Esterification (removal of water)	Amidases, amylases, cyanidase, esterases, hemi-cellulases, glucanases, lipases, pectinases, phospholipases, proteases, saccharases, ureases
Lyases	Addition of groups to double bonds or formation of double bonds by removal of groups	Hydroperoxide lyases, phenylalanine ammonia lyase
Isomerases	Transfer of groups within molecules to yield isomeric forms. Isomerisation, racemisation	Glucose isomerase
Ligases	Formation of bonds by condensation reactions coupled to cofactor ATP or other nucleoside triphosphate cleavage. C–C, C–O, C–S, C–N, etc.	Glutamine synthetase, Acetate-CoA ligase, Urea carboxylase

8.6 Total flavour enhancement

Total flavour enhancement focuses on the development of a complex multi-component system that reflects the overall flavour of a food. A classical example would be enzyme-modified cheese, where the application of a lipase causes the hydrolysis of the triglycerides into fatty acids. The quantity and type of fatty acids produced will depend on the triglycerides present and the lipase used. For example, the enzyme from *Mucor miehei* is known to exhibit a high selectivity towards short-chain fatty acids, whereas other lipases may preferably liberate long-chain fatty acids. The position of the fatty acid residue in the triglyceride molecule may also be important [9]. The aim is to produce the correct ratio of fatty acids to reflect the type of cheese flavour desired, but at a much higher concentration. The principal flavour component released is butyric acid which has a characteristic mature cheese flavour profile. The higher the chain length, the more fatty and soapy the taste becomes. In addition to the lipase, a protease may be added to breakdown short-chain peptides that can be responsible for bitter off-tastes. Another starting material used to produce natural butter flavourings is purified butter oil. This is a simpler system and under controlled conditions liberates good quantities of the C4–C6 acids, namely butyric and caproic acids.

The chemical compounds responsible for the characteristic flavour of blue cheeses, such as Roquefort and Stilton, are methyl ketones. They are produced by a different mechanism which results in a mixture of C5–C11 alkanones, especially 2-pentanone, 2-heptanone, 2-nonanone and 2-undecanone, from the corresponding C6–C12 fatty acids. In cheese, this microbial fermentation is caused by the deliberate addition of a mould, such as *Penicillium roqueforti*. Proteolytic enzymes hydrolyse the casein present, thereby producing precursors and substrates for mould growth. Lipolytic enzymes produce free fatty acids that are then oxidised to the equivalent α-keto acids and then decarboxylated to form methyl ketones, as shown in Scheme 8.1. Concentrated blue cheese powders are manufactured by the microbial fermentation of milk solids, caseinates and milk fats in an aqueous environment followed by spray drying. Methyl ketones are manufactured using the spores of *Penicillium roqueforti*, which still retain the required enzyme activities, despite being dormant [10]. This makes the production scale use of spores very easy, as once grown, they are easy to store until required and then require no growth media. Using this method, lipolysed fats or oils from a variety of sources can be transformed directly into methyl ketones. These may be used as a mixture or fractionated into the individual ketones.

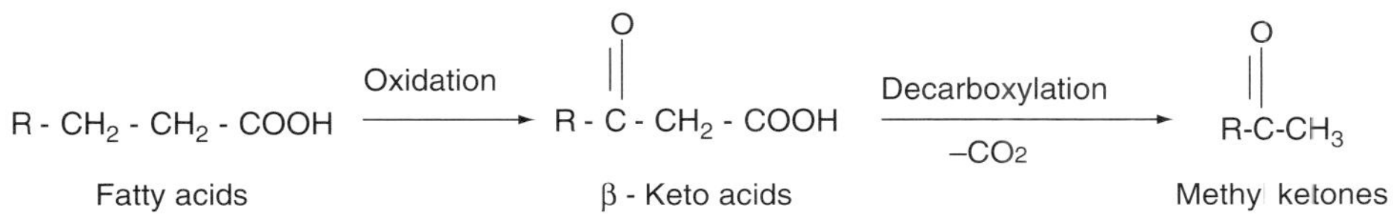

Scheme 8.1 Methyl ketone production in blue cheese.

8.7 Individual flavour chemicals

8.7.1 Alcohols

Short-chain alcohols are primarily obtained by classical fermentation techniques. They are either products or by-products of industrial ethanol production or from the alcoholic beverage industry. Fusel oil is the principal by-product recovered during purification or rectification. It is a complex mixture of secondary alcohols, primary alcohols, some esters and trace amounts of pyrazines. The actual composition is highly dependent on the base raw materials used as the feedstock for the fermentation such as cereals, potatoes, grapes, whey, etc. These alcohols may be used as mixtures, or separated and purified. Some alcohols can be used directly as flavour materials, e.g. isoamyl alcohol, butanol and isobutanol. They are all used as starting materials for further reactions to form aldehydes, acids or esters.

During conventional yeast fermentations (*Saccharomyces cerevisiae*, *Kluyveromyces marxianus*) low levels of phenyl ethyl alcohol are produced. This is an important flavour and fragrance compound which can be manufactured synthetically or concentrated from rose oil at high cost [11]. During conventional fermentation the amino acids present in the fermentation broth can undergo transamination to form the α-keto acids followed by decarboxylation to form the corresponding aldehydes. These are then reduced to the corresponding alcohols, which have one carbon atom less than the substrate amino acid. In the case of phenyl ethyl alcohol, the substrate is L-phenylalanine and the route is shown in Scheme 8.2. This reaction has been specifically targeted for investigation, and significant improvements in the yields obtained are reported using *Saccharomyces cerevisiae* and *Kluyveromyces marxianus* mutants, *Aspergillus niger* strains and *Hansenula anomala* with yields up to 3.4 g/l [3, 11].

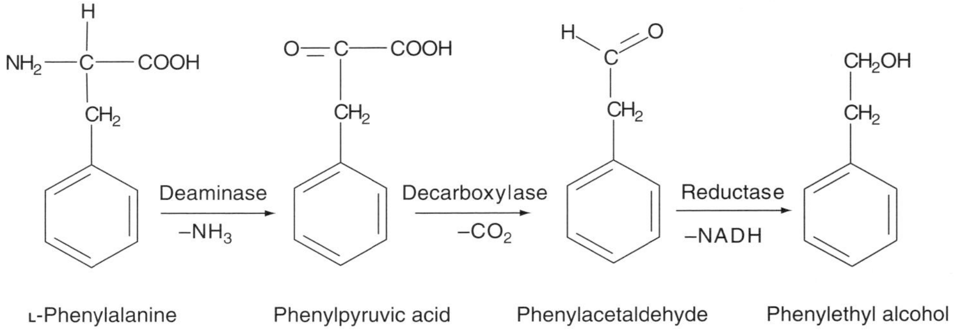

Scheme 8.2 L-Phenylalanine to phenylethyl alcohol.

8.7.2 Carbonyls

Chemicals within this group have a wide range of extremely different aromas. Simple aldehydes range from acetaldehyde, which provides a very fresh note, to hexanal, which is green in aroma, through to longer-chain aldehydes which become more soapy or waxy. The aromatic aldehyde benzaldehyde is characteristic of bitter almonds whereas phenylacetaldehyde is floral and honey-like. Methyl ketones, as described earlier, are characteristic of blue cheese whereas the short-chain diketone, diacetyl, is sweet and buttery. Investigations and manufacture of a number of these types of compounds are described.

Acetaldehyde is very important for fruit flavourings, particularly orange juice type flavourings. It is also a principal component of yoghurt flavourings. A process was patented in 1984 for the conversion of ethanol to acetaldehyde using the enzyme alcohol dehydrogenase [12]. However, the process suffers from a common problem associated with isolated enzyme oxidation reactions that rely on the need for and the regeneration of cofactors. When the ethanol is oxidised to acetaldehyde the cofactor (nicotinamide adenine dinucleotide) is reduced. For the reaction to continue, the cofactor must be regenerated. This is achieved by light-catalysed oxidation with a second substrate (flavin mononucleotide). This second substrate then has to be regenerated by oxidation with molecular oxygen which produces hydrogen peroxide as the by-product. This is decomposed to water and oxygen by the addition of a catalase. The process is shown in Scheme 8.3. As a general rule, this level of complexity should be avoided when developing commercial processes. An easier approach for producing a natural aroma chemical is to find a microorganism that has the appropriate metabolic pathways and/or that will produce the required enzymes in situ. Ethanol, propanol, butanol, but-2-en-1-ol and benzyl alcohol may be oxidised by the production in situ of methanol oxidase and catalase using certain microorganisms such as yeast of the genera *Pichia*, *Candida*, *Hansenula* and *Torulopsis* [13].

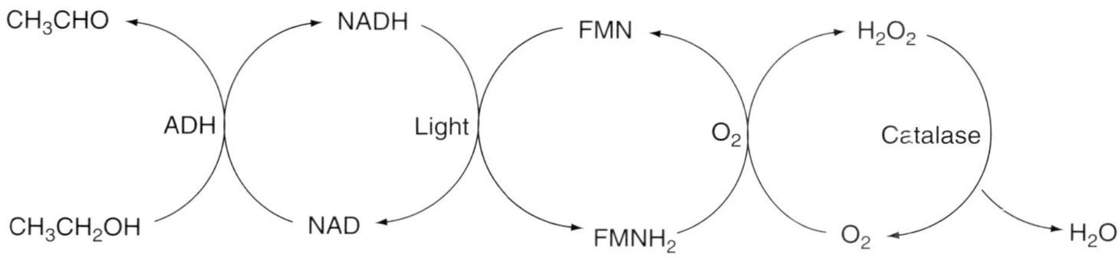

ADH – Alcohol dehydrogenase

NAD – Nicotinamide adenine dinucleotide

FMN – Flavin mononucleotide

Scheme 8.3 Oxidation of ethanol to acetaldehyde.

Diacetyl is formed during the production of cultured dairy products. It is the main compound responsible for the characteristic creamy buttery flavour of fermented dairy products such as sour cream and buttermilk [14]. A major product made in this way is 'starter distillate', the main component being diacetyl. It is produced by distillation from lactic dairy cultures. The starting material for the biosynthesis is citrate and a key intermediate is α-acetolactic acid which is decarboxylated to form diacetyl, as shown in Scheme 8.4. The level of diacetyl in such products is still relatively low.

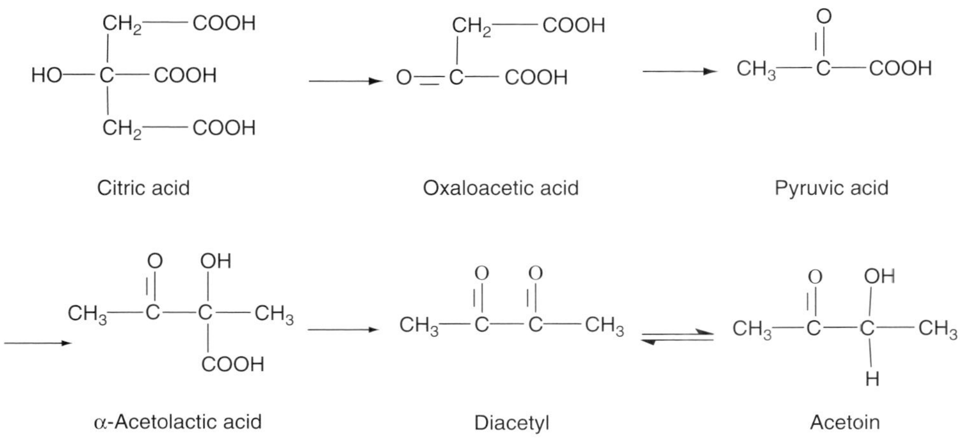

Scheme 8.4 Biosynthesis of diacetyl.

A number of industrial processes for the production of diacetyl have been reported, including the production from citric acid using *Streptococcus cremoris* and *Streptococcus diacetylactis* [15].

4-Hydroxy-phenyl-butan-2-one (raspberry ketone) is a characteristic flavour compound found in raspberries at levels of approximately 4 ppm [3]. It would not be practical to consider extraction of such low levels of a component from a complex natural source. Betuloside may be extracted from birch bark (also from rhododendron and maple) and hydrolysed with, for example, a β-glucosidase from *Aspergillus niger* to form betuligenol. This may then be converted with a microorganism containing a secondary alcohol dehydrogenase, such as the yeast *Candida boidinii*, into raspberry ketone (Scheme 8.5). An alternative approach is the conversion of 4-hydroxybenzalacetone. This is formed from 4-hydroxybenzaldehyde of botanical origin and acetone from sugar fermentation. A wide range of microorganisms has been evaluated and the most effective conversion to raspberry ketone was found using *Mucor subtilissimus* [16].

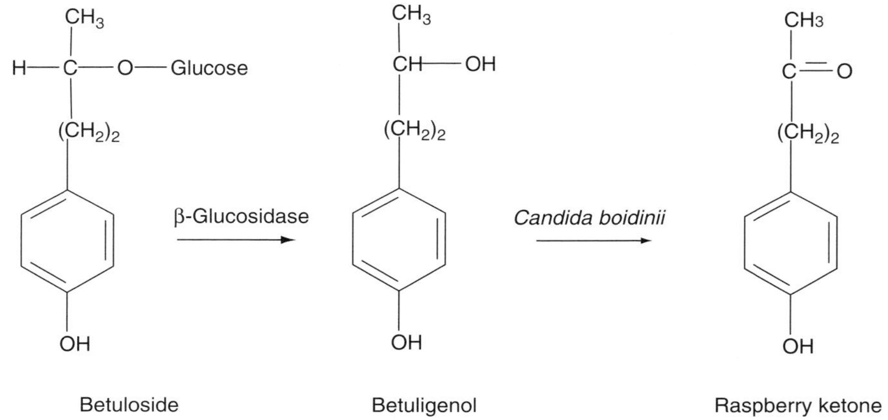

Scheme 8.5 Raspberry ketone formed from betuloside.

8.7.3 Acids

As discussed earlier, microbial transformation of ethanol into acetic acid, the production of vinegar from wine, dates back to ca. 2000 BC. This has to be the first individual flavour chemical to be produced through biotechnology and uses *Acetobacter* species. The basic process remains the same today with improvements to the production technology, the scale of operation and the strains used for the bacterial fermentation. This type of biotransformation has been extended to other natural substrates in order to produce the corresponding natural acids. Fusel oil, either as a mixture or separated into individual alcohols, can be converted by certain *Acetobacter* strains into the corresponding natural acids such as butyric, isobutyric, 3-methylbutyric and 2-methylbutyric acid [17], as shown in Scheme 8.6. Propionic acid is important for Emmental cheese type flavourings and for fruit flavourings. It can be produced by first forming lactic acid with strains of *Lactobacillus lactis* or *Streptococcus thermophillus* and then metabolising this lactic acid into propionic acid by *Propionibacter* strains [10].

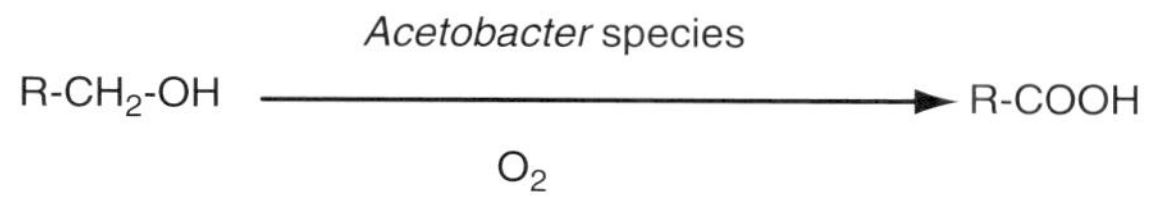

Scheme 8.6 Biotransformation of fusel alcohols to the corresponding acids.

Table 8.6 Sources of fatty acids from oils and fats

Fatty acid	Source
Saturated	
C4–C12 Butyric–lauric	Milk fat, butter oil
C8–C14 Caprylic–myristic	Coconut oil
C12–C14 Lauric–myristic	Palm kernel oil
C16 Palmatic	Palm oil, cottonseed oil
C14–C18 Myristic–stearic	Animal fats
Unsaturated	
C18:1 Oleic	Vegetable oils, animal fats
C18:2 Linoleic	Vegetable oils
C18:3 α-Linolenic	Linseed oil, rapeseed oil
C18:4 Arachadonic	Animal fats
C20:1 Gadoleic	Herring oil, whale oil
C20:5 Eicosapentaenoic (EPA)	Fish oils
C22:1 Erucic	Herring oil, whale oil
C22:6 Docosahexaenoic (DHA)	Fish oils
Hydroxylated	
C16:0–OH 11 Hydroxypalmatic	Jalap root
C18:1–OH Ricinoleic	Castor oil

Another important source of fatty acids is the degradation of triglycerides from oils and fats through the action of lipases which catalyse the hydrolysis of lipids to fatty acids and glycerol. Various sources such as butter, coconut, palm, soy, sunflower, linseed and caster oil may be used, depending on the free fatty acid chain length and degree of unsaturation and/or hydroxylation required. Table 8.6 outlines the main commercial sources of these fatty acids. Following the action of the lipase, the acids may be separated by, for example, steam distillation and further refined by fractional distillation. Alternatively, the mixture may be used in situ for further reactions with, e.g., alcohols to produce esters.

8.7.4 Lactones

Both γ- and δ-lactones are used widely as flavour and fragrance chemicals. They occur in foods such as dairy products, fruits, nuts and vegetables. An industrial example of lactone production is the microbial conversion of ricinoleic acid (from castor oil) via partial β-oxidation into γ-decalactone, which has a peach-like aroma, by yeasts such as *Sporidiobolus salmonicolor* and *Yarrowia lipolytica* [3]. The ricinoleic acid is hydrolysed from the triglyceride and then, via β-oxidation, degraded to 5-ketodecanoic acid. The yeast then reduces the keto acid to the hydroxy acid which can be lactonised by dropping the pH and heating (Scheme 8.7). Other methods are reported, such as the conversion of ethyl decanoate by *Mucor circinelloides* [18] and the fermentative β-oxidation of ricinoleic acid using the yeast *Candida lipolytica*, but it is not known if these methods have been commercialised. *Mucor circinelloides* is also reported to convert other fatty acid esters such as ethyl caprylate (octanoate) to the corresponding γ-octalactone. In peaches, the *R*-enantiomer of γ-decalactone is dominant whilst it is the *S*-enantiomer that occurs in mango varieties. Chemical synthesis results in a racemic mixture. Similarly, coriolic acid is the precursor hydroxylated fatty acid for the production of δ-decalactone with *Cladosporium suaveolens*. This lactone can also be obtained by yeast fermentation of 11-hydroxy palmatic acid present in Jalap root (sweet potato). The α,β-unsaturated lactone, 2-decen-5-olide, a main component of massoia bark oil, can also be efficiently converted by baker's yeast into δ-decalactone. Massoia bark oil also contains 2-dodecen-5-olide which, in particular, can be converted into δ-dodecalactone by the yeast *Saccharomyces delbrueckii* [19] and *Clostridium* strains [20].

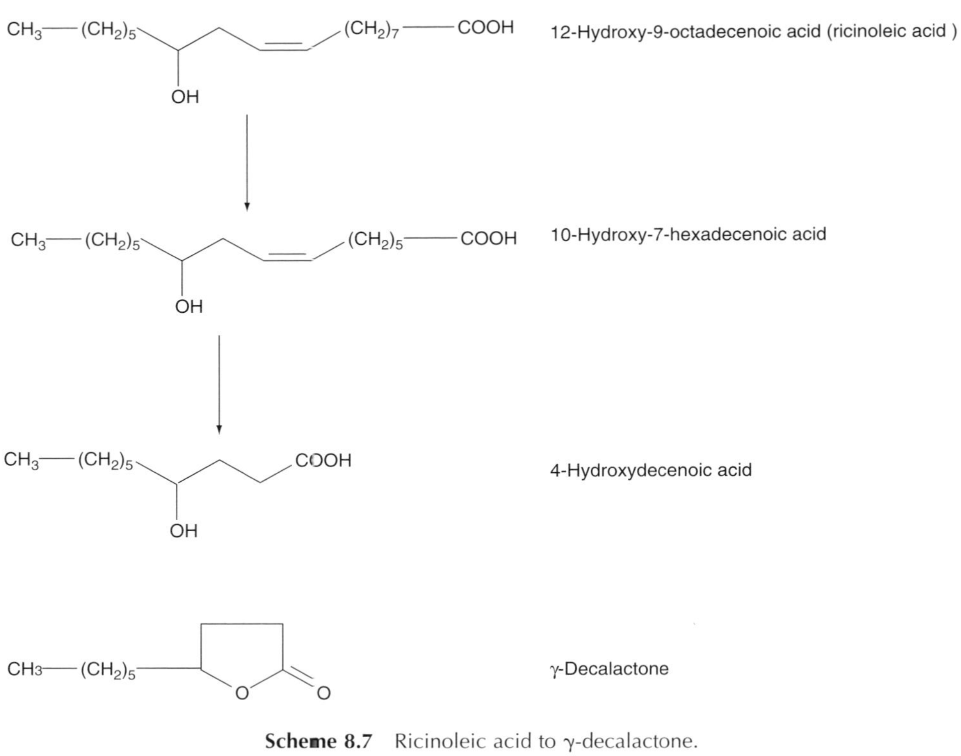

Scheme 8.7 Ricinoleic acid to γ-decalactone.

8.7.5 Esters

These compounds provide the flavour and fragrance industry with a wide range of characteristic sweet, fruity and, sometimes, floral aromas. Their synthesis through biotechnology is well documented and the principal method of manufacture is esterification of the appropriate alcohol and acid by the action of a lipase. Most lipase-based reactions are reversible to a greater or lesser extent, the enzyme being capable of hydrolysis or esterification depending on the specific enzyme used and the environment in which the reaction takes place (Scheme 8.8). The major requirement for ester production is the removal of water to drive the reaction forward. This can be achieved by a variety of methods, for example, the reaction may be performed in an organic solvent. Suitable solvents include hydrocarbons such as hexane or even super-critical fluids such as carbon dioxide. The organic phase may be a vegetable oil containing a mixture of fatty acids, or indeed the pure vegetable oil may be used to perform transesterification, where the acid residues are 'swapped' between the glycerol of the fatty acid triglyceride and the alcohol. An example is the use of the lipase from *Candida cylindracea* with butter fat and ethanol [12]. A mixture of ethyl esters was produced with the highest concentration being the butyrate and caproate. The enzymic esterification takes place at the solvent/water interface, the product of the reaction being more soluble in the organic phase. The process requires very little water to be present for the enzyme to function. Further techniques may be required to remove excess water such as distillation, membrane separation or a combined technique such as pervaporation.

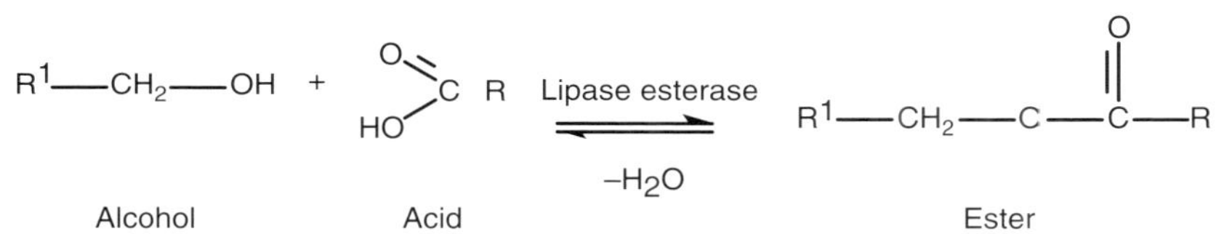

Scheme 8.8 Enzyme synthesis of esters.

Enzymes may be highly substrate-selective. For example, with alcohols of chain length C2–C10, the fungal lipase enzyme from *Mucor miehei* has a distinct selectivity towards C4–C6 fatty acids which are converted in yields of up to 95% whereas C2 or C3 fatty acids are poor substrates which do not yield significant levels of esters [17]. Other enzymes reported to have high specificity are lipases from *Aspergillus niger* and *Rhizopus arrhizus*, whereas *Candida rugosa* lipase has been reported to give high yields at low selectivity. Manufacture of iso propyl palmitate (C3 alcohol/C16 fatty acid) can be carried out using a lipase from *Candida antarctica*. Water is removed from the reaction by either azeotropic distillation or pervaporation [8]. This enzyme is also reported to convert acyclic monoterpenols such as citronellol and geraniol to their acetates in high yield using a microaqueous hexane reaction solvent [21].

Natural sources of the alcohols and acids for these reactions may come from natural isolates or through biotechnology. Provided that a suitable source can be found, virtually any ester can be produced. Esters of acids from acetic to hexanoic and alcohols from methanol to hexanol, citronellol and geraniol have been synthesised by lipases from *Mucor miehei*, *Aspergillus* sp., *Candida rugosa* and *Rhizopus arrhizus*. 2/3-Methylbutanol (commonly called isoamyl alcohol) esterification is reported by *Candida* sp. lipases, and geraniol and citronellol esters of C6–C12 fatty acids have been achieved in high yield using immobilised lipases of *Rhizomucor miehei* [22].

8.7.6 Green notes

This collective term refers to the aroma of certain organic chemicals, rather than to them being environmental friendly. Green notes are present in nature in a wide variety of leaves, fruits and vegetables. The aroma of 'green' chemicals is reminiscent of newly cut grass, under-ripe fruits and green leafy vegetables. This green aroma is sometimes referred to as a fresh note and the chemicals responsible are predominantly C6 alcohols and aldehydes, preferably unsaturated. These include compounds such as hexanol, hexanal, *trans*-2-hexenal, *trans*-2-hexenol, *cis*-2-hexenol and *cis*-3-hexenol. The fatty green notes more associated with the cucumber and melon family are typically C9, preferably the aldehyde and the double-unsaturated, i.e. *trans*-2-*cis*-6-nonadienal (violet leaf aldehyde). The natural biosynthetic pathway to these chemicals in many plants is the degradation of unsaturated fatty acids. Interestingly, many of these notes are also the cause of off-tastes formed during the oxidation of fats and oils. For many plants the biosynthetic pathway follows the lipoxygenase-catalysed oxidation of fatty acids containing a 1,4-pentadiene sequence to form hydroperoxides, followed by cleavage with a hydroperoxide lyase to produce a C6 unsaturated aldehyde. Aldehyde isomerase catalyses the formation of other unsaturated aldehydes, and the presence of alcohol dehydrogenase results in the formation of alcohols [23]. The biosynthetic pathway is shown in Scheme 8.9. Depending on the plant source, the enzymes differ in the reactions that take place both in terms of the isomers produced and chain length of the resulting aldehydes and alcohols. Thus, for cucumbers, melons, etc., the dominance of 9-hydroperoxides results in a predominance of C9 aldehydes and alcohols.

A number of proposed processes for producing green notes using homogenised vegetable materials and crude enzyme preparations are described in the literature [17, 23]. These include the use of strawberry leaves, radish leaves, soya flour, watermelon and alfalfa to provide lipoxygenase and hydroperoxide lyase enzymes, yeast (e.g. *Saccharomyces cerevisiae*) to provide alcohol dehydrogenase acting on substrates such as linoleic acid, safflower oil, linseed oil, sunflower oil and linolenic acid. Violet leaves are reported as the vegetable material of choice for the production of C9 aldehydes and alcohols [24]. However, within the published literature there appear to be very few processes that could be realised on an industrial scale for the production of green notes or individual 'green' chemicals. There appears to be much more that needs to be done before these materials become widely available from a biotechnological source.

8.7.7 Pyrazines

These compounds provide the characteristic aroma of nuts. They occur in a wide variety of foods, including roasted products such as cocoa, coffee and meat. They also provide the distinctive aroma of bell peppers. *Bacillus subtilis* and *Brevibacterium linens* have been used to produce 2,5-dimethyl pyrazine and tetramethylpyrazine, respectively, by fermentation of soy beans or flour enriched with precursors such as L-threonine, acetoin and ammonia. (Certain alkylpyrazines can also be isolated from fusel oil residues.) Bell pepper pyrazine (2-isobutyl-3-methoxy-pyrazine) is produced by a mutant strain of *Pseudomonas perolens* [10] (Figure 8.1).

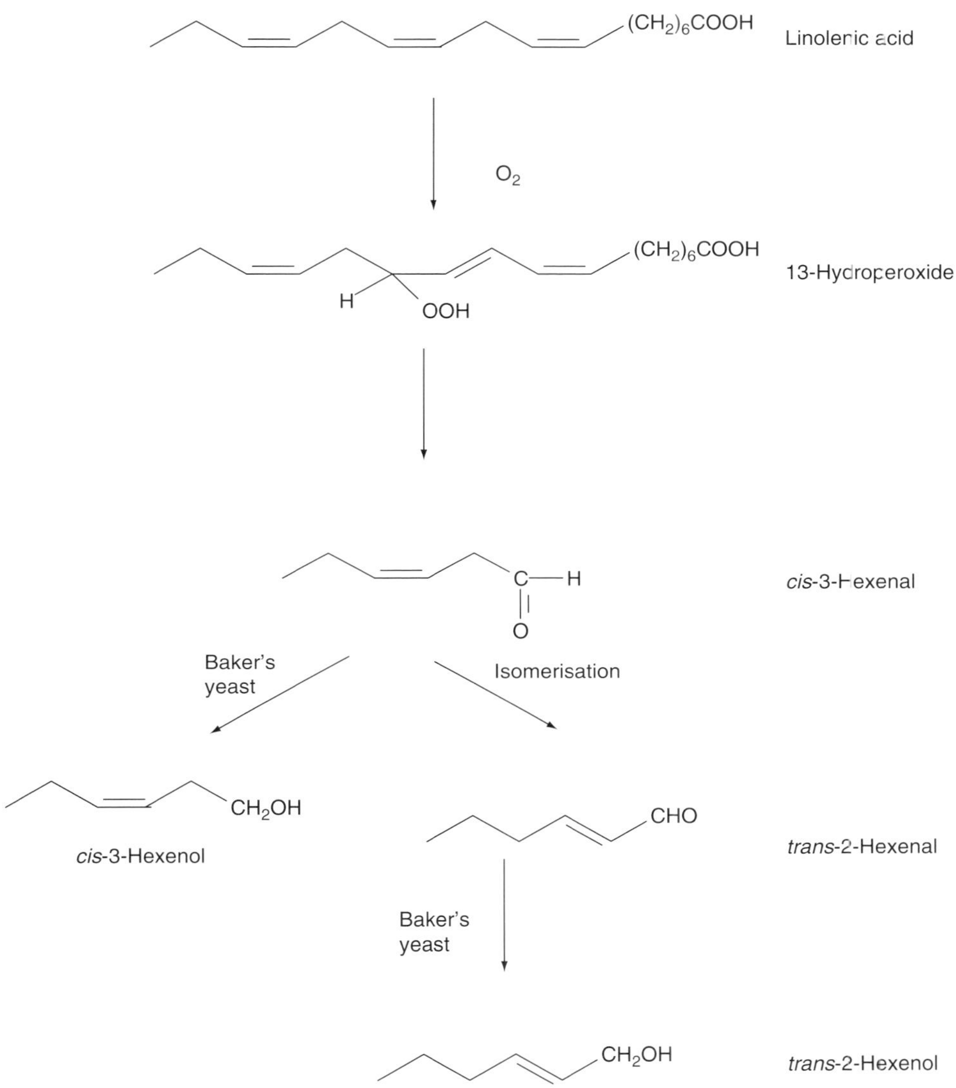

Scheme 8.9 Formation of green notes.

8.7.8 Sulfur compounds

A number of important flavour chemicals can be derived from L-methionine. These include methional (3-(methylthio)-propanal), methionol (3-(methylthio)-propanol) and the corresponding methyl and ethyl esters. Depending on concentration, methional can have a pleasant meaty, onion-type character and occurs widely in nature. Methionol is found in fermented products and is a particularly important compound in soy sauce. The esters have powerful fruity, sulphurous aromas and occur naturally in

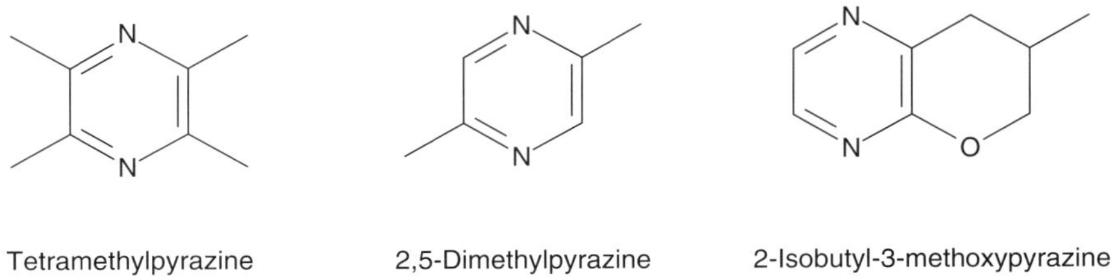

Figure 8.1 Examples of pyrazines.

pineapple and fermented products. Methods have been developed for the biofermentation of methionol and 3-(methylthio)-propionic acid from natural L-methionine [25]. Under conditions of nitrogen limitation, growing yeast will deaminate L-methionine to produce 2-oxo-4-(methylthio)-butyric acid. This α-keto acid is then decarboxylated to form methional, which in turn is rapidly reduced to methionol. This product can be recovered from the fermentation or, by altering the fermentation conditions, converted into 3-(methylthio)-propionic acid. This may then be esterified to form the esters (Scheme 8.10).

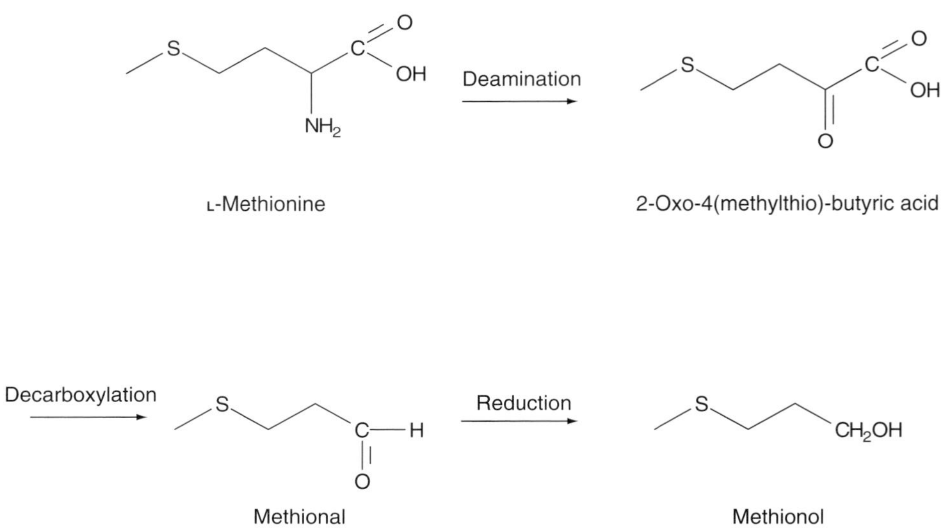

Scheme 8.10 Formation of methionol from L-methionine.

p-Mentha-8-thiol-3-one is a highly potent flavour molecule with a characteristic blackcurrant aroma at low concentrations. Natural cysteine and natural pulegone ((*R*)-(+)-*p*-mentha-4(8)-en-3-one), when mixed together in water at room temperature, react to form a precipitate of a cysteine–pulegone conjugate. The conjugate is incubated with any number of bacteria (e.g. *Eubacterium limosum*) which possess an enzyme belonging to the group known as β-lyases. These enzymes are able to hydrolyse the bond between the β-carbon atom and a sulfur atom in a relatively broad range of substrates. In this case, the conjugate is converted to the desired *p*-mentha-8-thiol-3-one. The same approach has also been reported for the conversion of furfural to furfuryl mercaptan (2-furfurylthiol), an important flavour chemical found in roast products, particularly coffee, together with other thiols from the corresponding aldehyde [25, 26]. An outline of the reaction is given in Scheme 8.11.

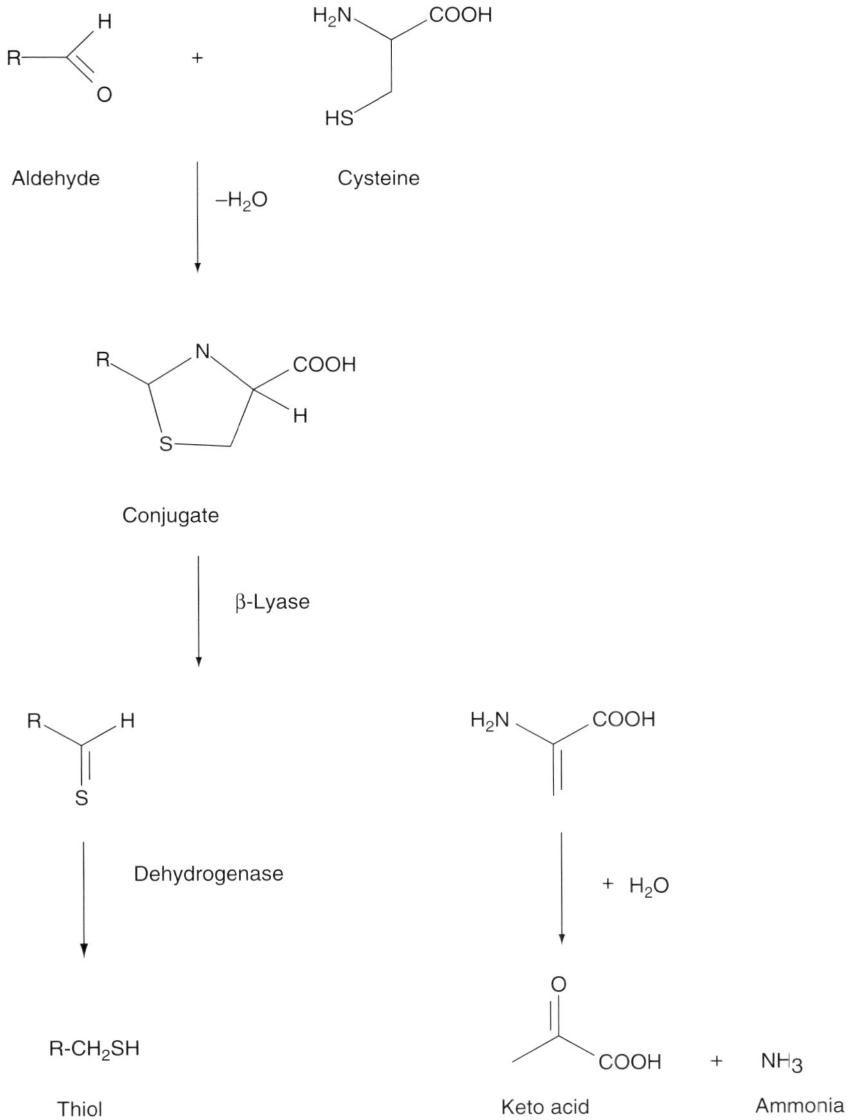

Scheme 8.11 Formation of thiols from a cysteine–aldehyde conjugate.

8.7.9 Terpenoid and related compounds

The essential oil industry produces large quantities of terpene hydrocarbons as a by-product of the purification, rectification and concentration of essential oils. They have little aroma, low water solubility and are unstable. It is estimated that 36 000 t of the monoterpene *R*-(+)-limonene (D-limonene) is separated from cold-pressed citrus oils during concentration, and that 160 000 t of α-pinene and 26 000 t of β-pinene are obtained from pine oils [21]. Naturally, these terpenoid by-products of the essential oil industry are seen as a potentially rich source of starting materials for biotransformation into more valuable natural chemicals. Whilst there is a body of literature on the research being carried out in this area, there are very few examples of commercially produced products due to the complex nature of the chemistry and the toxicity of terpenes to microorganisms.

(*R*)-(+)-limonene has been converted to 1*S*,2*S*4(*R*)-*p*-menth-8-ene-1,2-diol using *Corynespora cassiicola* in high purity, with few side products. Using a second organism, *Penicillium digitatum*, the (*R*)-(+)-limonene of racemic limonene was transformed into α-terpineol [10], a key component in lime flavourings (Scheme 8.12). Work on pinene transformations spans more than four decades [21]. Recent work using strains of mutated *Aspergillus* has resulted in the bioconversion of α-pinene to verbenol or verbenone.

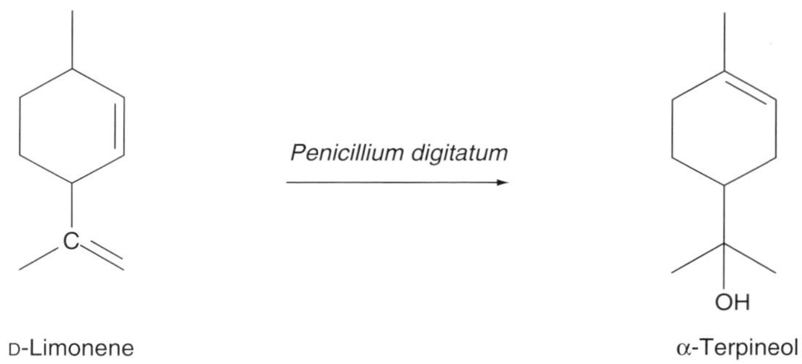

Scheme 8.12 D-Limonene to α-terpineol.

Nootkatone is one of the main characterising compounds found in grapefruit. A large number of synthetic methods have been described, including the oxidation of the sesquiterpene valencene to nootkatone (Scheme 8.13). The same reaction can be accomplished through biotechnology. Whole cell transformations of valencene to nootkatone have been reported using both bacteria and fungi [27]. A simple oxidation of valencene

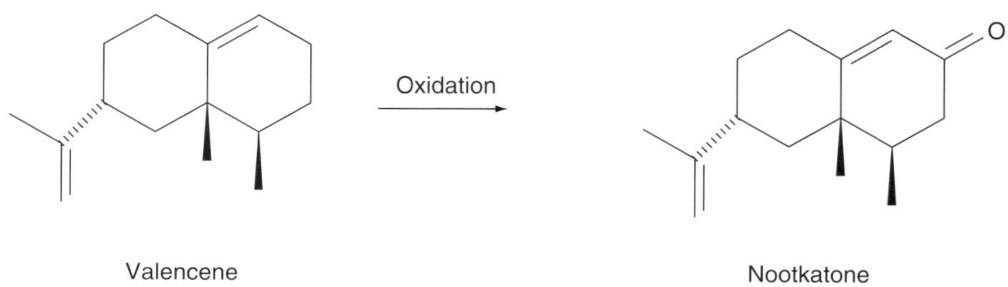

Scheme 8.13 Oxidation of valencene to nootkatone.

to form nootkatone and nootkatol is achieved in the presence of a hydroperoxide of an unsaturated fatty acid. The hydroperoxide may be formed in situ by the action of a fatty-acid lipoxygenase on certain fatty acids [28]. A further process describes the use of valencene, with a composition having laccase activity, to form valencene hydroperoxide, which is subsequently degraded to form nootkatone. Examples of microbial sources of laccase are quoted as *Botrytis cinerea* and *Trametes versicolor* [29].

Many important aroma chemicals are formed by the degradation of carotenoids in plants. One group of compounds of particular importance is the C13-norisoprenoids such as the ionones and damascones. A group of closely related compounds are the C14 irones (Figure 8.2). The chemistry of these compounds is highly complex due to the number of stereo centres, and work on degradation of carotenoids by microorganisms has tended to result in broad mixtures of related compounds. Using α- or β-ionone or damascone as substrates, regioselective insertion of oxygen was achieved with various fungal species [21].

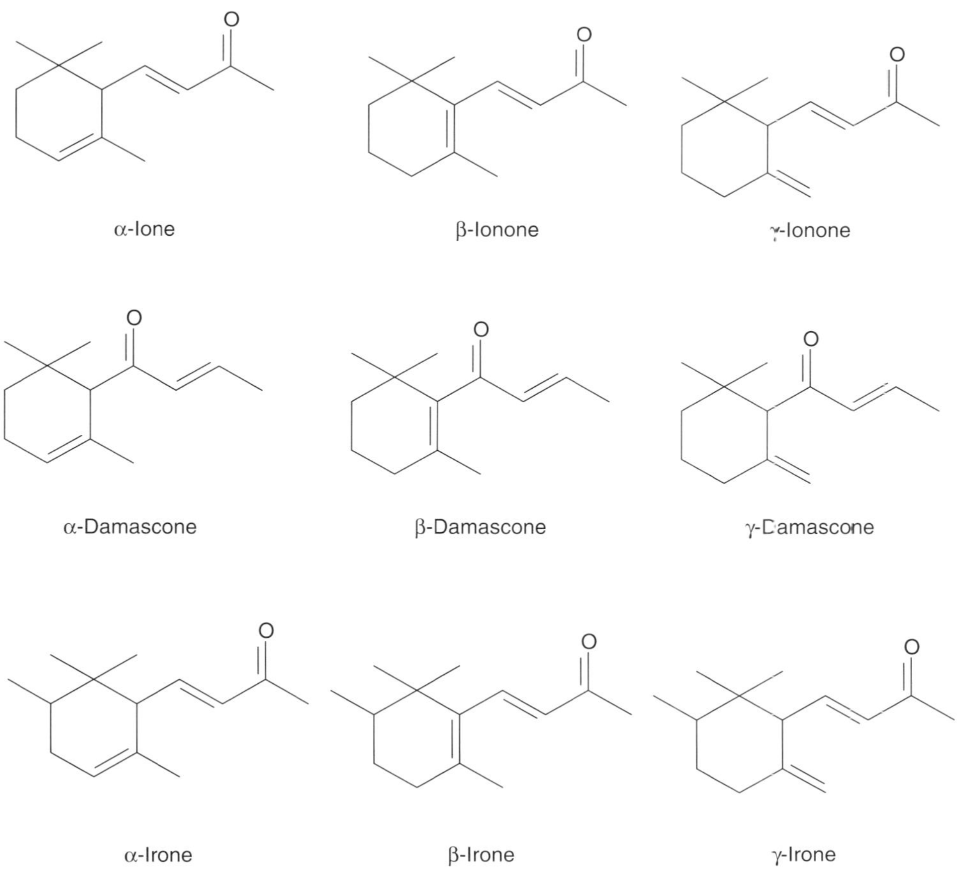

Figure 8.2 Ionones, damascones and irones.

As described under natural isolates, irones may be separated from orris root at an extremely high cost. Traditionally, the rhizomes have to be stored before extraction. Using microbial strains isolated from rhizomes such as *Serratia liquifaciens*, a method has been developed to accelerate this process. A yield of 1 g/kg of mixed irones was obtained after just eight days, compared to 0.4 g/kg for rhizomes stored for three years, as per the traditional method [16]. The mixed irones were produced in a similar ratio to their natural occurrence and were mainly *cis*-γ-irone, *cis*-α-irone and *trans*-α-irone [30].

8.7.10 Vanillin

Synthetic vanillin is produced on a scale of some 15 000 t per annum. Alternative biotechnological routes have been extensively investigated, and an excellent review appeared in 2001 summarising research, patents and potential production processes [31]. Many different microorganisms, starting materials, pathways and potential production methods are discussed. The most readily available starting materials are eugenol from clove oil and ferulic acid from degradation of cereal by-products. Microorganisms suggested for the conversion of these starting materials are *Streptomyces setonii*, *Amycolatopsis* sp. and *Pseudomonas* strains. Products are principally vanillyl alcohol, vanillin and vanillic acid. If the alcohol and acid are produced, these may be converted into vanillin using additional organisms. An outline of the most relevant structures is given in Scheme 8.14.

8.7.11 Precursors

Both for chemical synthesis and biotechnology, the availability of suitable precursors for conversion into the desired end-products is crucial. Biotechnology can provide a suitable route for the economic generation of these precursors. Furaneol® (2.5-dimethyl-4-hydroxy-3(2*H*)-furanone) is an aroma chemical which is key to strawberry flavourings. The synthesis of furanones, via chemical synthesis or 'soft chemistry', requires 6-deoxyhexose sugars and, in particular, 6-deoxy-L-mannose (L-rhamnose) for Furaneol®. These sugars are relatively rare in nature, but it has been found possible to obtain L-rhamnose from a number of natural sources by hydrolysis. Citrus peel, particularly bitter orange, contains flavanoid glucosides such as naringin and hesperidin. Hydrolysis by rhamnosidase, which is present in commercial preparations of pectinases, can be used to liberate L-rhamnose. Since pure L-rhamnose is needed to ensure good yields, other sugars liberated (e.g. glucose) can be selectively removed by fermentation or bioconversion to gluconic acid. The process is expensive and research into alternative biosynthetic approaches is reported [3].

Very recently, a new source of natural L-rhamnose has become available. D-Xylose is manufactured from birch/beach wood hydrolysate. The product is rich in other monosaccharides that are released during the processing. During purification these rare sugars are separated and recovered using column separation technologies. L-Rhamnose, together with L-ribose, L-fucose, L-arabinose and D-mannose, is now available in high purity.

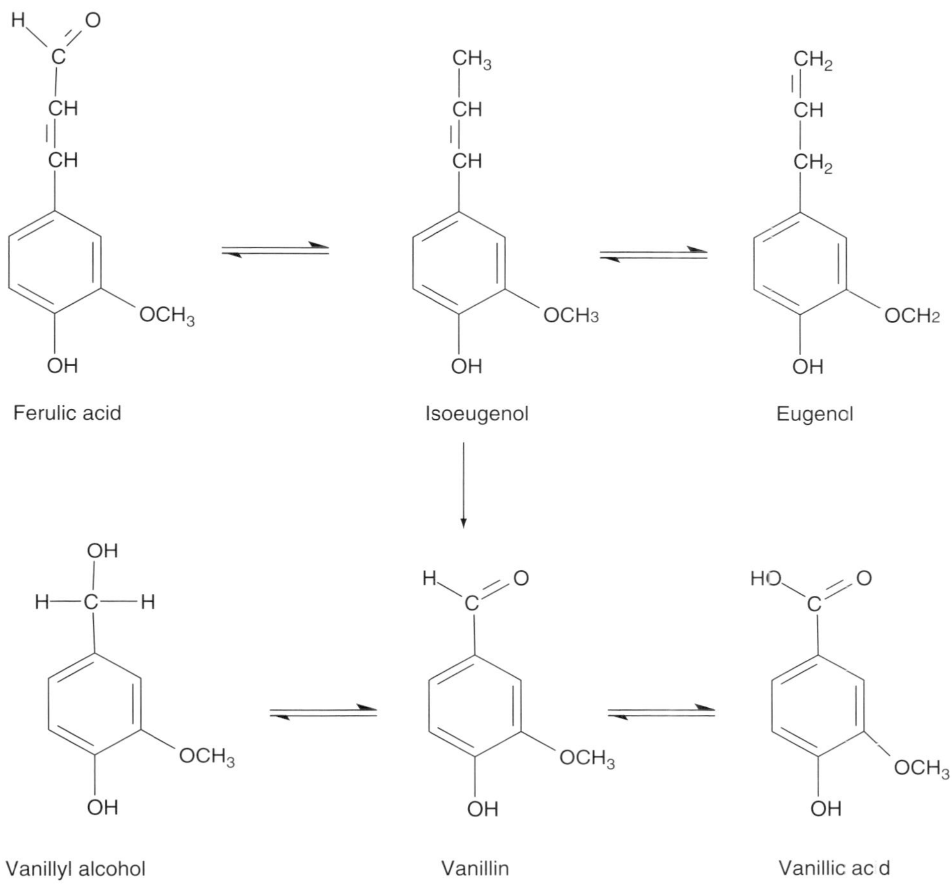

Scheme 8.14 Biotransformation to vanillin.

8.8 Soft chemistry

Natural flavour chemicals may be formed using so-called soft chemistry. This technology does not use biocatalysts and does not use hard or aggressive chemicals. It is based around the expression 'any product of roasting, heating' in the US definition of natural and the 'traditional food-preparation processes' described earlier in the EU definition of natural. The technology of 'soft' or 'cooking' chemistry has been generally accepted in the US whilst in Europe it is viewed with much suspicion and is widely thought of as a non-natural approach. The types of chemical reactions that can be performed include hydrolysis, oxidation, condensation or addition, rearrangements and the Maillard reaction. Reaction conditions may be adjusted, e.g. pH, reactions may be performed in organic solvents, oxygen or other atmospheres may be introduced and the processes can be performed at elevated temperature. Clearly, the big question is when does 'soft chemistry' become 'hard chemistry'? I leave that for the reader to consider with the following examples.

Perhaps, the simplest example of soft chemistry is ester formation. Many short-chain esters can be formed without the use of biocatalysts, synthetic catalysts or aggressive chemicals. Esterification with formic acid can be performed readily without catalysts or any other special conditions [32]. Other esters may require the continuous removal of water to drive the reaction forward. This can be achieved by boiling the mixture of reactants and condensing the vapour. Water plus some of the reactants or products is removed. The non-aqueous layer recovered is returned to the reaction mixture. An alternative approach would be to perform the reaction in an organic solvent.

Bitter almond oil is derived from steam distillation of the dried kernels of bitter almonds, peaches and apricots. Following the removal of fixed oil, the powdered cake is macerated in water to split the glucoside amygdalin. Enzymes may be added to assist with the breakdown of the glucoside. The resultant slurry contains benzaldehyde and hydrocyanic acid. The hydrocyanic acid is removed and the benzaldehyde is then distilled and dried [33]. Natural bitter almond oil is in short supply, and alternative methods to manufacturing natural benzaldehyde have been investigated. One method that has been commercialised in the US is the production of benzaldehyde from cinnamon bark, cinnamon leaf and cassia oil. These oils contain large amounts of cinnamaldehyde (cinnamic aldehyde). Under mild basic conditions, with heating, the cinnamaldehyde breaks down to form benzaldehyde and acetaldehyde in a retro-aldol reaction (Scheme 8.15). Both benzaldehyde and acetaldehyde can be recovered from the reaction in high purity [34].

Furanones occur widely in many fruits such as strawberry, pineapple and mango. They are also formed as a result of certain microbial fermentations, as in the production of soy sauce. Furanones are formed during the roasting or cooking of a wide range of foods by the reaction of amino acids with sugars, the Maillard reaction, which was discussed in Chapter 3. 6-Deoxyhexose sugars (rhamnose, quinovose, fucose, etc.), when reacted with amino acids such as L-proline and L-alanine, can lead to the formation of a wide range of compounds including furanones. Of specific interest is Furaneol® (2,5-dimethyl-4-hydroxy-2*H*-furan-3-one), which can be formed by soft chemistry in the reaction between L-rhamnose and L-proline (Scheme 8.16). Furaneol® is used in strawberry, pineapple, caramel and cooked or roasted food-type flavourings.

The smoking of foods, such as fish and meat, has been used as a method of preservation for thousands of years. Smoke flavourings can be produced by the controlled pyrolysis of various wood and cellulose sources. The resultant smoke flavourings contain a wide range of chemical compounds such as acetic acid, 2,6-dimethyl phenol, ethyl cyclopentenolone, eugenol, furfural, guaiacol, maltol, methyl cyclopentenoloned (cyclotene), 5-methyl furfural, vanillin and many more. A number of these compounds are now being separated from the mixture and purified, e.g. maltol and cyclotene.

The hydrolysis of curcumin from turmeric root to form vanillin was patented in 1990 [35]. Turmeric root, extracts of turmeric root or purified curcumin was subjected to the action of heat and pressure in the presence of water, where upon vanillin and other reaction products were formed (Scheme 8.17). The ratio of vanillin and other products was controlled by adjusting the pH of the reaction mixture. Temperatures of

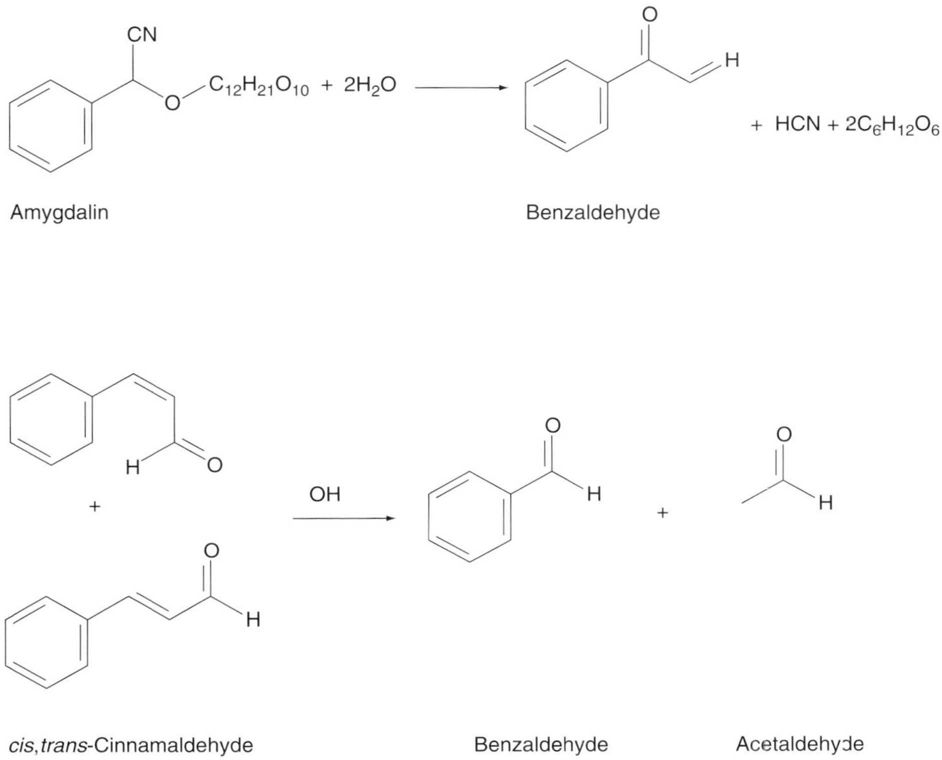

Scheme 8.15 Benzaldehyde formation.

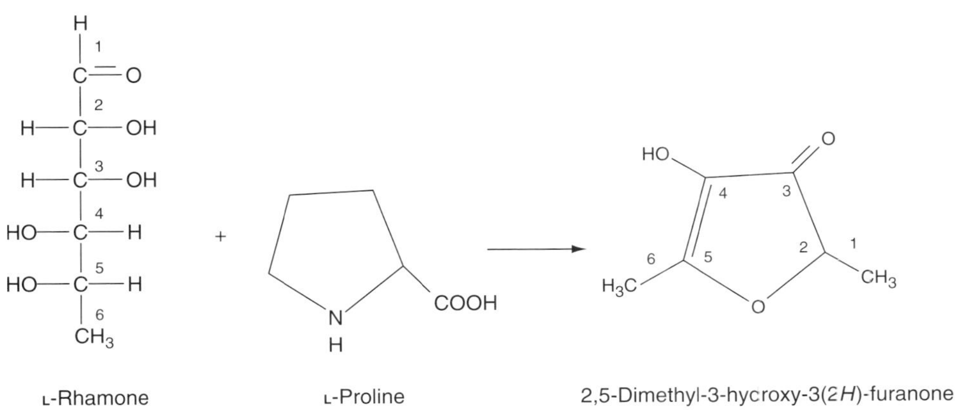

Scheme 8.16 2,5-Dimethyl-4-hydroxy-3(2*H*)-furanone.

up to 350 °C, pressures of up to 110 atmospheres and pH adjustment have to be used to achieve the required reaction rates which yield the economic production of flavour compounds, principally vanillin. Current market price for vanillin produced by this, or similar methods, is in the low hundreds of dollars per kilogram.

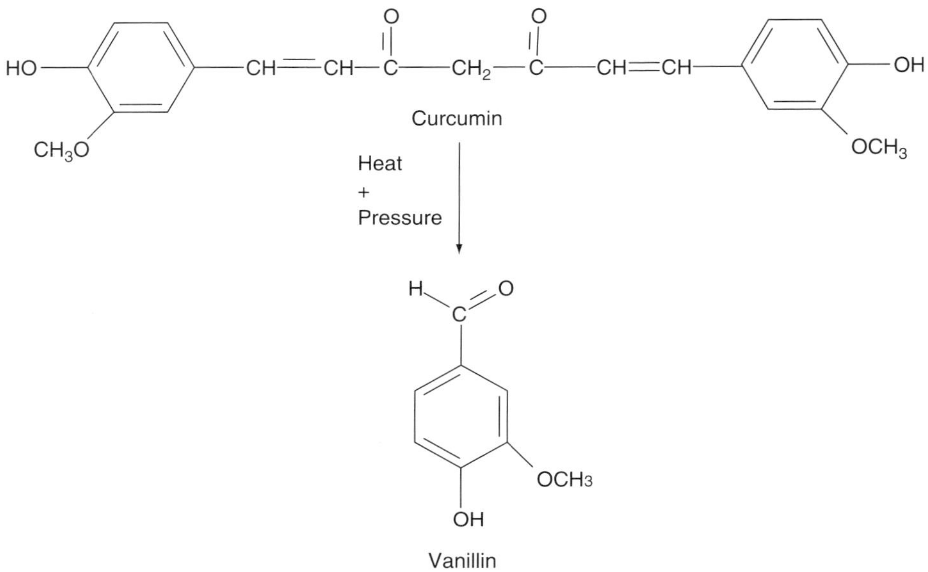

Scheme 8.17 Conversion of curcumin to vanillin.

8.9 Conclusion

This chapter has explored the types of natural chemicals now available through isolation, biotechnology and soft chemistry. It is by no means extensive, and much research effort continues with many reports appearing in the literature in all the three areas discussed. Two topics deliberately not covered are genetically modified organisms and plant cell tissue cultures. At present, the use of genetic-modified organisms to produce foods or food ingredients is not acceptable to European consumers. This may or may not be logical, but without a market there is little point in manufacturing a product. Whilst research is ongoing and the technology shows much promise, there are no industrial examples of genetically modified organisms being used to produce natural chemicals for the flavour and fragrance industry. For now, the industry must continue with mutation, rather than manipulation. To date, plant cell tissue cultures have not yielded commercial products. There is a large body of information in the literature demonstrating that plant cell tissue cultures can produce a wide range of natural chemicals. However, this is typically in complex mixtures and at very low concentrations. Much more research still needs to be done in this field.

Many important natural flavour chemicals are now available through isolation, biotechnology and soft chemistry. Whilst there is an obvious commercial pressure to accept natural products produced by soft chemistry, many questions remain unanswered. In particular, when does soft chemistry become hard chemistry? The range of natural chemicals available by isolation is limited. Nature's resources are finite and the costs of recovery can be prohibitive. Therefore, the flavour and fragrance industry continues to devote much effort into the development of natural chemicals through biotechnology for its ever-increasing demand for natural flavourings and fragrances.

References

1. Tyrrell, M. (1995) Advances in natural flavors & materials. *Perfumer & Flavorist*, **20**, 13–22.
2. Leffingwell, J.C. (2003) Chirality and odour perception. www.leffingwell.com.
3. Vandamme, E.J. and Soetaert, W. (2002) Bioflavours and fragrances via fermentation and biocatalysis. *J. Chem. Technol. Biotechnol.*, **77**, 1323–1332.
4. Keville, K. and Green, M. (1995) *Aromatherapy: A Complete Guide to the Healing Art*, Crossing Press, USA, p. 2.
5. Arctander, S. (1960) *Perfume and Flavor Materials of Natural Origin*. Published by the author.
6. Wright, J. (1995) Essential oils. In P.R. Ashurst (ed.), *Food Flavourings*, 2nd edn, Chapman & Hall, London, pp. 25–57.
7. Margetts, J. (1989) The search for natural flavour ingredients. In A. Turner (ed.), *Food Technology International Europe*. Sterling Publications, London, pp. 231–234.
8. Liese, A., Seelbach, K. and Wandrey, C. (2000) *Industrial Biotransformations*. Wiley-VCH, Germany.
9. Schreier, P. (1997) Enzymes and flavour biotechnology. *Biotechnol.*, **55**, 51–72.
10. Cheetham, P.J.S. (1999) Fermentation, Biocatalysis and Bioseparation. In M.C. Flickinger and S.W. Drew (eds), *Encyclopedia of Bioprocess Technology*. John Wiley & Sons, New York, pp. 1004–1029.
11. Fabre, C.E., Blanc, P.J. and Goma, G. (1998) 2-Phenylethyl alcohol: an aroma profile. *Perfumer & Flavorist*, **23**, 43–45.
12. Raymond, W.R. (1984) US Pat. 44481292.
13. Gabelman, A. and Luzio, G.A. (1998) US Pat. 5783429.
14. Gatfield, I.L. (1988) Enzymatic generation of flavour and aroma components. In R.D. King and P.S.J. Cheetham (eds), *Food Biotechnology*. Elsevier Science Publishing, New York, pp. 59–82.
15. Shukunobe, K. and Takato, S. (1989) Japanese Pat. 01168256.
16. Fuganti, C. and Zucchi, G. (1998) Product distribution in the microbial biogeneration of raspberry ketone from 4-hydroxybenzalacetone. *J. Mol. Catalysis B: Enzymatic*, **4**, 289–293.
17. Gatfield, I.L. (1995) Enzymatic and microbial generation of flavors. *Perfumer & Flavorist*, **20**, 5–14.
18. Kumin, B. and Munch, T. (1998) US Pat. 5849551.
19. Rosanna, C., Claudio, F., Massimo, B. and Gianna, A. (1996) US Pat. 5527693.
20. Bretler, G. and Dean, C. (2000) US Pat. 6025170.
21. Berger, R.G., Krings, U. and Zorn, H. (2002) Biotechnological flavour generation. In A.J. Taylor (ed.), *Food Flavour Technology*. Sheffield Academic Press, pp. 60–104.
22. Chatterjee, T. and Bhattacharyya, D.K. (1998) Synthesis of terpene esters by an immobilized lipase in a solvent-free system. *Biotechnology Letters*, **20**, 864–867.
23. Fabre, C. and Goma, G. (1999) A review of the production of green notes. *Perfumer & Flavorist*, **24**, 1–8.
24. Hausler, A., Ehret, C. and Binggeli, E. (2000) US Pat. 6150145.
25. Whitehead, I.M. (1998) Challenges to biocatalysis from flavor chemistry. *Food Technology*, **52**, 40–46.

26. Huynh-Ba, T., Matthey-Doret, W., Fay, L.B. and Bel Rhlid, R. (2003) Generation of thiols by biotransformation of cysteine–aldehyde conjugates with baker's yeast. *J. Agric. Food Chem.*, **51**, 3629–3635.
27. Sime, J.T. (1997) Biotransformations applied to plant chemicals: a natural combination. In S. Wrigley, M. Hayes, R. Thomas, and E. Chrystal (eds), *Phytochemical Diversity: A Source of New Industrial Products*. The Royal Society of Chemistry, p. 190.
28. Muller, B., Dean, C., Schmidt, C. and Kuhn, J. (1998) US Pat. 5847226.
29. Huang, R., Christenson, P.A. and Labuda, I.M. (2001) US Pat. 6200786.
30. Belcour, B., Coutois, D. and Ehret C. (1990) US Pat. 4963480.
31. Priefert, H., Rabenhorst, J. and Steinbuchel, A. (2001) Biotechnological production of vanillin. *Appl. Microbiol. Biotechnol.*, **56**, 296–314.
32. Schmidt, C. (2002) Esters. *Perfumer & Flavorist*, **27**, 36–39.
33. Shaath, N.A. and Benveniste, B. (1991) Natural bitter almond oil. *Perfumer & Flavorist*, **16**, 17–24.
34. Pittet, A.O., Muralidhara, R. and Liberman, A.L. (1988) US Pat. 4727197.
35. Dolfini, J.E., Glinka, J. and Bosch, A.C. (1990) US Pat. 4927805.

Chapter 9
Molecules of Taste and Sensation

Mark L. Dewis

9.1 Introduction

This chapter aims to give the reader an overview of all the important compounds in the areas that can be described as 'functional', rather than those traditionally described as 'organoleptic'. In an attempt to outline this a little more clearly, this 'functional' area will include compounds that may (in some cases) exhibit properties of taste, but elicit little or no overall flavour. In strict terminology 'organoleptic' is defined as the effect or impression produced by any substance on the organs of touch, taste or smell, and also on the organism as a whole. However, for the purpose of outlining the scope of this chapter, I shall define the 'functional' chemicals as being chemicals that impart taste, act on taste fibres, or stimulate trigeminal or other receptors in the oropharyngeal cavity (pertaining to the oropharynx, the area directly behind the mouth). That is to say that they impart taste or chemical irritation, but do not, for the most part, possess the properties of olfaction (evoke a sense of smell).

The compounds discussed will therefore fall into one of the following categories:

- Sensory molecules – tingle, cooling, warming or astringent;
- Taste molecules – sweeteners, bitter agents, sour compounds, salt, flavour enhancers and Umami compounds.

During recent decades, the advances in analytical techniques and instrumentation (most notably GC/MS and a variety of techniques for the trapping of volatile components in food) have enabled the identification and subsequent synthesis of the vast majority of the key volatile chemical components in nature. This has enabled the creative flavourist to access the chemicals required to make flavours acceptable to the public. In recent times, industry has looked more critically at non-volatile components in food. Inevitably, into this category fall mouthfeel components, bitter agents, sweeteners, salt, proteins, monosodium glutamate (MSG), fats, starches and other 'active' ingredients. Compounds that cool have been in demand for the last 30 years or so, whilst warming agents (with the exception of traditionally used spices) have been a more recent focus. The subject of tingle has seen great research activity up to, and including, the current time. However, despite the trends coming and going, research in most of these areas remains active and current, to at least some degree, as competition in the marketplace drives individuals to find the next small point of difference. Indeed, such is the importance of trigeminal stimulation and taste quality, that since 1992 salsa has outsold ketchup.

It is important to familiarise the reader with the terms used in the text; this will be covered in the next section of this chapter. The different categories of functional molecules, as defined by the author, will be covered individually. Within each category, a review will outline the important compounds and pertinent information the reader requires as background to the category. This will be followed or interspersed, where appropriate, by some recent advances in the area and an insight into possible future developments.

9.2 The trigeminal nerve system, taste and oral receptors

Since the vast majority of the discussion of this chapter relates to the interaction of chemical moieties with receptors in the oral cavity (mouth, oropharynx, tongue and nerves and the receptors therein) and in the skin, it is worth devoting a little time to explain in simple (and hence probably not totally scientifically correct) terms, what these interactions are. I will also attempt to provide an overview of the terms, and what each of these encompasses. For those readers wanting more elaborate detail, entire books have been written on these subjects.

There are two good primary texts covering the areas of nerves, taste, receptors and chemesthesis. The first [1] covers the biological aspects of taste and smell, whilst the second [2] deals specifically with chemical irritation.

As one would expect from the wide variety of functions it serves, the oropharyngeal cavity is immensely complex. There are three aspects of the oral cavity that require further discussion: the trigeminal nerve system, human gustation (the ability to taste) and the related neurons, receptors and taste fibres.

9.2.1 The trigeminal nerve

The head has twelve major nerves, the fifth cranial nerve is known as the trigeminal nerve. It is the largest cranial nerve and is the primary nerve responsible for sensation in the head and the face. It is also the motor nerve of the muscles of mastication. The nerve has a small motor root and a large sensory root.

The trigeminal nerve has three branches the ophthalmic, the maxillary and the mandibular (Figure 9.1). The ophthalmic and maxillary branches consist exclusively of sensory fibres; the mandibular branch is joined by a motor root.

The maxillary branch divides and contains the superior alveolar and the nasopalatine nerves. The mandibular branch is located in the lower part of the mouth in the lingual area, roughly following the line of the lower jaw. This too divides to include the lingual nerve, inferior alveolar and buccal nerves.

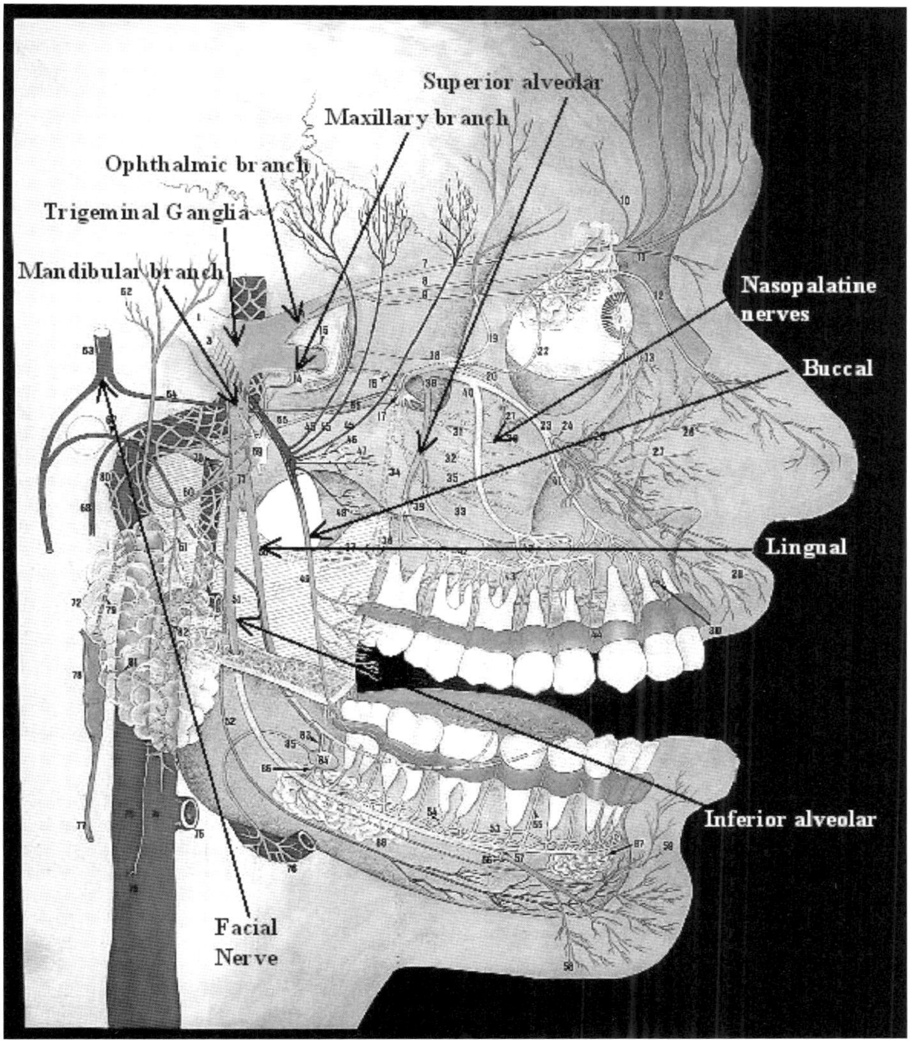

Figure 9.1 The primary cranial nerves responsible for the perception of taste and oral and nasal sensation. Reproduced by kind permission of Prof. Fausto Berzin, Unicamp, Sao Paulo.

Most trigeminal sensory nerve fibres lie in the trigeminal ganglia, and it is at the trigeminal ganglion that the three main trigeminal nerve branches come together. Chemicals can stimulate trigeminal nerve endings causing rise to a range of sensations, for example, stinging and tearing. However, in the oral cavity they are associated with pain, touch and temperature.

9.2.2 Gustation

Breslin discusses human gustation in detail [3]. Taste is elicited by chemical stimulation of 'taste receptor cells' in the mouth. The signalling pathways of taste reception, like olfaction, rely on G-protein-coupled receptors. Taste was traditionally described as being characterised by four descriptive senses (basic taste qualities), which when combined can create all taste sensations and are the basis for all gustatory sensation (see Figure 9.2). The traditional four basic tastes were salt, sweet, bitter and sour. Currently, Umami (of Japanese origin), which is the taste described as meaty, brothy or savoury is recognised by many as the fifth basic quality of taste [3]. Until recently, taste has not been so extensively studied as olfaction. It is therefore known that mammals have hundreds of olfactive receptors that bind odorants selectively to allow the mammal to distinguish smells. In the case of taste, how the stimulants bind to and are recognised by the taste cells, how this is selective, and how the taste cells innervate the epithelium are still largely unknown subjects.

Colloquially (though incorrectly), taste is often described by words that do not describe taste qualities but sensations such as hot, cool, prickling and pungent (see Section 9.3), mouthfeel such as gritty, greasy, astringent or olfactory sensations, for example, fruity, floral, green or sweaty. It is clear that distinguishing gustatory sensations from somatosensation and olfaction is not always trivial. Indeed, these are often difficult sensations for even expert tasters to distinguish between. Therefore, it is often flavour, the total sensory combination of taste (see Sections 9.2.3 and 9.5), olfaction and the common chemical sense (see Section 9.3) that is described as taste. Flavour, in the strictest definition, is a combination of oral chemoreception (taste – split into chemical irritation and gustation) and nasal chemoreception (smell – split into chemical irritation and olfaction).

9.2.3 Oral receptors

As outlined in Section 9.2.2, special cells in the epithelium of the mouth and larynx elicit human taste.

The seventh (facial, see Figure 9.1), ninth (glossopharyngeal) and tenth (vagal) cranial nerves are responsible for the perception and sense of taste. There are two sensory

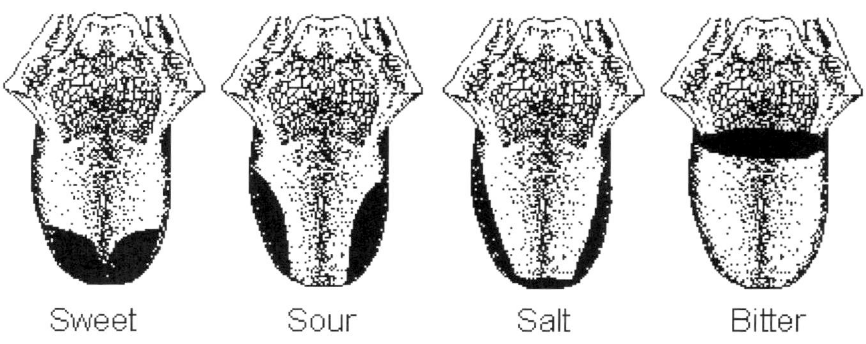

Figure 9.2 The regions of the tongue responsible for the perception of the basic taste qualities.

components of the seventh nerve related to taste. (The facial nerve contains the chorda tympani and the greater superficial petrosal nerves.) The first leads to taste buds in the front of the tongue and the sides of the back of the tongue. The second innervates the soft palate and the nasoincisor ducts. The ninth nerve serves taste buds in the vallate and the back of the tongue, whilst the tenth leads to the larynx and the upper oesophagus. The ninth and tenth nerves also have somatosensory function, and all the three nerves serve motor function in addition to the functions described above. Note – the somatosensory system is the sensory system of the body which is not included in the five senses of taste, touch, smell, sight and hearing. For example, thermal perception and irritation.

The oral mucosa and lingual epithelium also carry fibres mediating touch, pain and thermal sensation carried in the trigeminal nerve. It is estimated that 75% of fibres in the fungiform papillae are trigeminal in nature.

Somatosensory nerves contain different sizes of fibres, large fast-conducting mechanoreceptive fibres, along with smaller slower-acting and small slow-acting thermoreceptive and nociceptive fibres. Humans have no taste sensation in the hard palate but thermal (among others) sensations are still observed.

The oral receptors can be summarised as follows:

Relayed primarily by the trigeminal nerve

- Pain receptors – sensation of discomfort, can be due to polymodal, thermal or mechano reception;
- Mechano receptors – sense of texture, mouthfeel, astringency;
- Thermal receptors – sensation of heat and cold, can be hot/cold fibres or trigeminal;
- Nociceptors – provides sensations such as tingle, prickling, can be trigeminal (painful thermal, mechanical and/or chemical irritation) in origin;
- Polymodal – multimodal sensations for example, thermal heat and pain.

Relayed by the gustatory nerves

- Taste – senses of sour, sweet, salt, bitter and Umami.

Whilst some chemicals provide irritation and stimulate combinations of olfaction, somatosensation and taste, it is important to point out that many chemicals also act on combinations of somatosensory receptors. Menthol is the classic example of this phenomenon, as it possesses olfaction in its mintiness, somatosensory stimulation in a feeling of irritation, cool and/or heat, and taste usually described as bitterness. Other examples include carbon dioxide, fatty acids and citric acid which all stimulate cold nociceptors, as well as cool/acid-sensitive lingual nerve fibres [4].

The fact that many chemical substances interact with such a wide variety of receptors in the mouth, makes molecule design very difficult for the chemist, when looking for compounds that exhibit taste or irritation effect. There is considerable overlap between chemical sensitivity, pain, thermal modalities and tactile modality. The picture is even more complex for olfaction and fragrance molecule design, with several or many areas of the olfactory bulb being stimulated by a single chemical entity. In case the challenge were not already difficult enough, the position is further complicated by people exhibiting specific anosmia or super sensitivity – 'super tasters'. Indeed, especially when talking about chemicals providing somatosensory effect, the threshold level for two individuals can be different by a 100-fold (or more) concentration.

The message (as one might easily suspect) is that taste and olfaction are incredibly complex, involving several major nerves, multiple receptor types in multiple locations and that these are also very individual.

9.3 Chemesthesis

The compounds discussed in the remaining sections of this chapter are used primarily to provide or enhance a taste or flavour, or in flavours and fragrances to stimulate nerve fibres and receptors in the oral cavity or on the skin. It is therefore important to outline one further term; that of 'Chemesthesis'.

Chemesthesis [5] is a word (from Chem = chemical and esthesis = ability to perceive or feel) that became adopted [2] to describe what has been long referred to as the 'common chemical sense'. The 'common chemical sense' was coined [6] to describe the sensory system responsible for the physical detection of irritating chemicals. The chemosensitive fibres are a subset of pain- (nociceptive) and temperature- (thermal) sensitive fibres, which occur throughout the body, in the skin and the mucous membranes, as part of the somatic sensory system. The coolness or menthol, the warming of chilli peppers and the tingle of carbonation are examples of the sensation of chemesthesis. Chemesthesis is a major part of oral perception.

An interesting feature of chemesthesis is that, for many irritants, sensitisation occurs. This is the effect of repeated stimulation (exposure) leading to a progressively stronger perception of the sensation. However, an interruption to the exposure leads to desensitisation, a decrease in perceived intensity following a break in exposure.

In order to reach trigeminal nerve endings in and below epithelia, compounds must pass through the lipid phase of the cell membrane or through the aqueous phase to arrive at free nerve endings. For this reason, chemesthetic sensations have a longer onset time before perception and linger longer than taste perception or olfactory perception. Since lipid solubility is a pre-requisite for most trigeminal stimuli and the primary route of access for hydrophobic irritants, compounds that evoke a 'chemesthetic effect' have a hydrophobic part to their structure. Since the compounds generally require at least some binding at the nerve ending or receptor and an ability to partition the fat/water boundary, a hydrophilic functionality is also present. There is evidence in the literature that increased lipophilic chain length leads to an increase in trigeminal stimulus. However, since this is in competition with time of onset, longevity of experience and, in some flavour and fragrance systems, a hydrophobic base, it is not necessarily borne out when one puts the theory into practice.

9.4 'Sensates' – compounds which provide a sensory effect

Olfaction is covered elsewhere in the book and is deliberately not tackled in this chapter. Indeed, whilst olfaction is a physical response to a chemical exposure, and could therefore be classed as an irritation, it is not generally considered as a chemoreception under chemesthesis. To explain this by example, early studies on the psychophysical aspects of chemesthesis in the nasal cavity asked test subjects to describe irritation and odour separately. For example, menthol would be described separately as minty in odour and irritating to the nose. In contemporary studies, carbon dioxide is used because it stimulates trigeminal receptors in the nose but is odourless. Therefore, a sensation is observed, without the subject having to mentally ignore any olfactive aspect.

The compounds discussed have different intensities. Even a single irritation type can be experienced in different parts of the mouth. It is not known why this occurs and this is still an unanswered question. Similarly, it is uncertain why synergies, cross-desensitisation and cross-potentiation occur.

For the purpose of this chapter, I have classified the chemical compounds and compositions according to the type of sensation or physiological response perceived by the 'exposure' to the flavour or fragrance ingredient. However, the materials are sub-classified according to the site of action (in this case largely mouth and nose, but also skin, eyes and respiratory tract), where practicable. The type of response may be trigeminal irritation, taste reception, olfaction or reception on skin, or as previously alluded to, combinations of these response factors.

9.4.1 Tingle compounds

9.4.1.1 What is tingle?

Compounds used in the flavour industry to impart a tingle sensation all exhibit a similar type of characteristic sensation in the mouth. The sensation can be described in a number of different ways but the vocabulary most commonly used to describe the sensation ranges from a numbing anesthetic effect or a tingling paresthesia, an attachment of 9 V battery terminals on the tongue effect, to pins and needles or buzzing, or in extreme cases pin prickling or pain (a pricking pain rather than the cold irritation of menthol or the burn or capsaicin). The feeling is generally perceived on the tongue and lips, but may also be experienced on the gums, teeth, cheeks and roof of the mouth.

There are several reasons why tingle has value in flavour applications. Firstly, because it synergises with other attributes of the product. Carbonation and alcoholic strength are both enhanced, and other sensory effects, for example, cooling, warming and sweetness are perceived differently, usually perceived as enhanced and preferred. Secondly, the effect can be perceived as refreshing or cleansing, whether on the skin or in the mouth. Thirdly, anti-microbial properties of the tingle chemicals could be beneficial, for example, in the mouth and the underarm. Finally, the sensation can be marketed as a sensory trigger for a product claim that the consumer might miss. An example might be a numbing sensation on the scalp to demonstrate to the user the skin-soothing effect of an anti-dandruff shampoo.

9.4.1.2 Where are the compounds found?

Compounds possessing tingle character are found in nature in a large number of botanical species. However, these are concentrated into five plant families: dicotyledonous, Piperaceae, Aristolochiaceae, Rutaceae and Asteraceae (Compositae), and monocotyledonous, Poaceae. The compounds of interest primarily occur in the following genera: *Achillea* (Asteraceae–Anthemideae), *Acmella* (Asteraceae–Heliantheae), *Ctenium* (Poaceae), *Echinacea* (Asteraceae–Heliantheae), *Heliopsis* (Asteraceae–Heliantheae), *Piper* (Piperaceae), *Spilanthes* (Asteraceae–Heliantheae) and *Zanthoxylum* (Rutaceae). The compounds in question fall into two chemical types: olefinic alkamides and acetylenic alkamides.

Acetylenic alkamides occur only in the Asteraceae family and principally in two tribes, Anthemideae and Heliantheae. At the time of writing, around 100 of these acetylenic amides are known to occur in nature, and these are comprised from the combination of a wide range of amine (around 15) and acid (over 80) moieties [7]. Although acetylenic amides are structurally very closely related to, and are found alongside the purely olefinic alkamides, they are not known to be taste- or sensorial-active. As a result of this finding, the acetylenic amides will not be further discussed in this text. I discussed them at all because some are undoubtedly active but the effect is as yet undiscovered and because as they are in occurrence, the larger set of the two chemical types. In short, they might one day prove very important.

Purely olefinic alkamides (with no triple bonds) are also widely reported, with over 50 being known to occur across the five plant families. A number of these are known to exhibit the tingle character that the creative and application functions of the flavour industry so much desire. Indeed, the area of tingle compounds (both natural extracts and synthetic compounds) has been the subject of much industry research over recent years. Since the historical focus of plant research (identification of components and phytochemistry) was in the area of anti-microbial activity, significantly more is known in the scientific literature about the anti-fungal, anti-bacterial and insecticidal activities of these compounds and plant extracts than about activity in the oral cavity. Therefore, a surprisingly small number of these alkamides are reported to be active in the oral cavity. Thus, a handful have some current use, a few more are known to be active but have no current utility, while a good number more are likely to be active compounds (based on their structural similarity to known actives) whose activity is not yet reported.

The compounds listed below, along with some references, are the naturally occurring purely olefinic alkamides that are currently used (in some form) in flavour or edible compositions, or will likely be used in the near future. Additionally, the compounds have reported tingle activity in the oral cavity, or activity is strongly implied. The use of the known active compounds in flavoured products, particularly in oral care, is extensively documented in the literature.

Spilanthol/Affinin (**1**) found in the species *Spilanthes* [8–10], *Heliopsis* [8, 11], *Acmella* [8, 10, 11] and subspecies thereof.

α-Sanshool (**2**) is found in five species of *Zanthoxylum* [7, 12, 13]. γ-Sanshool (**3**) is found in six species of *Zanthoxylum* [7, 14–18], Hydroxy-ε-sanshool (**4**) occurs in *Zanthoxylum* (*Szechuan pepper*) [19]. Hydroxy-α-sanshool (**5**) is known to occur in four species of *Zanthoxylum* [13, 18–21] and two species of *Echinacea* [20]. Hydroxy-γ-sanshool (**6**) occurs in four species of *Zanthoxylum* [7, 13, 17, 18].

Isoaffinin (**7**) has been found in *Ctenium aromaticum* [22] and *Achillea millefolium* [7].

Pellitorine (**8**) is found in two species of *Zanthoxylum* [14, 16], *Ctenium* [22], *Echinacea Pallida* [23], *Artemisia dracunculus* [24], along with a number of *Piper* species including *Piper guineese* [25] and *Piper Longum* [26].

N-(2-methylbutyl) *E*,*Z*,*E*-2,6,8-decatrienamide (**9**) is found in *Spilanthes* [10], *Acmella* [10, 11] and *Heliopsis* [11].

N-isobutyl *E*,*E*-2,4-undecenamide (**10**) is reported in *Piper guineese* [25] and is an active component in *Piper longum* [26].

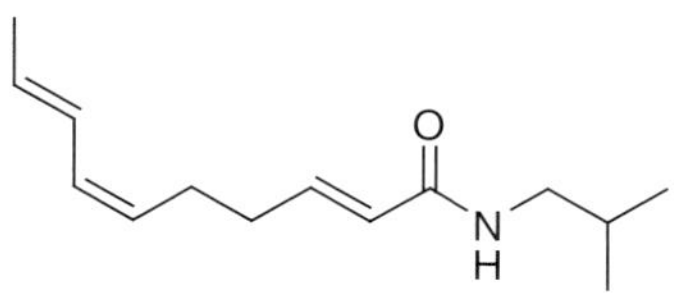

(**1**)

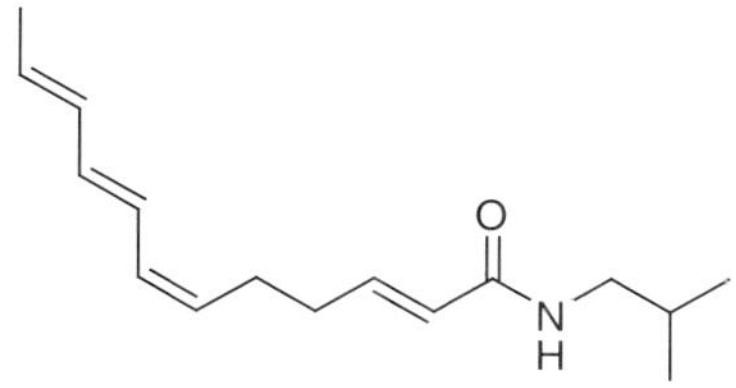

(**2**)

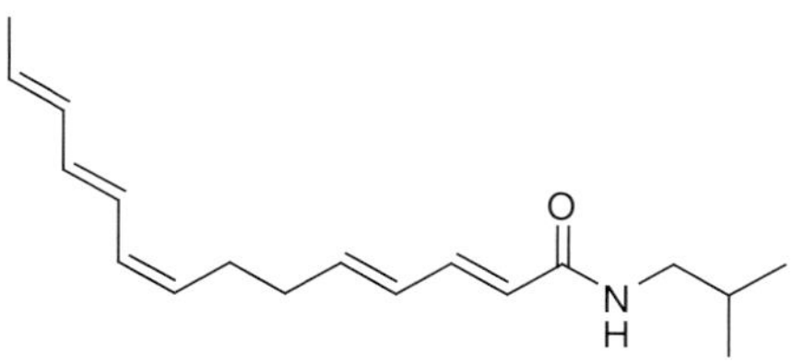

(**3**)

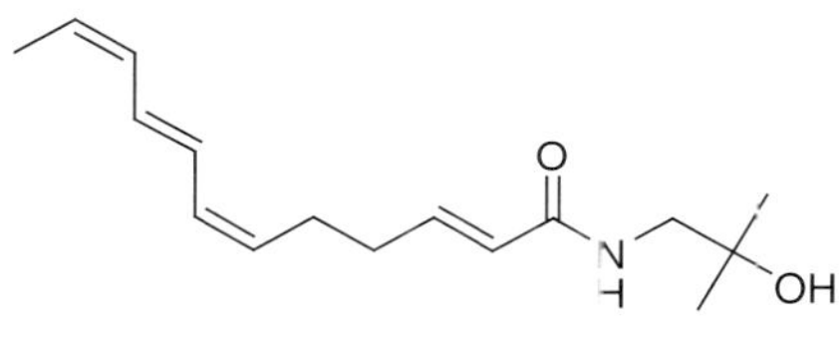

(**4**)

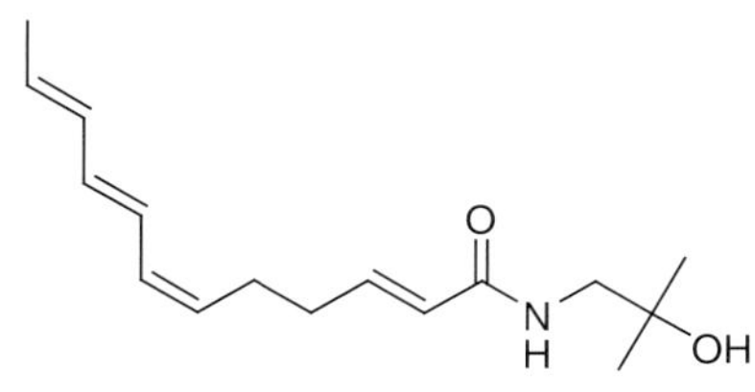

(**5**)

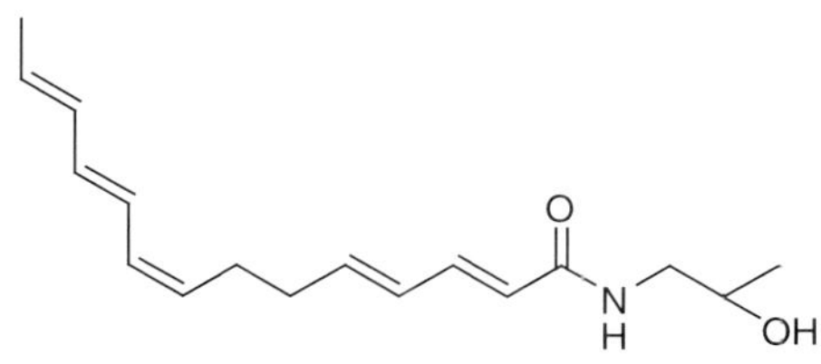

(**6**)

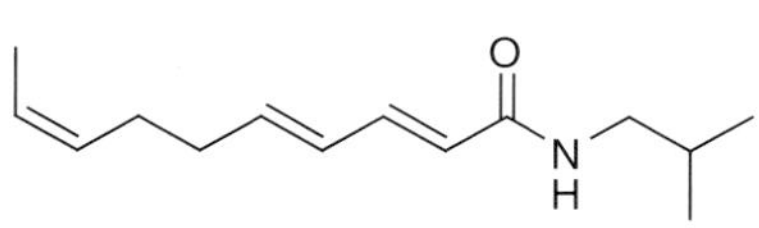

(**7**)

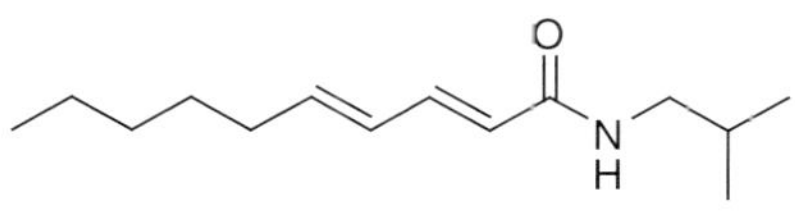

(**8**)

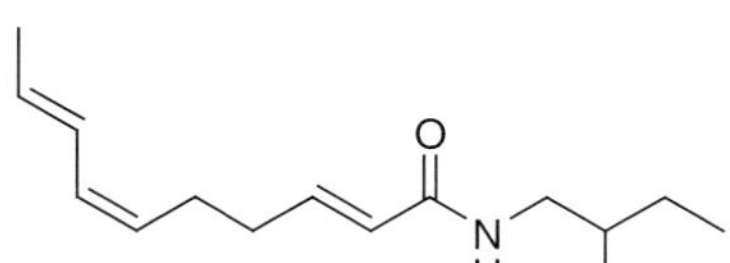

(**9**)

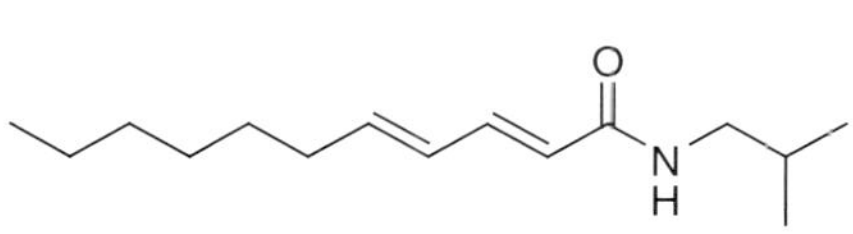

(**10**)

9.4.1.3 Structure and activity

It is not certain what the relationship is between structure and activity of the tingle molecules. These would also be difficult to model since the structures possess a large degree of flexibility and, hence, spatial conformations. Some general conclusions can be drawn when one compares the structures of inactive molecules with active ones.

An amide function with double bond in conjugation appears essential, as does the presence of a second double bond. This second double bond must be in further conjugation in the 4 position or be a z-double bond, somewhere remote in the chain. The conjugated amide is separated from the remote double (if present) by a carbon chain spacer unit of 2 or more carbons. These features are illustrated in Scheme 9.1. Some examples have all of these features. One or more additional double bonds are of moderate importance. The size of the R′ group does not appear so important, nor does the size or functionality of the R group (although most active examples contain isobutyl). The important aspect of the R and R′ groups is that they must allow the molecule to have sufficient polarity to cross the oral mucosa, therefore and for this reason, a one-carbon difference can make a significant change to the threshold and the activity. All of the molecules in (**1**)–(**10**) include these features.

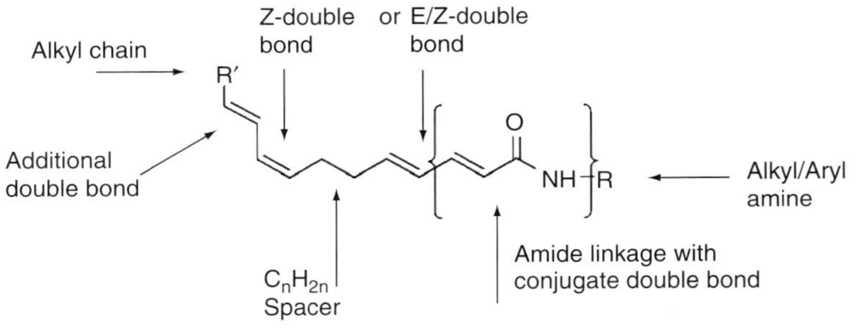

Scheme 9.1

These structural conclusions are supported by the structure of the inactive molecules hydroxy-β-sanshool [19] (**11**), β-sanshool [13], *N*-isobutyl *E*-2-decenamide and *N*-isobutyl *Z*-8-decenamide, and may be responsible for the weak activity of the aromatic containing fagaramide [8], a component of *Zanthoxylum* and *Piper* families.

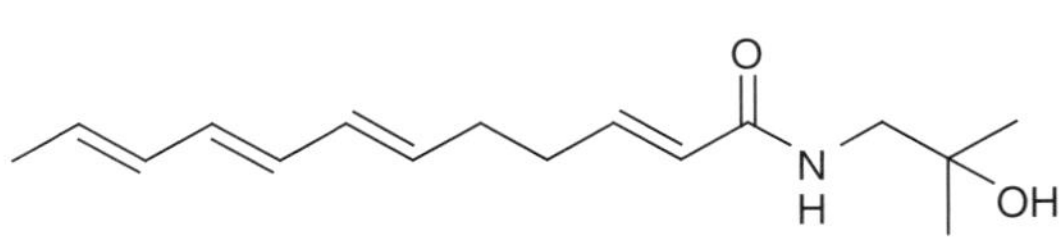

(**11**)

The best reviews covering the effects of these molecules and describing the tingle effect 'tingling paresthesia' and the activation of tactile and thermal trigeminal nerves are referenced [7, 19, 20].

Spilanthol/Affinin (**1**)
As the name implies, this occurs most widely in the Spilanthes family 'toothache plant'. The compound was first isolated in by Gerber in 1903 [9], and the plant which has flowers and leaves possessing a pungent taste accompanied by tingle and numbness, has long been used as a herbal remedy for a variety of complaints, for example, stammering, toothache, throat complaints. The flowers have also been used for the flavouring of dentifrices and gum [8]. Occurrences in Indian long pepper, *Heliopsis* (a plant used for toothache possessing sharp flavour and used to add to spirits), *Acmella* 'Spotflower' (the source of the oleoresin Jambu which is used in flavour applications) and mountain atlas daisy are also reported in the literature. The synthesis is also reported [7, 27]. As the active component of Jambu oleoresin, this compound is registered as a GRAS-flavouring ingredient in the USA under FEMA number 3783. The pure compound and most extracts that contain it are found to be somewhat salty-tasting to most people. At higher concentrations it has been found to be metallic and is described by some tasters as woody.

α-Sanshool (**2**), γ-sanshool (**3**), hydroxy-ε-sanshool (**4**), hydroxy-α-sanshool (**5**) and hydroxy-γ-sanshool (**6**)
α-Sanshool has a reported synthesis [17]. α-Hydroxy sanshool is described as altering the spontaneous activity in cool-sensitive fibres as well as inducing activity in tactile fibres, cold nociceptors and silent fibres insensitive to thermal and tactile stimuli. The activity for these molecules was recorded as being a mild electric shock 5–7 V at 25–50 μg having a delayed onset of about 60 seconds and lasting for 10–20 minutes. This is common for chemesthetic irritation (see Chemesthesis, Section 8.3), and, in addition, a sensation of evaporative cooling. At 100 μg the perception changed to a painful stimuli.

γ-Sanshool has stronger activity than the hydroxy-γ-sanshool derivative [7] and also has a reported synthesis [13] while α-hydroxy sanshool and hydroxy-ε-sanshool are equally active components. The sanshools cause enhanced salivation effects in the mouth in addition to the tingling paresthesia. The 'sanshools' often co-exist in botanicals but the most important is the α-hydroxy sanshool because of occurrence in *Echinacea* which has wide utility in edible products in many areas of the world. Although *Echinacea* is used in herbal preparations it is also added to vitamin supplements, including those prepared for children. *Echinacea* is known as 'purple coneflower', and plant extracts are regarded as a non-specific stimulant of the immune system.

Isoaffinin (**7**)
This active tingle molecule is reported to occur in *Ctenium*, a species prevalent in the south-eastern USA and known commonly as 'toothache grass' and is additionally reported to occur in the perennial Yarrow. Yarrow oil is used in China and is FDA-approved for food use where it is used in limited quantity. The activity is described as numbing on the tongue and gums [22]. The synthesis of isoaffinin and two higher homologues of unknown activity is reported [17].

Pellitorine (**8**)
Pellitorine is one of the most widely occurring alkamides, occurring in prickly ash, toothache grass, *Echinacea*, some pepper species and Russian tarragon (dragon sagewort) [24]. In addition to a number of reported syntheses it is reported to have pungency [14–16] and to be tasteless [22]. Pellitorine is one of the strongest tingle agents and,

like spilanthol, has a little salt taste at typical tasting levels. This 'saltiness' is not as pronounced an effect as in the case of spilanthol (**1**). The 'tingle profile' is quite different to most other compounds in the class and is usually best described as clean. Although probably currently used as a component of both *Echinacea* and tarragon, the clean tingle profile and strong activity will likely lead to use of the material specifically as a flavouring compound in the near future.

N-(2-methylbutyl) E,Z,E-2,6,8-decatrienamide (**9**)
This compound often co-occurs with spilanthol and hence is also found in *Spilanthes*, *Acmella* and *Heliopsis* species. The material is active but the activity is weak [10], as compared to spilanthol. Use probably results only because of the co-occurrence with spilanthol in jambu.

N-isobutyl E,E-2,4-undecenamide (**10**)
This alkamide has only recently been found in nature. It has some tingle effect at 25 ppm in aqueous solution. The tingle is relatively weak but very long-lasting. As a pure chemical it has very poor aqueous solubility and this is perhaps responsible for the delayed, weak but long-lasting effect. In most botanical species it occurs alongside pellitorine and so will be used where natural pellitorine extracts are used, probably synergising to enhance the tingling effect.

9.4.1.4 The future

With the obvious application of the tingle molecules in food preparations, particularly for use in gum, mouthwash, toothpaste, mints, candy and a variety of beverages, it seems likely that more natural 'tingle' extracts, nature identical and purely synthetic molecules will become commonplace in the flavour industry in the coming decade. Until now, the molecules have been in the form of natural extracts and oleoresins which possess much taste, can have a pungency perceived as warming, can be somewhat salty, have limited aqueous solubility and can have stability issues. With time these hurdles, along with reduction of the relatively high price are likely to be overcome.

Clearly, there are consumer benefits to drive the application of such effects. This can be manifest in the market place in a variety of different ways. Examples of such perceived benefits are as a fresh brushed feeling from gum or toothpaste, a numbing or refreshed feeling from mouthwash, an effect of increased alcohol or increased carbonation in a beverage, a feeling of astringency or dry mouth in a tea, coffee or wine flavour.

A time will come when the interaction of these molecules used in combination with cooling, warming and other flavour molecules is better understood. This interaction coupled to taste, smell, touch and temperature, will create great consumer benefit. This should cause a change in the taste and experience of products we consume in the market place. Since most of these components have demonstrated anti-microbial effects, including anti-bacterial and anti-fungal activity, it is likely that the scope of these materials, in combination with other ingredients, in oral care will be explored to the full, especially if anti-microbial activity in the absence of negative taste attribute is successfully demonstrated.

The industry is ever keen to offer new claims and dimensions to the quality of the products that it produces. There is evidence in the literature of the usefulness of these compounds to provide non-traditional attributes to flavour and fragrance applications. Examples include, but are not limited to, *N*-isobutyl *E,E,Z,E*-2,4,8,10-dodecatetraenamide (**12**) as an antioxidant component of *Echinacea* [28]; the anti-bacterial properties on spilanthol (**1**), isolated from Heliopsis [11], and the use of the same extract in oral hygiene [29]; the use of spilanthol as a refrigerant and its use in toothpaste and oral compositions [30]. Spilanthol is claimed as a component diminishing the acidic and astringent effects of dentifrices [31], and for fragrance as a useful component in bath preparations [32]. Fagaramide, pellitorine (**8**), and hazaleamide (**13**), a component of prickly ash [26], are reported to be modest, natural, anti-malarial compounds [33]. Additionally, there are several reports on the anaesthetic properties of pellitorine.

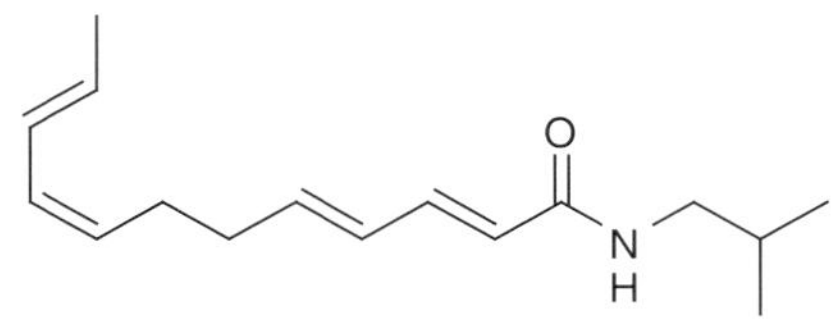

(**12**)

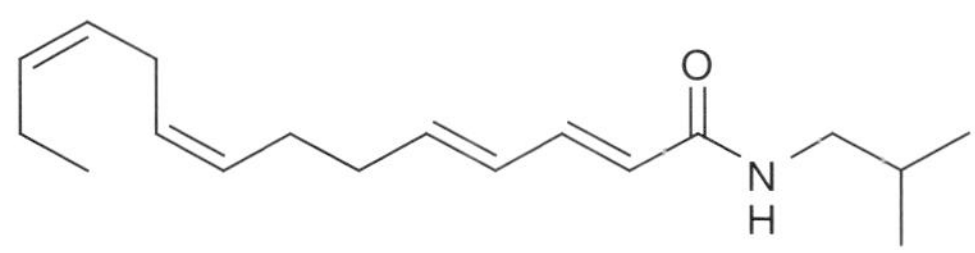

(**13**)

Since the use of synthetic tingle compounds offers (in addition to not being reliant on a crop) the luxury of easily controlling many key parameters such as purity, in-use concentration, stability, solubility and flavour, this has been a key focus area of research activity in the last few years. A number of new synthetic compounds have been identified by the author as potent tingle chemicals, although other compounds may be known to other researchers. Some of these potent compounds have been reported to occur in nature, but many are thus far not reported to occur in nature. The best four synthetic compounds identified are compounds (**14**)–(**17**). *N*-Isobutyl *E,E,Z*-2,4,8-undecatrienamide (**14**) is structurally related to *N*-isobutyl *E,E,Z,E*-2,4,8,10-dodecatetraenamide (**12**), Isoaffinin (**7**) and a number of other known naturally occurring compounds. It has extreme potency as a tingle agent [34], and at the dose level required to exhibit tingle effect, has little saltiness and none of the heat character associated with some of the other compounds and extracts. The heliotropyl derivative (**15**) and the *N*-isobutyl dodecadienamide (tetrahydro-α-sanshool) (**16**) are milder tingle compounds but exhibit additional interesting characteristics [34]. A similar effect is experienced with (**17**), the dihydro derivative of compound (**14**).

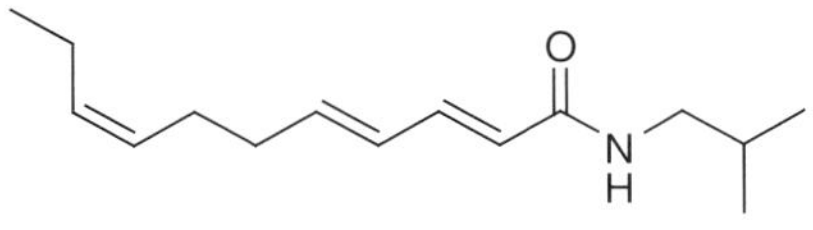

(**14**)

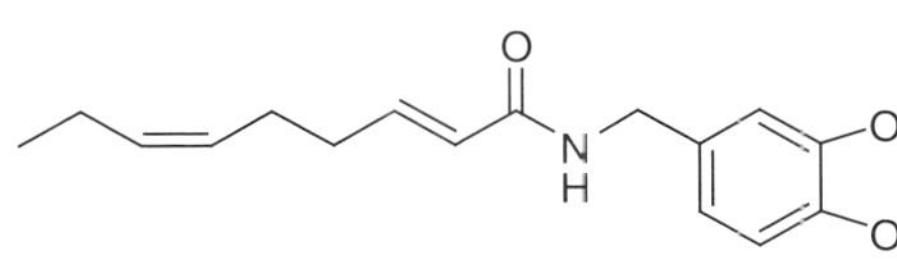

(**15**)

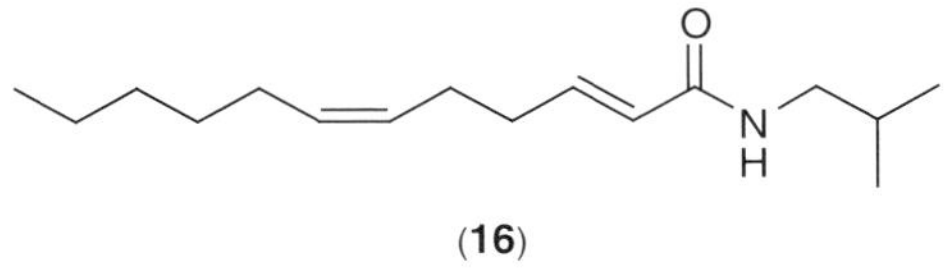

(**16**)

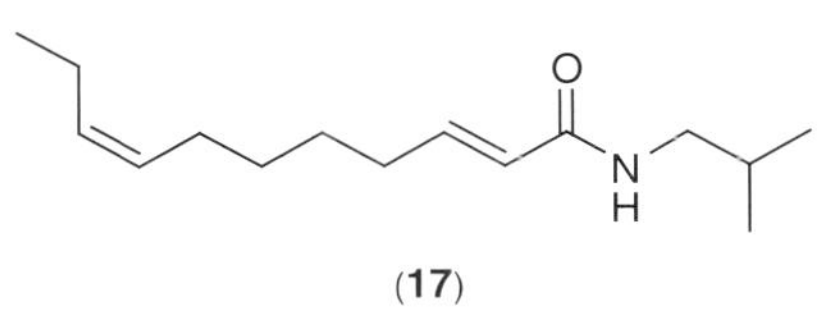

(**17**)

9.4.2 Cooling compounds

9.4.2.1 The use of cooling compounds

For many years, the natural cooling agents such as menthol, menthone, methyl acetate and peppermint oil have been used to provide a cooling effect, often along with providing a minty flavour and nasal clearing. Although use of these compounds in products such as body wash and shampoo was (and still is) employed, the irritating effect to the eyes and nose, along with the relatively high dose level required to achieve effect, has focused use mostly in the area of flavour application. Historically, chewing gum and mints were the main end-application for cooling compositions, as in these applications the 'non-cooling effects' exhibited by the components were acceptable.

These applications remain the major users of coolers today; however with the desire to branch cooling effect into other product areas, along with the desire to make gum (and other products) with flavours other than mint, the search for alternative coolers became a creative and application need. Since the 1970s synthetic compounds providing the physiological sensation of cooling have been known and utilised in the flavour and fragrance industry. The pioneering work of Watson and co-workers [35] was done by Wilkinson Sword Ltd [36] and interest in this area has been researched more keenly ever since. The value of coolers (other than menthol) is that they can give enhanced cooling without the 'burn' associated with higher levels of menthol. Additionally, due to lower or absence of taste and bitterness, they can be used in non-mint flavouring systems. These coolers give synergy if used with menthol and exhibit novel effects with other sensates (see Section 9.4.5).

Cold receptors on skin are in the basal layer of the epidermis, 0.1–0.2 mm below the surface, and are about 1 mm in diameter. When temperature cools to ~32°C the discharge is background (starts to become perceivable) and this discharge (perception) increases as the temperature lowers. Below 13°C thermal pain is felt, but this is through a different receptor system. The mouth is far more sensitive to cooling than is the outer skin. Presumably, this is due to the thinner barrier for penetration. Cooling of menthol is believed to be caused by reduced calcium^{2+} ion conductance.

Ambient temperature and molecular synergies have a profound effect on the perception of cooling. For example, menthol cooling is enhanced by cold but may be enhanced or inhibited by warming. Another problem with the perception of cooling is individual genetic sensitivities. Taking menthol as an example, individual oral thresholds vary from 0.02 to 10 μg, with the majority of people being in the more-sensitive category.

The cooling agents reviewed are generally soluble in flavour and fragrance oils and bases. Therefore, they are best mixed with the flavour or fragrance prior to loading in the end-application. The in-use levels for products of this type vary widely depending on the application. For flavours, levels in chewing gum and toothpaste are the highest, followed by candies and tablets, with soft drinks and other highly water-based products having the lowest dose levels. Additionally, these agents deliver best from polar systems making use in beverage, for example, much easier than from gum. Therefore, in these difficult applications, use of the cooler in the coating of the gum (or similar product) is often employed as a solution to overcome the lipophilic nature of the base. When this technique is used alongside other components such as menthol, an effective 'quick hit' is delivered. Examples of common uses of cooling agents include shampoo, cosmetics, oral care (mouthwash/toothpaste), chewing gum, pressed tablets, chewy candy, soft drinks and alcoholic beverages. Consumer research suggests that a cool feeling in the mouth equates to a sense of freshness or cleanness. Obviously, this is a desirable attribute in oral-care products.

9.4.2.2 Cooling molecules

Cooling compounds are generally believed to require a hydrogen-bonding group, a compact hydrocarbon skeleton, a correct hydrophilic/hydrophobic balance and a molecular weight between 150 and 350 [36]. A general feature of chemical cooling agents is that very little fatigue in cooling sensation is perceived, so long as the stimulus remains in contact with the oral/dermal receptors. For some reason, carboxamides generally provide a more rapid onset of cooling than do those coolers not containing nitrogen. Menthol, having very rapid onset of cooling, is an exception to this general rule.

In general, structure–activity relationships in the cooler area are not understood. As discussed above, there are some recognised guidelines regarding the molecular attributes required for cooling. These factors help in molecular design, however they are not comprehensive in terms of coverage of the active compounds. The best H-bonding groups are hydroxy, *N*-alkyl carboxamides, sulfoxides and phosphine oxides. Since the lipophilicity and H-bonding properties are fundamental to activity, most synthetic cooling compounds possess these functional groups. I have divided the molecules with a cooling effect into three classifications, menthol and its direct derivatives, carboxamides and miscellaneous coolers.

9.4.2.2.1 Menthol and direct derivatives

Considering compounds directly based on the menthol structure, those derived from L-menthol are significantly more cooling than those derived from D-menthol. Eccles has published a good review article on menthol and related cooling compounds [37].

DL-Menthol, (±)-menthol – stereochemistry relative, L-(−)-menthol – stereochemistry absolute (**18**).

Menthol is listed on the EFFA list of European flavour ingredients and is FEMA 2665.

Menthol is of botanical origin occurring in mint, specifically in mint oils from *Mentha piperita* (peppermint oil) and *Mentha arvensis* (Cornmint oil). Steam-distilled cornmint oil contains about 75% L-(−)-menthol [37] whereas the L-(−)-menthol level in peppermint oil is about 50%. Isolation from cornmint is the most economically viable source of natural menthol (but others exist), and natural isolation competes with synthesis of, from among other things, thymol. The cost of natural menthol fluctuates, whereas the cost of the synthetic menthol is stable. Which is more expensive varies year to year but in general the natural version is cheaper. The synthesis, physical chemical properties and isolation of menthol, in particular the synthesis of (−)-menthol by various routes, is discussed in Bauer [38].

Traditionally, menthol has been the most widely used cooling agent. This is particularly the case for fragrance applications where the high cooling activity of menthol can be perceived on skin at the <1% level. The effect is dependent on concentration, with cooling at low concentrations and irritation and local anaesthesia at higher concentrations of 2–5%. Unlike flavour applications, in fragrance applications most of the negative olfaction and taste of menthol is not perceived. This makes menthol a valuable cooling agent; however, some irritation and lachrymatory (irritating to the eyes, causing extreme tearing) effect, especially on the eyes, is still observed at higher concentrations.

The threshold for menthol is about 0.3 μg for oral cooling. L-Menthol is 45 times more active than D-menthol in terms of its cooling effect. The fact that menthol is relatively volatile (boiling point 212°C) and that menthol has a strongly irritant property, inherent odour and flavour, limits its use as a cooling agent in the oral cavity. That said, in flavour applications, menthol is still valuable in end-products where a minty taste is acceptable, such as chewing gum, breath fresheners and mouth wash. In these applications menthol can be used to provide a 'quick hit' of cooling, often in combination with other cooling agents with less taste and longer-lasting sensory effect, for example, monomenthyl glutarate.

Menthol isomers and related structures:

- D-(+)-Neomenthol, the hydroxy epimer of L-menthol

This compound is listed on the EFFA list of European flavour ingredients and is FEMA 2666.

- (±)-Menthone

This compound is listed on the EFFA list of European flavour ingredients and is FEMA 2667.

- L-(−)-Isopulegol, 'Coolant P' (**19**)

This compound is listed on the EFFA list of European flavour ingredients and is FEMA 2962.

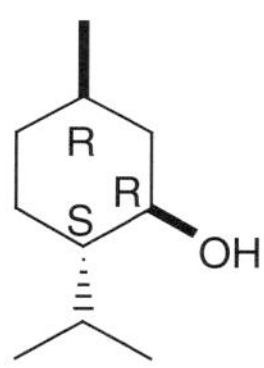

(**18**)

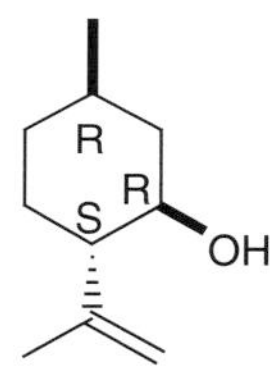

(**19**)

Isopulegol is normally considered to provide a cooling sensation, possess olfaction (minty, herbaceous) and to be somewhat bitter. However, Takasago International have found that very high purity (−)-isopulegol (>99.7 enantiomeric purity) is odourless and can provide a freshness, crispness and coolness to citrus fragrances. (−)-Isopulegol is sold as 'Coolant P' by Takasago International.

Menthol derivatives:

- Menthyl acetate

This compound is listed on the EFFA list of European flavour ingredients and is FEMA 2668. This compound has occasional use as a cooler. Literature reports include use for breath freshening, dentifrices and in combination with molecules such as monomenthyl succinate.

- L-Menthyl lactate, 'Frescolat ML' (**20**)

This compound is listed on the EFFA list of European flavour ingredients and is FEMA 3748. L-Menthyl lactate has long been GRAS-registered as a flavour ingredient, although the compound possesses only a slight minty odour and faint earthy taste. It does, however, provide a long-lasting cooling effect in the mouth. It was sold under the name 'Frescolat ML' by Haarmann & Reimer and has been proposed as a fragrance ingredient in cosmetics.

- L-Menthol ethylene glycol carbonate, 'Frescolat MGC' (**21**)
- L-Menthol propylene glycol carbonate, 'Frescolat type MPC' (22)

Both of these compounds are listed on the EFFA list of European flavour ingredients and are FEMA 3805 and 3806, respectively.

The ethylene derivative is listed under the CAS number [156679-39-9] on the European list of flavouring substances; however, the CAS number for the absolute L-isomer is [156324-78-6]. In the case of menthol propylene glycol carbonate it is the L-isomer of the 1- and 2-propylene glycol isomers that is listed as GRAS. These compounds were the subjects of a patent application [39] describing the preparation of the compounds and them as causing 'a cooling sensation on skin and on the mucous membranes while being taste and flavour neutral'.

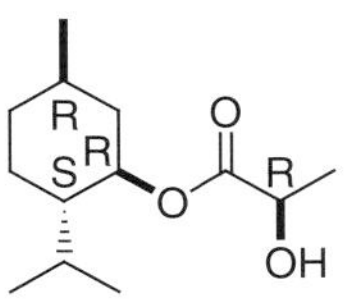

(**20**)

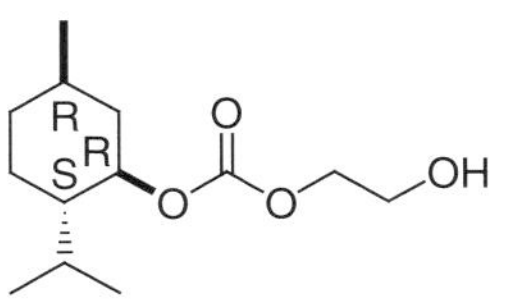

(**21**)

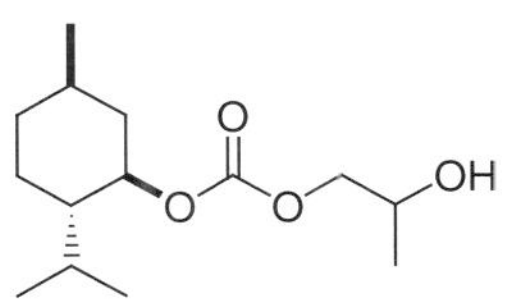

(**22**)

- Menthone 1,2-glycerol ketal, 'Frescolat MGA' (**23**)

This compound is listed on the EFFA list of European flavour ingredients and is FEMA 3807/3808.

Menthone glycerin ketal is sold as Frescolat MGA by Haarmann & Reimer (now Symrise) and there is substantial literature reporting and proposing its use in a range of edible and topical applications. Much of the literature reports usage with other agents in applications, but there are a couple of interesting articles specific to this compound. The first [40], by Greenberg of Wm Wrigley Jr Co, details the preparation of this and other ketals for use as flavour enhancers. The second investigates consumer acceptance of cooling agents (Frescolat MGA and Frescolat ML) in three different cosmetic formulations in four different regions of the world. The study shows general acceptance with variation from product to product and region to region [41]. The racemate L-isomer is FEMA 3807 and the DL-isomer is FEMA 3808, but the latter appears to be the item of commerce.

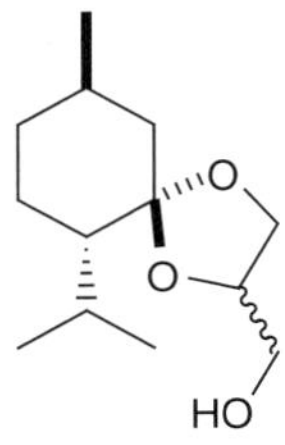

(**23**)

- L-Monomenthyl succinate (**24**)

This compound is listed on the EFFA list of European flavour ingredients and is FEMA 3810. Monomenthyl succinate is one of the better synthetic coolers that are a current item of commerce, although it has been a known molecule for over 100 years. The synthesis and use (as a cooling agent) in a consumed product was first reported by the British-American Tobacco Co Ltd in 1962 for use in the manufacture of tobacco, but the potential use as a flavouring ingredient was reported in 1965 for use as a menthol-releasing compound in menthol cigarettes. Several syntheses are known, both published and unpublished, for example Jabloner [42] and Shimizu [43], but in general, synthesis is easily accomplished by reaction of menthol with succinic anhydride under pressure or in solution with basic catalysis (Scheme 9.2).

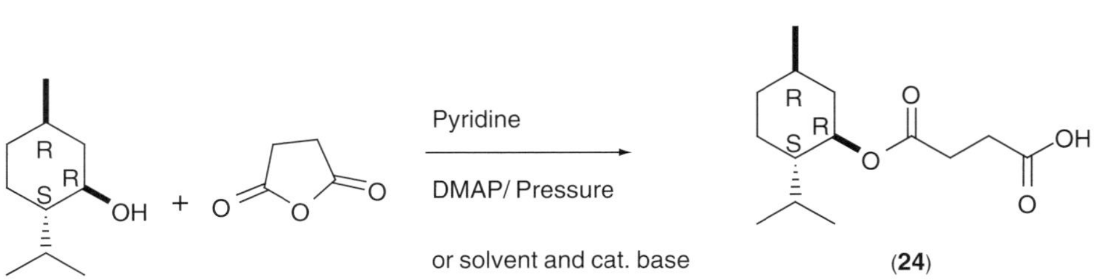

Scheme 9.2

The use of monomenthyl succinate in cooling compositions has been patented by V. Mane Fils SA. [44]; however, there are several subsequent patents, for example Grainger *et al.* [45], detailing use in both flavour and fragrance applications. The compound is essentially tasteless, provides a good balance of cooling onset and length of cooling profile at a very economic cost in use.

- L-(−)-Monomenthyl glutarate, 'Cooler 2' (**25**)

International Flavours & Fragrances sell this close relative (FEMA 4006) to monomenthyl succinate as Cooler 2. The sensory profile is similar to the succinate, but the onset is slower at about 40 seconds (25 ppm in water) and the cooling is tenaciously long. It is probably the longest-lasting oral cooling agent that is currently commercially available. Despite being a known molecule, for many years, the application in flavours is a recent adoption of the compound. It has an uncommon attribute among molecules of this type, in that it is also nature-identical, bringing a significant commercial advantage in the traditional European markets. The other nature-identical cooling agents are menthol, menthyl acetate and menthone, all of which are comparatively short-lasting and have olfactive properties which are usually undesirable. The compound was first synthesised back in 1928 from the dimenthyl ester by alcoholysis, but like the succinate can be readily synthesised from corresponding anhydride.

- 3-(L-Menthoxy)propane-1,2-diol, 'TK-10', 'Cooling agent 10' (**26**)

This compound is listed on the EFFA list of European flavour ingredients and is FEMA 3784, sold by Takasago International as TK-10. As with the other menthol-derived coolers, the compound with absolute stereochemistry in the L-menthol configuration is probably the preferred cooling ingredient in terms of activity [44]. It has been reported by the synthesis patent inventors that the 1-acetyl-2-substituted-3-L-menthoxypropane derivatives, the L-menthyl-2*S* isomer (**27**) and the 1-acetyl-2-substituted (H or Ac) derivatives thereof might be more greatly preferred embodiments [46] although none of these appear to be current items of commerce. TK-10 is a potent cooling agent with a threshold of about 1 ppm in the mouth. Although reported to be odourless [46] to most tasters it possesses some menthol type olfactive properties, has some irritation and is perceived as bitter by some tasters. Long cooling duration is possible at higher concentrations >50 ppm, but to myself at least, a substantial menthol-like taste is observed. There are numerous reports of use in both flavour and fragrance applications, in, for example, antiperspirant cosmetics, body-refreshing compositions, oral compositions, bath preparations, mint flavours, non-alcoholic mouthwashes and anti-dandruff shampoo. The strength of the cooling agent, along with its relatively high taste, may explain the prevalence of reports of use by the fragrance fraternity.

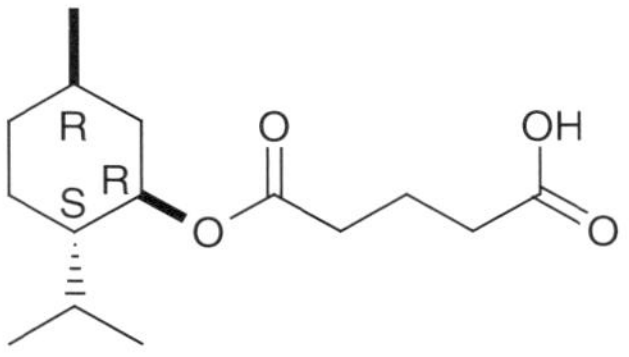

(**25**)

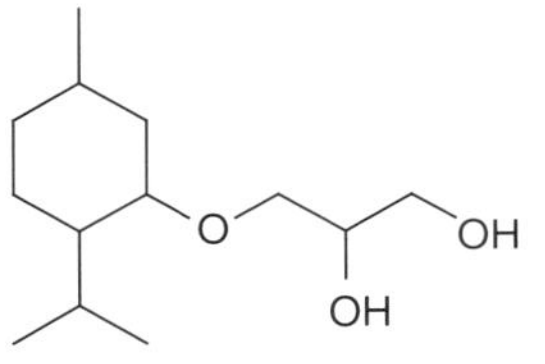

(**26**)

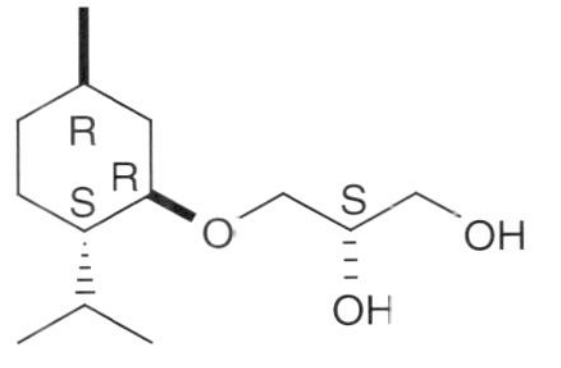

(**27**)

The original synthesis of TK-10 was reported via formation of the sodium salt of menthol, using sodium or sodium hydride and subsequent alkylation, with allyl halide, to form 3-L-menthoxy-1-propene. This is oxidised to the epoxide using organic peroxide and then hydrolised to the product. A number of more economic and safer synthesis of TK-10 and the other active derivatives are reported [46] with the preferred routes being outlined in Scheme 9.3.

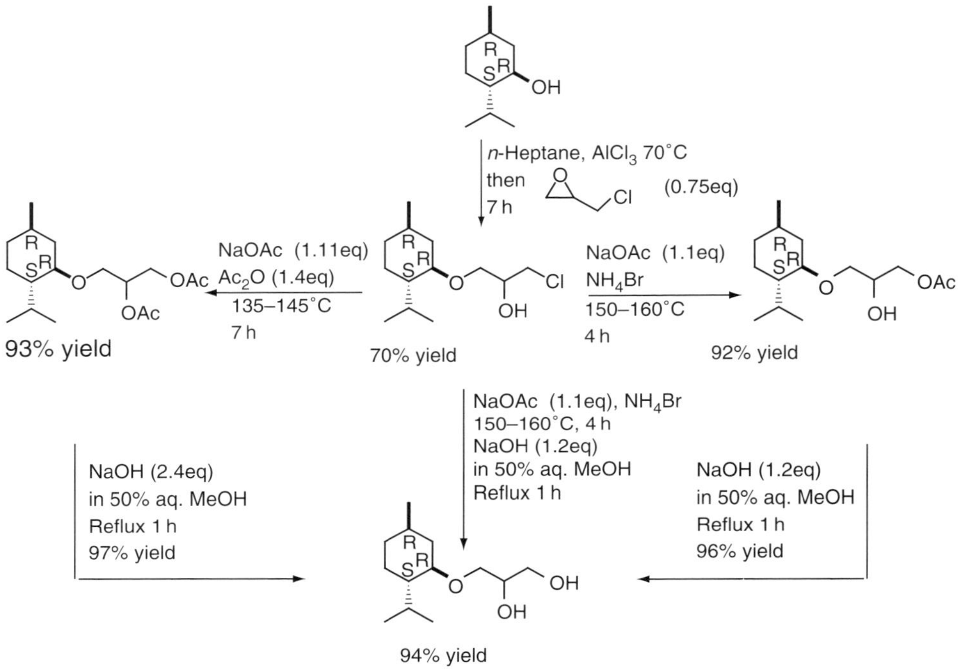

Scheme 9.3

The most efficient route appears to be synthesis of the 1-chloro derivative from epichlorohydrin and L-menthol, followed by in-situ preparation of the 1-acetoxy derivative and in-situ hydrolysis to the product, the overall yield being over 60%.

The 2*S* derivatives can be similarly prepared from (2*R*)-(−)-epichlorohydrin.

Two derivatives of 3-(L-menthoxy)propane-1,2-diol are listed as being FEMA. The first, 3-(L-menthoxy)-2-methylpropane-1,2-diol is FEMA 3849 but appears to have little, if any, commercial volume. The second, vanillin 3-(L-menthoxy)propane-1,2-diol acetal is FEMA 3904 and is patented for flavour and flavour applications [47].

9.4.2.2.2 Carboxamides

As previously discussed, carboxamide cooling compounds tend to be rapid onset coolers and, in some cases, can also be very long lasting. Certainly those carboxamide coolers discussed in this section are among the best coolers in the market today.

Among the carboxamides, the two most commonly used items of commerce are *N*-ethyl-*p*-menthane-3-carboxamide, WS-3® and 2-isopropyl-*N*,2,3-trimethylbutanamide WS-23® (WS-3® and WS-23® are registered trademarks of the Warner-Lambert Company).

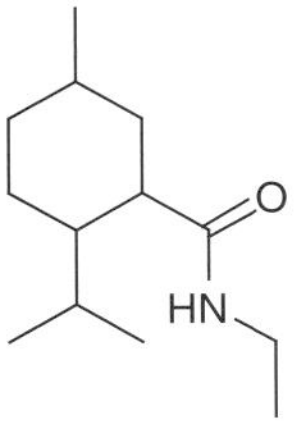

(**28**)

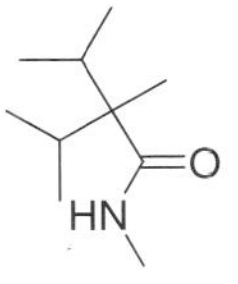

(**29**)

N-ethyl-*p*-menthane-3-carboxamide, 'WS-3' (**28**)
This compound is listed on the EFFA list of European flavour ingredients and is FEMA 3455. WS-3® is probably the quintessential cooler for flavour applications having a cooling threshold in the region of 0.2 ppm. The compound is virtually odourless, has only the faintest minty flavour character and provides rapid onset cooling (comparable to menthol), whilst providing a good duration of physiological cooling. One of the first coolers to be known, it was reported (along with many other active and never commercialised compounds) in the pioneering work of Wilkinson Sword; hence the 'WS'. WS-3 is the subject of many application patents, but the original literature [35, 36] and patents are from the inventors at Wilkinson Sword [48]. Among the many references referring to the use of this product in isolation, or in combination with other cooling agents, recent examples of uses in edible flavouring compositions have been published by Warner-Lambert Company [49]. The compound is currently marketed my Givaudan and Millennium Chemicals. Although not specified, the L-menthyl configuration is almost certainly the preferred cooling structure. Synthesis is generally accomplished by conversion of menthol to menthyl chloride, preparation of the corresponding Grignard agent and quenching with carbon dioxide. The acid is then converted to the ethyl amide via acid chloride or similar means, Scheme 9.4.

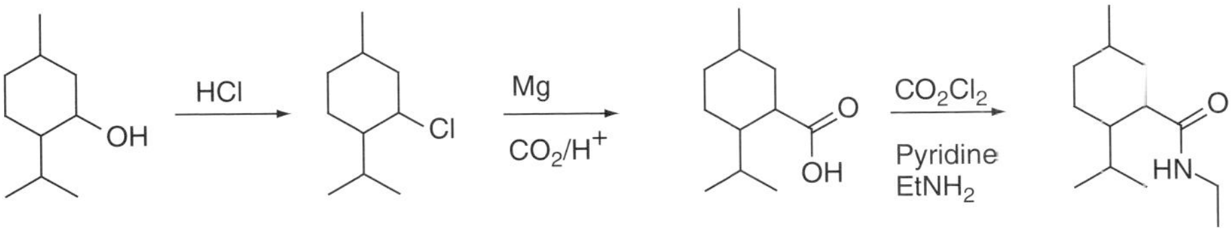

Scheme 9.4

2-Isopropyl-*N*,2,3-trimethylbutanamide, 'Cooler 3', 'WS-23' (**29**)
This compound is listed on the EFFA list of European flavour ingredients and is FEMA 3804. When called by its chemical name the structure is confusing, indeed a rather more sane chemical name would be *N*-methyl 2-isopropyl-2,3-dimethyl propionamide.

In any event, it is sold by some companies as WS-23 and by International Flavors & Fragrances as Cooler 3. Like WS-3, it too was born out of Wilkinson Sword and is described as an acyclic carboxamide with cooling effect [50]. The original patent and indeed recent subsequent literature discuss use in cosmetics, especially after-shave lotions, fragrance aerosols and reduced skin sting. However, primarily due to cost restraints, contemporary use of the cooler has been more successful in flavour applications such as gum, medicinal preparations and breath fresheners. The chemical has clean cooling with no taste, bitterness, burn, sting or tingle. The cost of the cooler has significantly reduced in recent months and is now in the region of $70/kg.

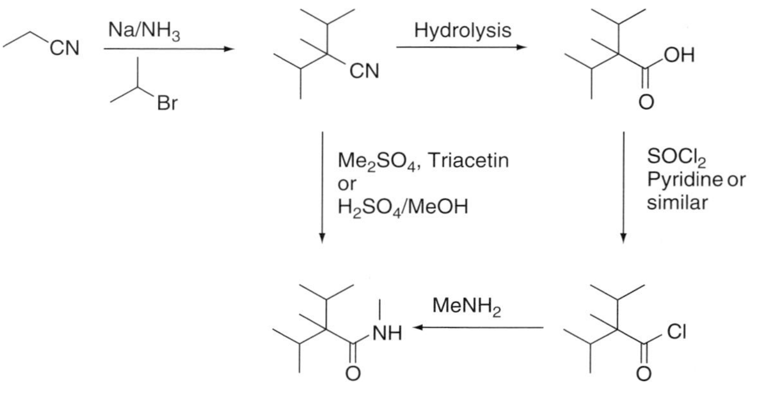

Scheme 9.5

Synthesis of WS-23 (Scheme 9.5), can be envisioned a number of different ways, but key to the synthesis is the diisopropyl propanenitrile (DIPPN) (**30**), known affectionately to folks at International Flavors & Fragrances as 'little man nitrile' due to its 2D structural appearance on paper. The nitrile, formed by alkylation of propanenitrile with isopropyl bromide can be hydrolysed to the acid and then converted to the methyl amide. Two elegant syntheses have been recently patented demonstrating direct conversion of nitrile to methyl amide. The first, by International Flavors & Fragrances [51], involves a novel alkylation of the nitrile by dimethyl sulfate in triacetin. The second, by Millennium Chemicals [52], is a clever variation on the chemistry using sulfuric acid and methanol.

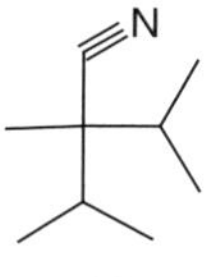

(**30**)

Menthyl 2-pyrrolidone-5-carboxylate, 'Questice' (**31**)
As the trade name 'Questice' suggests, this is a cooling agent marketed by Quest International. Both of the absolute L- and D-proline isomers of 'L-menthyl' Questice are

registered; however it is sold having the absolute stereochemistry of L-menthol and D/L-proline, and is as such an epimeric mixture. Thus far, it has only been sold and registered as a material for use in fragrances and as a cosmetic ingredient. Given the relative sensitivities of skin and the oral mucosa, and since it is a carboxamide of L-menthol, it is highly probable that it is a cooling agent in the mouth and will one day see flavour application. Recently, Quest International patented Questice as an insect repellent for use in cosmetic and other applications [53].

9.4.2.2.3 *Miscellaneous coolers*

Other miscellaneous cooling compositions exist.

AG-3-5, 'Icilin' (**32**)

Reported as a chemical producing sensations of cold [54], AG-3-5 is in a family of compounds known as tetrahydropyrimidine-2-one derivatives. The appearance of 'wet shakes' in mammals such as rats was reported as long ago as 1972 when low doses (<1 mg/kg) were injected intraperitoneally. When the compound was first (accidentally) contacted with humans, the compound was described as having sensations of coolness which lasted up to 15 minutes. On oral dosing at 5–10 mg in orange beverage, a sensation of 'coldness in the mouth, pharynx (the area that includes the oropharynx and the nasopharynx) and in the chest down to the xyphoid process' was observed. Cooling of the cheeks and inner surface of the arms and legs was also reported. The compound tested as fairly low toxicity LD50 is in the region of 1500 mg/kg but of suspect mutagenicity. At the time of writing, this cooling compound has not been adopted for flavour or fragrance use. I include it as it is reported in the literature, is extremely potent and is likely to be the subject of occasional industry consideration.

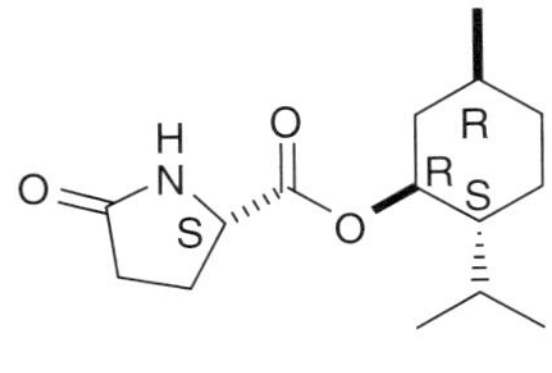

(**31**)

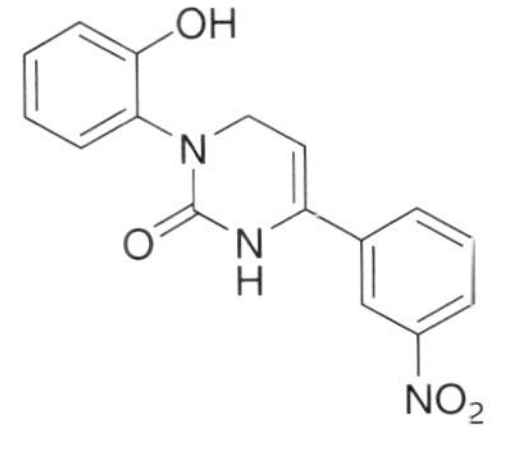

(**32**)

3-methyl-2-(1-pyrrolidinyl)-2-cyclopenten-1-one, '3-MPC' (**33**), 5-methyl-2-(1-pyrrolidinyl)-2-cyclopenten-1-one, '5-MPC' (**34**) and 2,5-dimethyl-4-(1-pyrrolidinyl)-3[2H]-furanone, 'DMPF' (**35**)

Three compounds found in malt, 3-methyl-2-(1-pyrrolidinyl)-2-cyclopenten-1-one (**33**), 5-methyl-2-(1-pyrrolidinyl)-2-cyclopenten-1-one (**34**) and 2,5-dimethyl-4-(1-pyrrolidinyl)-3(2*H*)-furanone (**35**), have recently been disclosed by Hofmann and co-workers [55] as potent cooling sensates in the oral cavity and on the skin. The compounds were first identified as cooling agents in Maillard reactions of glucose and L-proline; the compounds were first isolated and subsequently the identity confirmed by chemical synthesis. The cooling was described as 'cooling' and 'fresh' by a sensory panel.

The cooling thresholds are reported to be 5–100× higher, for these compounds than for (−)-menthol. However, when tasted at 50× threshold levels the cooling is 'instantaneous', more strongly perceived than menthol and significantly longer lasting than menthol. Although all three have a perceived odour, the odour thresholds are higher than menthol and the ratios of cooling threshold/odour threshold are lower, indicating a better chance (relative to menthol) of cooling without olfaction. It is not clear whether these materials are commercially viable, but they would provide the industry with a natural cooler.

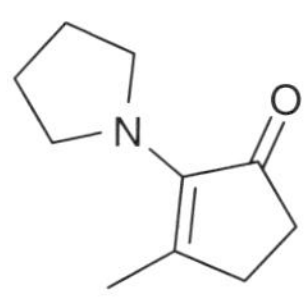

(**33**)

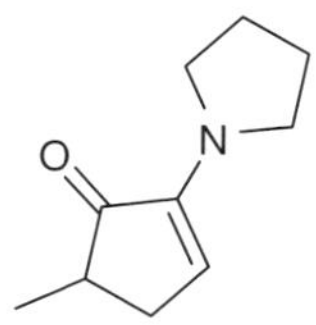

(**34**)

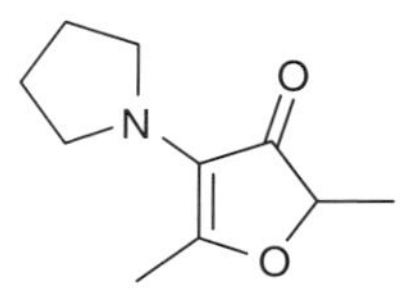

(**35**)

4-Methyl-3-(pyrrolidinyl)-2[5H]-furanone, '4MPF' (**36**)
Subsequently, but in the same year, Hofmann [55] reported some related compounds with higher activity. Again reported to occur naturally in malt, the most active cooling agent reported is 4-methyl-3-(pyrrolidinyl)-2[5*H*]-furanone, 4MPF (**36**). This is reported as 35 times more powerful than menthol in the mouth, and approximately 500 times more powerful on the skin. Like the original compounds, the cooling is longer lasting than menthol, additionally it is reputedly twice as long lasting.

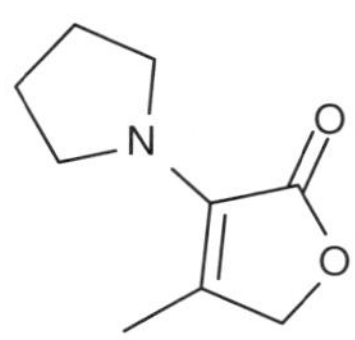

(**36**)

Mannitol and sorbitol
Both of these compounds exhibit cooling due to a negative heat of solution in the saliva. Both are widely used by flavour applications groups as a non-nutritive sweetener and anti-caking agent, for example, in chewing gum and soft candy.

Urea
Urea has a negative heat of dissolution in water and is used in personal care fragrance applications to provide a rapid but short-lasting cooling effect.

Methyl salicylate
Methyl salicylate, FEMA 2745, has a characteristic wintergreen odour and flavour. It does however provide some sweet cooling character when used at usual flavour application levels.

Camphor
Provides a cooling character when used in fragrances.

9.4.2.2.4 The future

With so much potential opportunity to make a positive impact on the mind of the consumer, coupled with such a wide variety of possible applications across flavours and fragrances, it is unlikely that the industry has seen the last of new cooling and refreshing compounds. There are still some unmet needs in the industry. Opinions regarding these needs are different depending on who you ask; however some general needs are clear. Fragrances desire coolers that lift some notes in fragrances or make them perceived as different or stronger on the skin. Flavours desire natural and water-soluble coolers along with rapid onset and longevity of cooling. Both disciplines would like the same or better performance at reduced cost.

The literature is still rich with work in the area of topical and oral cooling, and it has to be said that this is particularly true in the area of applications. Most coolers are no longer used individually but in combination with materials providing a synergic benefit. Examples of the bringing together of 'sensates' will be discussed later (see Section 9.4.5) but there are multitudes of examples of coolers used in combination. For example, a Proctor & Gamble patent describing a composition comprising a ketal, menthol and carboxamides [56], an International Flavors & Fragrances patent outlining the oral sensory perception using a menthol derivative, a carboxamide or a vanillin derivative in combination or as complexes and salts of dimethyl sulfoxide [45]. This patent cites many of the coolers discussed in this chapter as well as vanillin butyl ether (see Section 9.4.3). A recent Lion Corporation patent discusses use of menthol and guanidine surfactants, in combination with coolers such as isopulegol, methyl lactate, WS-3 and Frescolat MGA, for a long-lasting refreshing feeling in hair cosmetics [57]. A Proctor & Gamble patent describes the combination of Frescolat MGA and l-menthoxypropane-1,2-diol in combination with sweeteners to overcome undesirable aesthetics in oral cough treatments [58]. Many more examples exist, the point being that the industry is learning, and will continue to learn, new ways of applying these agents in traditional and non-traditional [59] end products.

Researchers, including the author, co-workers and others in the industry, are working to launch cooling compounds that fill the remaining unfulfilled needs of the creative and applications teams. Indeed, most of the leading companies in the industry have filed recent patents in the field. A range of new cooling compounds offering rapid onset cooling at a low threshold level, with a longevity of cooling, at an attractive cost are currently being developed [34]. These cooling compounds synergise very effectively with existing products such as WS-3 and monomenthyl glutarate and possess little or no taste, flavour or bitterness.

9.4.3 Pungent, warming and hot irritants

Foods having hot and painful irritation effects in the mouth have been enjoyed for hundreds, or even thousands, of years. Despite use of warming compounds in the form of spices the world over, only a very few individual chemicals are used to create a warming or hot sensation in the flavour and fragrance industry. Perhaps this is due to the abundance of available spices, or the cheap relative cost compared to materials synthesised or extracted from nature; but whatever reason, it remains so.

As was the case for cooling, ambient temperature and molecular synergies have a profound effect on the perception of warming. For example, warming enhances the effect of capsaicin, but cooling inhibits the effect. Also, as was the case for cooling, the physiology of warming will not be discussed here. The physiology of warming is even more widely covered else where in the literature than that of cooling. As a brief overview, warm and hot (pain) sensations are transmitted by thermal pain receptors (this perception can be inhibited by mastication or other mechanical stimuli of mechano receptors), hot/cold thermal fibres, trigeminal chemical irritation and polymodal receptors. Among the known heat-sensitive channels are the polymodal vanilloid receptors VR1 and VRL-1, so-called because the compounds that activate the channel, for example capsaicin, have a vanilloid moiety in their structure.

The enjoyment of 'pain when eating' has to be learned. This 'acquired taste' is partly genetic, and also results from desensitisation caused from regular exposure to the irritant. Removal of the stimulant will eventually cause recovery and the desensitisation is lost. Conversely the food 'gets hotter as we eat it' due to sensitisation effected by continued re-exposure to the irritant in a short time frame (much less than the time required for the recovery). In general, these hot irritants cause increased salivation, which is an oral defence mechanism. The same enhanced salivation is seen with most of the more potent tingle compounds (see (**1**)–(**6**) in Section 9.4.1).

Examples of edible foodstuffs that are used to provide warming in cooking, each of which contains at least one pungent chemical irritant, are pepper, chili pepper, ginger, Szechuan pepper, mustard, white mustard, horseradish, Wasabe (Japanese horseradish), cloves and onions.

The following section lists the more common ingredients that are used to exhibit a warming effect, whether individually as chemicals or as part of a natural foodstuff or extract.

Taking first those compounds with a vanillyl or heliotropyl moiety:

Capsaicin (chilli pepper) (**37**)
This compound is listed on the EFFA list of European flavour ingredients and is FEMA 3404. The active ingredient in capsicum pepper *Capsicum frutescens* or *Capsicum annum* causes a burning in the tip and middle of the tongue which then travels to the back of the throat. It is primarily used in the form of capsicum extract or oleoresin, in Indian and other Asian cuisines, but is permitted as an individual chemical ingredient under FEMA in the United States as well as being permitted in Europe. Capsaicin has also been used as a topical agent, for example, in rub-on lotions for muscle warming prior to sporting activity, or to help increase blood flow to aid healing.

Piperine (black pepper) (**38**) and Chavicine (**39**)
These compounds are the active ingredients in pepper oleoresin.

Piperine is listed on the EFFA list of European flavour ingredients and is FEMA 2909. It is the heat principle component of black pepper, *Piper nigrum*, although pipercide, small quantities of some of the tingle amides and chavicine may be reasonably considered as synergising pungent ingredients. The piperine concentration of pepper extract is typically about 50%, although the level can very somewhat and still be compliant with the standard of identity for the extract. Recent work has indicated that piperine, chavicine and similar combinations, can be used to remedy skin de-pigmentation disorders and therefore be used cosmetically to enhance natural skin colouration [60].

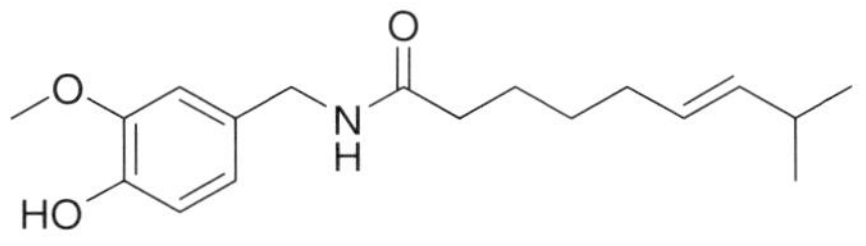

(37)

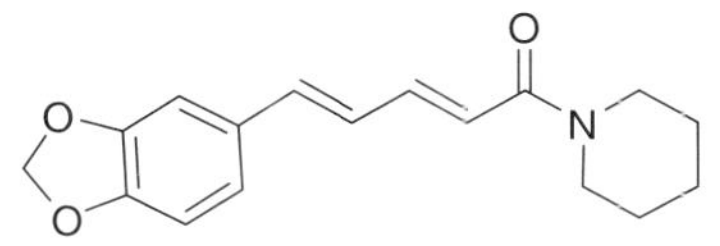

(38)

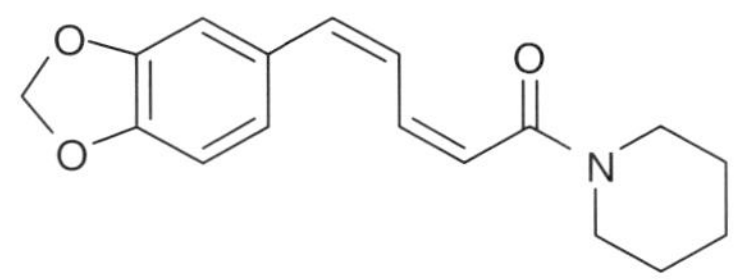

(39)

Vanillyl butyl ether, FEMA 3796 and vanillyl ethyl ether, FEMA 3815

Both of these ethers of vanillin are gentle warming agents. Although both have a pleasing sweet vanillic flavour, the useful characteristic of these molecules is that they provide an effect similar to the warming of ethyl alcohol.

N-Nonanoyl vanillylamide, 'Nonivamide' (**40**)

This compound is listed on the EFFA list of European flavour ingredients as *N*-Nonyl 4-hydroxy-3-methoxybenzylamide and is FEMA 2787.

A fully registered compound nonivamide can be used in either flavour or fragrance applications. The compound occurs in nature [61], has anti-inflammatory and analgesic properties [62], and modest warming potency on skin and in the mouth. There are reports of the use of nonivamide in both flavour and fragrance applications; however, the vast majority of activity appears to be in the area of cosmetics, particularly in products therapeutic to the skin.

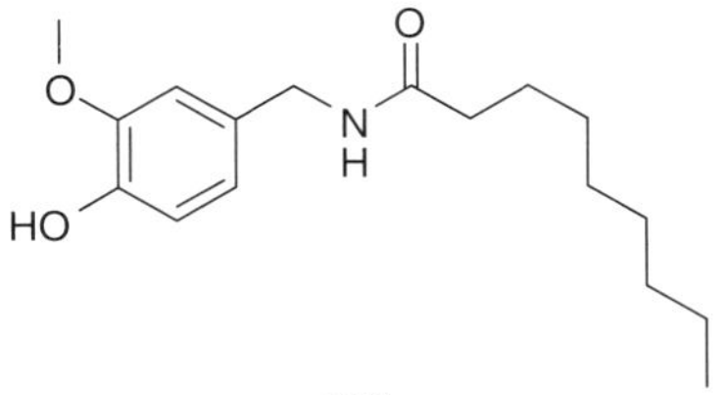

(40)

Gingerols and zingerone (ginger)

The mixture resulting from the natural botanical form of ginger is usually the preferred embodiment of the active components. Ginger, in the various natural forms, enjoys global registry status, for example FEMA 2520 (botanical), 2521 (extract) and 2522/2523 (oil). Zingerone, FEMA 3124 (**41**) is reported to be a key component as it provides the biting character. Ginger is widely used in cooking and is particularly prevalent in oriental cuisine, where it is often used in combination with citrus or other flavour types, for example, with lemongrass in Thai cooking. The 'gingerols', for example, 6-gingerol (**42**) are probably also heat-active components of the natural composition. There are a number of recent literature examples of the use of these compounds in flavour [63] and fragrance [64] applications.

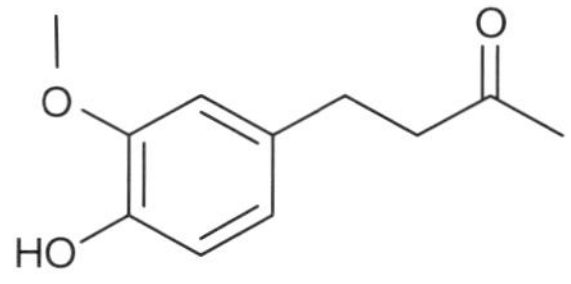

(**41**)

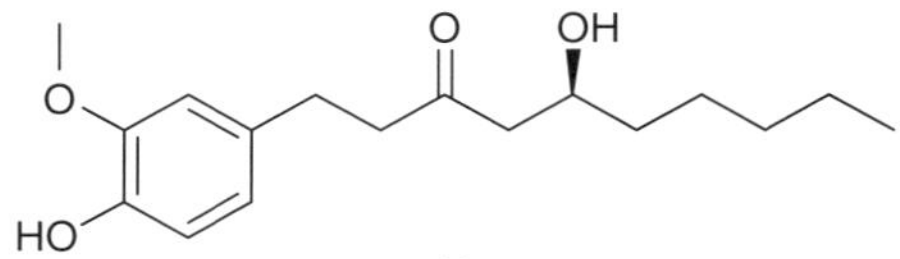

(**42**)

Eugenol (cloves)
This compound is listed on the EFFA list of European flavour ingredients and is FEMA 2467. Widely used either in clove oil, in which it occurs or as a pure material, *p*-eugenol has a desirable odour in addition to the mild warming properties it displays. Eugenol has been used in the treatment of dental pain for hundreds of years and is still used to the present day. Eugenol is also a key component of vanilla flavours.

Secondly, non-vanilloid compounds based on isothiocyanate moiety:

Allyl isothiocyanate (mustard and horse radish), FEMA 2034
This (often under the name of mustard oil) compound is the pungent and somewhat lachrymatory ingredient of Wasabe or Wasabi (Chinese horse radish) and mustard. Formed by enzymatic conversion of a glycoside by the action of the endogenous glycodase enzyme, in the form of mustard (brown or black mustard), it is widely used the world over as an accompaniment to cooked meats. In the form of Wasabe, it is an ingredient for sashimi, and an accompaniment to sushi. Allyl isothiocyanate is a powerful rubefacient, which means that it turns skin red, and usually hot, on contact. Although fully registered, the chemical is generally regarded as toxic, although the extent of the toxicity is uncertain. Most bodies consider the chemical a mutagen, and a toxic and an active fungicide and bactericide. The overall cancer evaluation for humans is 'Group 3, Unclassifiable as to carcinogenicity to humans', according to the International Agency for Research on Cancer. The anti-microbial qualities make it useful at low levels to increase product shelf life.

Phenethyl isothiocyanate (white mustard)
Although allowed as a flavour ingredient, this material is rarely used since white mustard does not yield an essential oil. Additionally, the chemical is listed as a cosmetic ingredient, but use as a cosmetic ingredient is not common place. A related hot component *p*-hydroxybenzyl isothiocyanate is also a known constituent of white mustard.

A variety of other active warming and heat compounds are known and these are listed with their FEMA numbers (and other regulatory numbers), where appropriate.

The Sanshool isomers (Szechuan pepper) (see (**2**)–(**6**) in Section 9.4.1), Corydalis (well known in Chinese and Tibetan herbal medicine as an analgesic), Galanga (Chinese ginger), 1-propenyl sulfenic acid (onion/lachrymator), diallyl sulfide FEMA 2042 and CoE 2174, methyl nicotinoate FEMA 3709 (another rubefacient chemical, also EINECS 202-261-8 and on the TSCA inventory) and cinnamic aldehyde FEMA 2286.

9.4.4 Astringency

Astringency is a chemoreceptive sensation since it occurs due to the irritation of mechano (tactile) receptors. It is experienced as a tactile sensation and is caused by tannins, polyphenols, aluminium salts and acids. The sensation is described as mouth drying, shrinking, drawing or puckering of the epithelium, roughing of the mouth and a drawing sensation of the cheeks.

Acids and phenols are the primary compounds responsible for astringency. Tannins (**43**) and (**44**) bind to salivary proteins delubricating the mouth. Wine is often described as having an astringent taste, even though it is a tactile sensation rather than a taste. Astringency is often considered a negative sensation, but some astringencies can be important, for example in wine, tea and coffee. The structures of the two types of tannins that are the principle contributors to astringency are shown below.

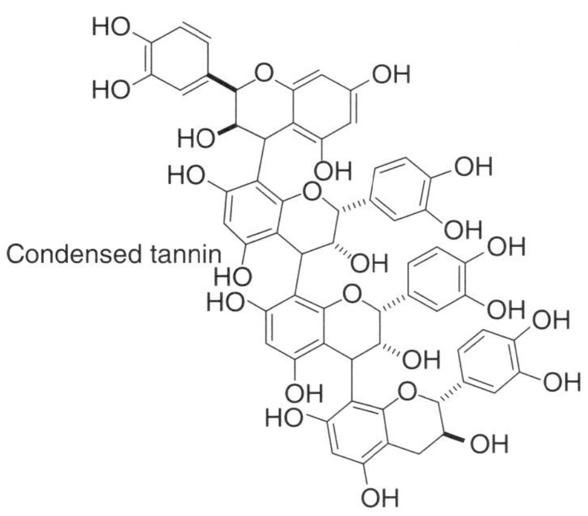

(**43**)

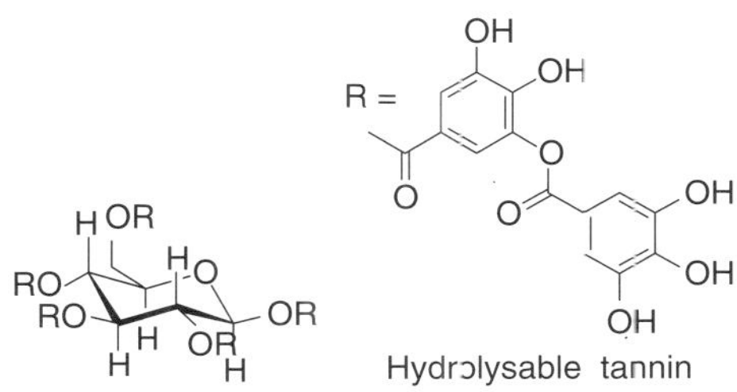

(**44**)

9.4.5 Synergies

It is apparent that in the case of chemical compounds and mixtures that exhibit tingle, cooling and warming, synergies exist between compounds possessing the particular sensation. Usually, the mixture possesses all the attributes of the individual components, and often, much more besides. It is a reasonable assumption that the compounds and extracts synergise across the sensory attributes too, be it a positive or negative perception of the molecular interaction with the receptor sites.

Recently, combinations of these sensory agents with different properties have been tested in the industry with positive effect. From the literature it appears that Takasago have taken the lead in this area. Kumamoto and co-workers cite the use vanillin (or a close derivative), in combination with a 'cooling agent', for example, those listed in this chapter but also L-menthylacetic acid *N*-ethyl amide, 3-(L-menthoxy)alkan-1-ols (recently patented by Takasago) [65], L-menthyl-4-hydroxypentanoate and L-menthyl-3-hydroxybutyrate and a 'warming agent', for example, vanillyl ethyl ether, vanillyl butyl ether, vanillin propylene glycol acetal (and ethyl ether thereof), a *N*-vanillyl carboxamide such as capsaicin or nonivamide, and Gingerone or other natural extracts. The composition is claimed as a 'warming composition' for fragrance and cosmetic applications [66]. Similarly, Nakatsu and co-workers [67] claim use of a sensate composition, 'A composition causing different skin sensations' constituting at least one cooling sensate, warming sensate and tingling sensate. The preferred mixture appears to consist of Spilanthol, WS-3 and vanillyl butyl ether, although a range of agents are cited including sanshool, *Zanthoxylum piperum, Piper nigrum*, piperine, chavicine and menthyl lactate. Kazutoshi and Kenichiro patented the use of warming and cooling agents in combination to provide an improved cooling sensation. The compounds used are similar to those used by Kumamoto, but the application focus, in this case, is hair-care products [68]. A great example of sensate combinations with flavours and other compounds used in food products has been demonstrated by Wm Wrigley Jr Company for chewing gum [63, 69]. In this example, not only are combinations of cooling agents, tingle agents and warming agents used, but it is claimed that in this application a hydrophobic sweetener can prolong the release of the flavour and the sensate agents.

9.4.6 Closing comments on compounds which provide a sensory effect

The search for new compounds providing the sensations of tingle, cooling and warming is likely to continue. It also seems likely that activity in ingredient discovery will be surpassed by actively exploring the application of the ingredients. This activity is likely to include exploration of mixtures within a sensate group, exploration of sensate mixtures with different sensory properties, interaction of sensates with flavour and fragrance compositions, and learning how to apply these findings across product categories. Therefore, in the opinion of the author, there is along way to go before the concept of chemesthesis and its application is understood, even empirically, let alone scientifically.

Another future development, still in its infancy in the marketplace, is how the effects and perceived benefits will be marketed to the consumer. Will sensation be used to link an experience to a performance in the product otherwise unseen, for example, anti-tangle in shampoo? Will attempts be made to explore the herbal healing attributes? Will the products target particular demographic groups for geographical regions? Or will traditional roads of increased consumer preference, long-lasting effect and increased impact be the preferred marketing ploy? The answer, which may not even be any of these strategies, should become apparent in the next few years.

9.5 Taste-active compounds

9.5.1 Sweeteners

Sweetness can be considered to be the most important taste quality as it is the taste category in which the most compounds have been researched and brought to market. This is probably due to the fact that a craving for foods that are sugar-sweet is among the strongest of human desires, and of the animal kingdom, in general. In recent decades, perhaps as a result of increased global wealth, consumption of empty calories in the form of sweet foods has caused a general trend toward obesity. In the USA this has been linked to an increase in diabetes to the extent that effects around 17 million of the population of the United States, with a further 16 million showing pre-diabetes. Additionally, of course, sugar is a factor in tooth decay. Whether driven by technology or health benefit there has been a demand for artificial sweeteners, or other sweeteners in the form of natural products, to replace nutritive sugars in the daily diet. Adoption of the 'high potency' sweeteners is a technically difficult business, and replacement of sugar in a foodstuff often requires a change in the design of the product to accommodate such an adoption. Several of these artificial sweeteners have come under scrutiny as to the long-term health benefits (and risks) associated with prolonged consumption.

An article outlining the chemistry of the sweet taste sensation, in theory and in practice, has been written by Shallenberger [70]. In principle, it is proposed that pairs of hydrogen-bonding functional groups such as oxygen, nitrogen and ether moieties are required to exhibit a sweet taste. These groups AH,B should be available and adjacent in the molecule to present activity.

Sweeteners fall into three categories: (a) nutritive, (b) non-nutritive (providing no significant nourishment) and (c) 'high potency'. Sugar replacements are often called 'high intensity sweeteners;' however, sweetness and taste are a function of appearance time and extinction time; hence the effect is non-linear, and a direct comparison of sweetness to sugar is difficult. Therefore I have preferred to use the term 'high potency' to describe sugar replacements. Sweeteners used in flavouring foodstuffs may be from any of these categories; additionally they may be natural sugars, non-sugars (corn sweeteners) or artificial sweeteners.

Nutritive sweeteners include the carbohydrates such as xylose FEMA 3606, sucrose, dextrose, honey, maltose, lactose, fructose, levulose and galactose. Some of these and other principal sweeteners used by the flavour and food industry, along with the class of the sweetener, are presented in Table 9.1.

Sugar, in the form of sucrose, still remains pretty much the standard against which all other sweeteners must be judged. Brief comments summarising some of the most important 'high potency, sugar replacement compounds are expressed below. The one thing that they have in common is controversy over the safety of their use and they are continuously under intense scrutiny.

Saccharin (45)

Considered to be the oldest non-nutritive sweetener, it has been in use since 1900. A zero-calorie product, before the introduction of cyclamate (1950s) and during the

Table 9.1 Comparative sweetness data for some common sweeteners and trademarked products

Sweetener	Class	Sweetness relative to sucrose	Comments
Sucrose	Sugar	1.00	The preferred standard
Fructose	Sugar	~1.20	Preferred in beverages as high fructose corn syrup in liquid form
Dextrose (crystalline glucose)	Sugar	0.70	
Molasses	Sugar	0.80	
Xylitol	Corn sweetener	1.00	Also provides cooling
Sorbitol	Corn sweetener	0.60	Also provides cooling
Mannitol	Corn sweetener	0.50	Also provides cooling. Laxative in excess
Maltitol	Corn sweetener	0.90	
High fructose corn syrup (55%)	Corn sweetener	1.10	The preferred sweetener for non-diet beverages
Saccharin (**45**) Saccharin sodium salt [128-44-9]	High potency	~300–500	Sweet N Low brand® name in the USA
Aspartame® (**46**)	High potency	~180–200	NutraSweet™ or Equal® (Merisant Company). Generic brand Natra-Taste® (Stadt Corporation, NY). Approved for use in over 100 countries
Sucralose, Splenda (**47**)	High potency	~ 600	Sucralose™ Tate & Lyle Splenda®, McNeil-PPC, Inc. FDA-approved in carbonated beverages
Acesulfame K (**48**) K represents potassium	High potency	~200	Sunett™ brand sweetener – Hoescht Celanese™. Approved in most countries >30
Cyclamic acid (**49**) and salts – Cyclamate	High potency	~30	Banned in the USA
Neotame (**50**)	Art. High potency	7000–13000 depending on the end application	Developed by Monsanto. Next generation Aspartame. NutraSweet Company product
Alitame, Thietane (**51**)	High potency	~2000	Approved in many countries. Discovered by Pfizer, brand name Aclame™

Table 9.1 (Continued)

Sweetener	Class	Sweetness relative to sucrose	Comments
Neohesperidin dihydrochalcone (**52**) [20702-77-6]	High potency	665 as free phenol	FEMA 3811. Listed in the register of Flavouring Substances in Europe
Glycyrrhizin ammoniated (**53**)	High potency		FEMA 2528. Enhances sweetness and masks bitter and acid off-notes. Tastes of licorice
Glycyrrhizin	High potency	~50	Extracted from Licorice root
Thaumatin (Talin) [53850-34-3]	High potency	~1600 for thaumatin I	FEMA 3732. Two isomers, each >200 amino acids. Approved in a number of countries including Japan, Canada, UK and USA

period 1970–1981, saccharin was the only approved sweetener in the USA. It is often perceived as having a bitter and metallic aftertaste, but is stable to heat, including baking. Saccharin and sodium saccharin have been the subject of extensive scrutiny and have been provisionally listed for use with a disclaimer. In 1977 the FDA proposed a ban on the product based on animal research regarding it as hazardous to health and has been determined to cause cancer in laboratory animals. It has been claimed that the doses of saccharin in the experiment were unrealistically high, equivalent to 750 cans of soda per day, every day, for life. In excess of 20 human studies have shown no overall association between saccharin consumption and cancer. It still remains (at least to myself) unclear whether or not the product is a possible carcinogen. The FDA's position appears to be to allow use as a tabletop sweetener, but not for use in foods and beverages. Many people, including the author and the editor, find the taste of saccharin bitter and obviously 'artificial'.

Aspartame™ (**46**)

Discovered in 1965, Aspartame™, is classified as a general-purpose sweetener. Unlike other artificial sweeteners, it is completely metabolised to its natural component parts. Being an amine it can react with aldehydes (to form Schiff bases among other things), and reaction with the common flavour aldehydes such as benzaldehyde, citral, cinnamic aldehyde and acetaldehyde is a problem. Additionally, breakdown at low pH, such as in beverages, occurs. The material breakdown is also observed on high heating, although it survives pasteurisation. An encapsulated form is available which is bake-stable. Since aspartame contains a phenylalanine component, people with the hereditary disease phenylketonuria should beware of its possible addition to foods. In the USA products containing it must be labelled 'This product contains phenylalanine'.

Sucralose (**47**)
Highly soluble, heat stable, stable at low pH (e.g. in a non-alcoholic beverage) and at high temperature storage for long periods sucralose is a zero-calorie sweetener (although the author is not a regular consumer of diet products and artificial sweeteners, sucralose is his sweetener of choice, though some bitter aftertaste persists). A US patent describing a method for improving the sweetness delivery of sucralose has recently been issued [71].

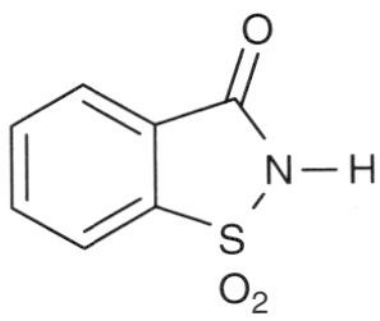
(**45**)

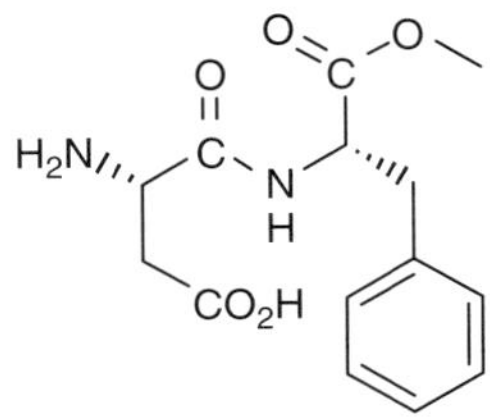
(**46**)

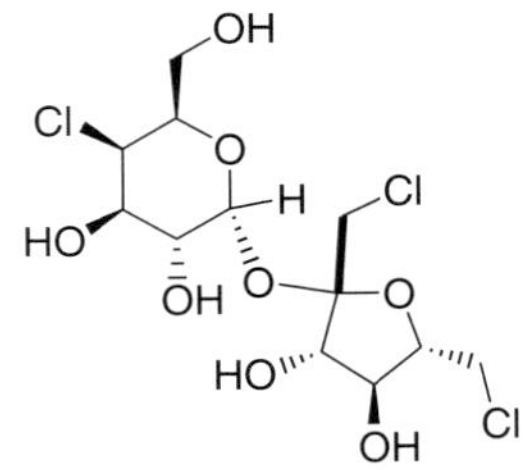
(**47**)

Acesulfame K, Oxathiazinone dioxide K (**48**)
Discovered in 1967 by Hoechst AG, Acesulfame K is marketed under the brand name Sunett® by Nutrinova Co. The product is stable under processing conditions, including high heat and acid media. Acesulfame K is reported separately to have a 'metallic aftertaste', 'bitter aftertaste' and 'no lingering aftertaste'. The bitter aspect appears to occur at high concentration and this is not universally observed. It is reported to synergise well with aspartame to reduce the overall level of artificial sweetener. The compound has been reviewed for safety by several organisations, and has been regarded as safe by most, if not all, studies to date.

Cyclamic acid (**49**)
The calcium and sodium salts are known as cyclamate. Cyclamate was discovered in 1937 and banned in the USA in 1970 due to safety concerns over bladder cancer in rats and mice, but that ban is currently being reconsidered. Several attempts to re-introduce the compound have been made over the years, but to date, without a successful re-introduction.

Neotame (**50**)
Neotame [72] is a new sweetener (next-generation aspartame) which is ~40 times sweeter than aspartame. Due to the hydrocarbon functionality on the nitrogen of the aspartic acid moiety, neotame should be more stable than the aspartame from which it is derived. Indeed it is claimed to be water-soluble, stable to storage and thermally stable. Since it is so much stronger than aspartame it is more cost effective in use. The taste is described as 'providing acceptable sweetness throughout the shelf life'. The taste of decomposition products when breakdown occurs at high temperature, or at the extremes of pH, is described as contributing 'no off flavours or odours', so that the taste is probably clean. Neotame was approved for use by the FDA as a general-purpose sweetener in 2002, for a wide variety of food products, but excluding meat and poultry. Studies indicate the safety of this sweetener for human consumption.

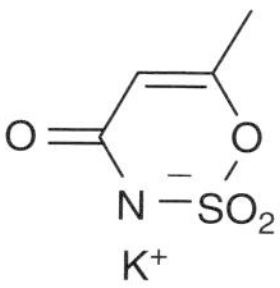

(**48**)

(**49**)

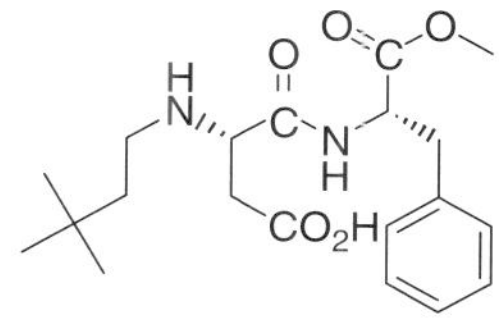

(**50**)

Alitame (51)

Alitame is freely soluble in water and alcohol, 12 times sweeter than aspartame and stable to processing at neutral pH. Some loss of sweetness occurs on baking, but in general the thermal stability over a broad pH range is good. Like the other peptide sweeteners, alitame should undergo Maillard reaction; this can lead to some browning, but no negative taste or flavour aspects have been reported. Additionally, alitame synergises well with other sweeteners (e.g. acesulfame K and aspartame) and is suitable for a wide range of applications. Studies in animals and humans have demonstrated good safety for human consumption. Approval in the United States and other countries are pending, but alitame enjoys approved status in Australasia, China, Mexico and other countries.

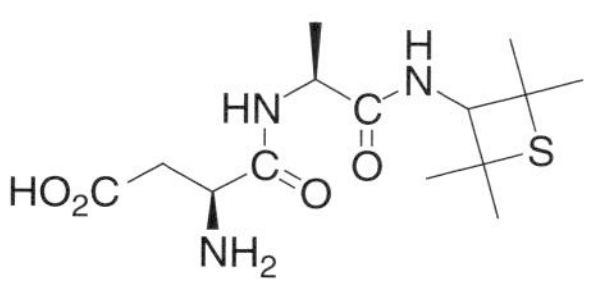

(**51**)

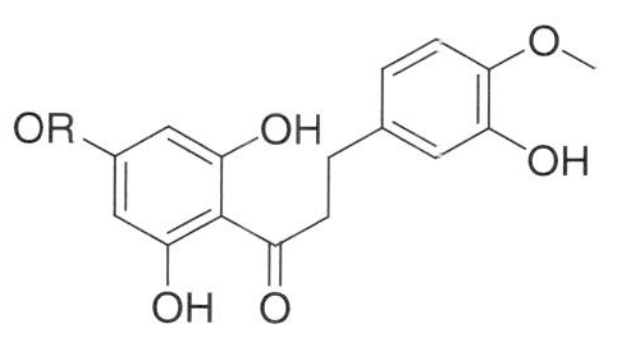

R = 2-O-α-L-rhamnosyl-β-D-glucopyranoside

(**52**)

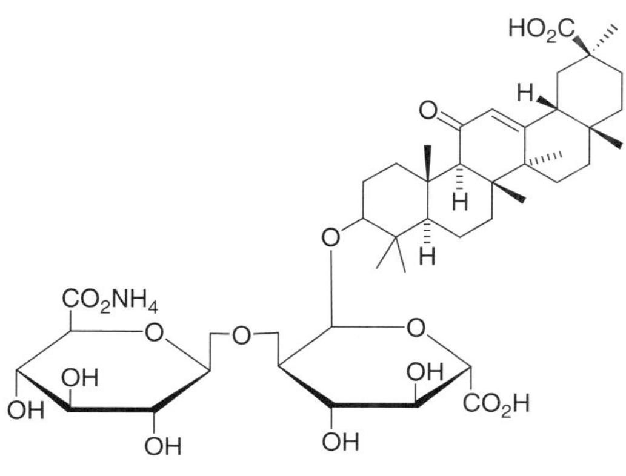

(**53**)

Neohesperidin dihydrochalcone (**52**)
Classified as a member of a large class of derivatives known as 'isovanillyl sweeteners', this is a semisynthetic sweetener. There exists a considerable amount of literature on hydrophobic sweeteners of this type [63, 69, 73], particularly referring to the structurally related compound 2-(3-hydroxy-4-methoxyphenyl)-1,3-benzodioxan [69, 73, 74], but neohesperidin appears to be the only compound that is an item of any commercial significance.

Thaumatin (talin)
A protein that occurs naturally in (and can be extracted from, in unmodified form by physical methods) the West African plant *Thaumatococous daniellii*, thaumatin is also known as Katemfe fruit or the miraculous fruit of the Sudan. Thaumatin is water-soluble to ~60% and is somewhat soluble in alcohols. In freeze-dried form it is stable to heat, UHT and pasteurisation. The sweetness is delayed and the extinction of sweetness is protracted. Thaumatin can be used as a flavour enhancer and as a masking agent for bitterness and metallic flavours, for example, of other 'high potency' sweeteners, with which it is reported to synergise well.

Miraculin
Miraculin is a product of the plant *Richadella dulcifica* that bears a fruit called miracle fruit. When tasted in the pure form it has no taste. However, all acids tested taste sweet if tasted after tasting miraculin. The sweetness is induced by acid, and a lemon will taste sweet if eaten after consumption of miraculin. The sweetness induced by a 0.1 M solution of citric acid after tasting a 1 μm miraculin solution is equivalent to a 0.4 M solution of sucrose. It is reported that the sweetness induced by the 0.1 M citric acid solution is 400 000 times that of a sucrose solution [75]. The effect is reported to last for around 2 h after the application of the miraculin. The referenced article also reports compounds possessing the ability to block or suppress sweet taste.

9.5.1.1 Other sweet offerings

It has been claimed that maltol, ethyl maltol and combinations of tartaric acid, peptides and gluconates can be used to mask the bitterness of artificial non-nutritive sweeteners.

Other compounds exist with sweet or 'brown' aroma that are used in flavour applications. These are not used as sweeteners *per se*, but none-the-less act as sweet flavour enhancers. Such compounds include maltol, cyclotene, maple lactone, iso-maltol and 4-hydroxy-2,5-dimethylfuran-3-one.

Sweetness and bitterness are closely related chemically, and this is perhaps why some sweet compounds, including mannose and indeed even sucrose, have a bitter aspect to the taste at high concentration. Compounds exhibiting bitterness (see Section 9.5.4) are closely related in terms of chemical functionality. As was the case for the sensates, where synergy allows the use of the individual sensate compounds at a lower level, to obtain a better profile, while maintaining the same level of overall sensate, sweeteners can be used in synergy to obtain about a 25% increase in intensity (for the same dose level) along with a better sweet profile. Actually, nutritive sweeteners synergise with each other and in combination with the artificial sweeteners too. Since sweeteners also effect flavour profiles, and since the flavour profile of a flavour and sweetener in combination changes with acidity, any flavour system must be designed specifically for the sweetener selected (or combination thereof), be the selection natural, artificial or a combination of the two.

Other sweeteners not discussed in any detail include, but may not be limited to, Stevia, Brezzein, Monellin, Mabinlin (all natural and protein-based), Lactitol and Erythritol.

9.5.1.2 Recent news

Recent literature in the field of sweeteners revolves largely around new applications, delivery and combinations of the known sweeteners. Hofmann [76] recently reported the use of (+)-(*S*)-alapyridaine (**54**), as having a sweet taste. Ley and co-workers of the Bayer Corporation have claimed use of hydroxyflavanones for masking bitterness [77]. These compounds were previously disclosed in the sweet vanillyl class. Additional aspartame derivatives and their salts are claimed to be up to 35 000 times sweeter than sugar and have recently been disclosed in the issued patent literature [78].

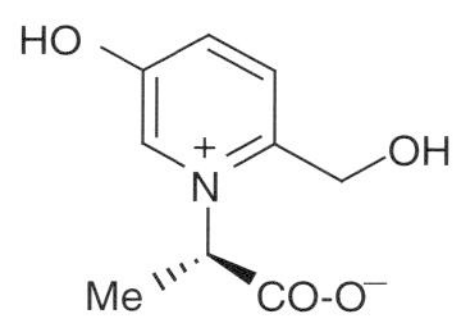

(**54**)

9.5.2 Salt and enhancers

In food use, salt is table salt (sodium chloride). This is the ingredient added to foods to provide the salt-taste quality. Salt is generally regarded as a general enhancer of most savoury flavour types. Salt replacement (to combat heart disease) is a challenge to the industry since it is nerve pathways transmitted by potassium–sodium ion exchange that results in the perception of the salt taste quality. These ion exchanges are fundamental systems in electronic nerve impulses of mammals and other organisms. Typically, replacement of sodium is the goal of low-salt (low-sodium) products and processes. Potassium, magnesium and calcium are the preferred replacement alkali metals, with chloride and sulfate as the counter ions. A recent US patent for salt enhancement with low sodium has been issued [79].

Whilst salt enhances savoury flavour types, it is itself enhanced by synergy with a variety of compounds that are either trigeminal stimulants, for example, tingle and warming compounds or Umami-effect compounds (see Section 9.5.5). It is not clear, at least to the author, whether these synergistic compounds are enhancing what salt is already there, or the salt enhances the molecules stimulation of the salt receptor. However, most people appear to believe that the compounds enhance the salt already present. A more detailed look into the theory of salt and salt-taste perception quickly becomes very complex.

9.5.3 Sour agents

The perception of sourness is activated by acidity. The commonly used compounds to provide sourness and to add acidity are lactic acid, acetic acid and citric acid.

9.5.4 Bitter agents

Bitterness is a taste quality perceived on the back of the tongue. It probably developed as a defence mechanism, since most bitter-tasting plants and chemicals are not good to eat. Indeed, many poisons and toxins are bitter, as are alkaloids and metallic compounds in food. Bitterness, or the masking of it, is a significant problem for pharmaceutical companies in the development of drugs. This is part of the reason why many drugs are coated or flavoured, particularly those targeted at the paediatric market.

In flavours, bitterness is generally a negative attribute and is often considered as an undesirable off-note in a food or flavour, for example, the bitterness of soy, on top of the already negative astringent character. However, in some products, such as coffee, dark chocolate and grapefruit, some bitterness is a desirable characteristic.

There are three principal types of bitter components: alkaloids, proteins and glycosides. Four bitter agents, in frequent use in the creation of flavours and in foodstuffs, are listed here, all are listed under both TSCA and EINECS:

(1) *Caffeine(55) FEMA 2224* Like most alkaloids this must be used sparingly due to the biological effect on the user.
(2) *Naringin (56) and naringin extract* Also known as naringoside, FEMA 2769.
(3) *Tannic acid, FEMA 3042* Also called tannin (see Section 9.4.4), tannic acid has bitterness secondary to the astringency and metallic aftertaste.
(4) *Quinine (57)* Another alkaloid, quinine is the traditional compound found in Indian tonic water. An insecticidal material, it was the reason for tonic water being used to aid in the prevention of malaria. Quinine is FDA approved for use at <83 ppm, however it is commonly used in salt form as the bisulfate FEMA 2975, sulfate FEMA 2977 or hydrochloride FEMA 2976.

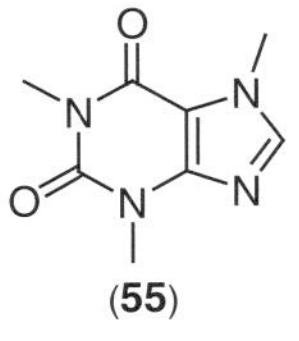

(55)

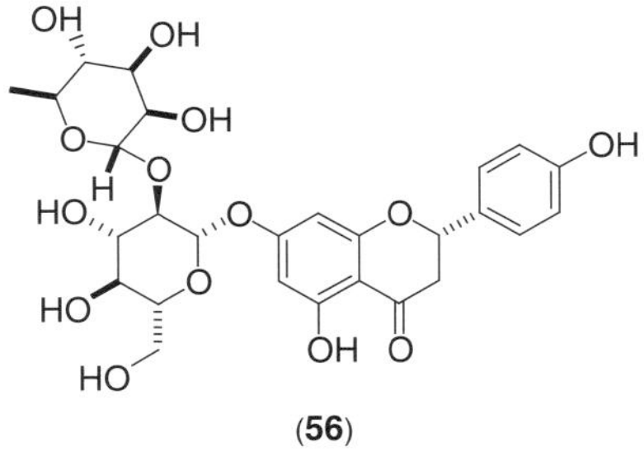

(56)

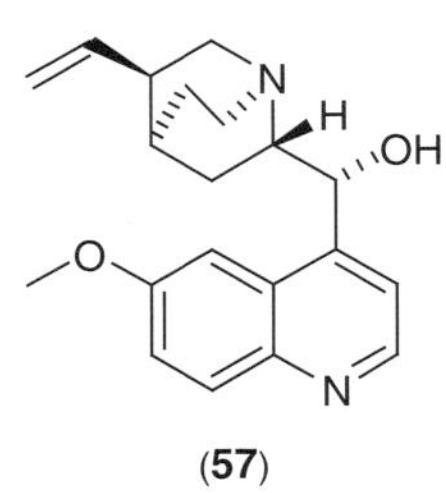

(57)

Other bittering agents exist as bitter principles in botanicals commonly used as foodstuffs. These include Aloe extract FEMA 2047, Chervil FEMA 2279, Chicory FEMA

2280, Cocoa leaf FEMA 2329, Hops (humulone 2579), Yarrow (see Section 9.4.1) FEMA 3117.

9.5.5 *Umami the fifth taste quality*

The term umami, from the Japanese word 'umami', which means delicious, best translates in english to savoury or meaty. It is the term used to describe the unique overall fullness, savoury or salivatory taste of food. Materials that exhibit this taste quality generally potentiate the intensity of glutamate solutions, and this is one important characteristic of umami taste [80]. Umami is increasingly becoming recognised as the fifth sense of taste, adding to the traditionally accepted sour, sweet, salt and bitter. Recent books on flavouring ingredients and taste tend to list only the four traditional qualities, or list Umami as a subject of controversy. Compounds traditionally described as possessing this character are MSG, protein hydrolysates, some amino acids, and certain nucleotides and phosphates.

MSG, FEMA 2756 [81], is the most widely used material as a 'taste enhancer' where it synergises the perception of 'savoury' ingredients. Currently 70% of the MSG market (1.1 million tonnes) is in the Pacific Basin, with the US and Europe accounting for only 15% of the market. However, MSG has also been alleged to cause an allergic reaction in a proportion of the population. Since MSG is widely used in Asian cuisine (especially Chinese), this has been referred to as the 'Chinese Restaurant Syndrome'. Free glutamic acid occurs naturally in food, but this, like MSG, is under scrutiny, being the subject of review by The Federation of American Society for Experimental Biology.

In addition to MSG, other chemical compounds, most notably several nucleotides, have been described to exhibit the Umami effect: adenosine 5′-(trihydrogen diphosphate), 5′-cytidylic acid (5′-CMP), 5′-uridylic acid (5′-UMP), 5′-adenylic acid (5′-AMP) (**58**), 5′-guanylic acid (5′-GMP) and 5′-inosinic acid (5′-IMP). The di-sodium salts of 5′-guanylic acid (**59**), FEMA 3668, and 5′-inosinic acid (**60**), FEMA 3669, are both FDA-approved flavour enhancers at GMP levels. Foods claiming to be MSG free may contain derivatives such as IMP and GMP salts.

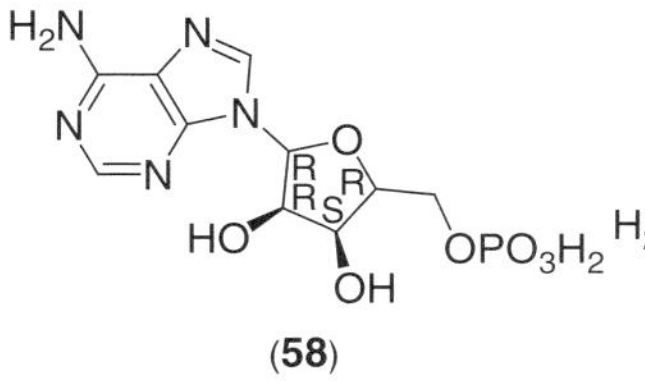

(**58**)

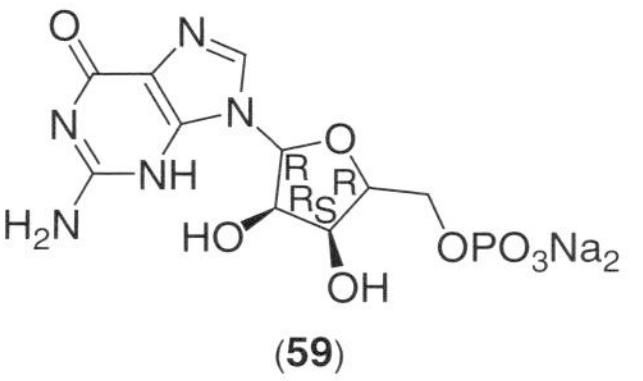

(**59**)

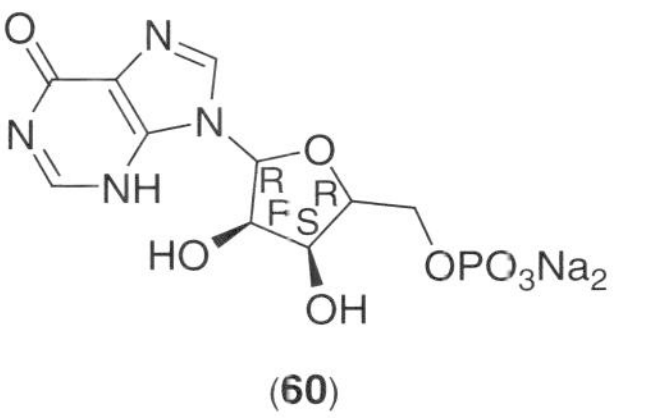

(**60**)

Recent literature [81, 82] cites an extensive range of other organic compounds which, as taste-active components of mixtures, have been implied to give the umami taste effect. This effect is probably the result of a synergy of the components in the mixture, with only a few (if any) really exhibiting umami taste quality.

These compounds include but are not limited to:

- Organic acids – succinic acid, lactic acid, saturated straight-chain aliphatic acids of six, eight, fourteen, fifteen, sixteen and seventeen carbon chain lengths, *Z*4,*Z*7,*Z*10,*Z*13,*Z*16,*Z*19-docosahexaenoic acid, *Z*5,*Z*8,*Z*11,*Z*14,*Z*17-eicosapentaenoic acid, *Z*9,*Z*12,*Z*16,*Z*19-octadecadienoic acid, *Z*9-octadecenoic acid, glutaric acid, adipic acid, suberic acid and malonic acid.
- Aminoacids – glutamic acid, aspartic acid, threonine, alanine, valine, histidine, proline, tyrosine, cystine, methionine, pyroglutamic acid, isoleucine, leucine, lycine and glycine. Dipeptides – Val–Glu and Glu–Asp.
- Other miscellaneous compounds including α-amino adipic acid, malic acid, α-aminobutyric acid, α-aminoisobutyric acid, *E*2,*E*4-hexadienal, *E*2,*E*4-heptadienal, *E*2,*E*4-octadienal, *E*2,*E*4-decadienal, *Z*4-heptenal, *E*2,*Z*6-nonadienal, methional, *E*3,*E*5-octadien-2-one, 1,6-hexanediamine, tetramethylpyrazine, trimethylpyrazine and *cis*-6-dodecen-4-olide.

In addition to the above, there exists a large number of recent patents that relate individual ingredients to possible umami taste effects. It must be said that the umami taste quality has, until now, been somewhat elusive. No real understanding of the quality, or its relation to compounds that cause the effect, either taken individually, or more usually in complex compositions and reaction flavours, appears to exist. What knowledge does exist is mostly in the area of ribotides, MSG and salt, along with the usage of these in combinations with each other, and other savoury flavour chemicals. Since research in this the area is focused mainly in Japan, and additionally, despite increased acceptance, there is still some doubt that umami is widely recognised as the fifth taste quality, it can be of no surprise that discovery in this area is lacking at the molecular level. Work is extensive on mixtures though, and it is likely that the creative and applications fraternity in the savoury field will, at least in the near future, continue with this 'Savoury Alchemy'.

9.5.5.1 The future

A new series of compounds belonging to the family of chemicals known as the 'alkamides' (see Section 9.4.1) have been investigated by researchers at IFF [34]. Surprisingly, a few examples from this series have been found by to exhibit umami taste, and other somatosensory, taste and flavour attributes, when dosed at the 1–10 ppm level onto the oral mucosa. There are no reports of amides having 'Umami' or other taste effects, and the most closely structurally related compounds, the aliphatic dienals and unsaturated acids, are not specifically reported to possess umami character when tasted in isolation. The odour of these acids and aldehyde compounds are fatty, herbal and melon/cucumber in character.

The author and co-workers at IFF are still exploring the full value of this discovery. Early results suggest that a significant enhancement to savoury foodstuffs is achievable with low levels of these flavour chemicals. Additionally, the Umami quality is not the only property of the chemicals, for example, they have olfactive properties like most flavour chemicals and enhance other taste qualities.

Additional work by Hofmann *et al.* [83] to explore the recently reported (+)-(*S*)-alapyridaine (**54**), has shown the molecule to have possible potential as a general enhancer, enhancing the umami and salt effects, in addition to possessing a sweet taste.

9.6 Conclusions

So this is the end of the whirlwind tour of the world of taste and sensation, and the molecules that effect them, from the humble viewpoint of an organic chemist in the flavour and fragrance industry. When I look back over the sections reviewed and discussed, a couple of general comments and opinions remain to be written.

On the sensory side of the coin, opportunities for business growth are largest in the area of tingle. Why? The area is still relatively undiscovered when compared to most aspects of the flavour and fragrance chemistry world, and its consumer products. Some small gains can still be made in the area of cooling, although this is nearer to being a mature field. Good, clean, natural coolers are still somewhat elusive. Warming is an area where the chemist must compete with some very potent molecules already easily available in nature. Some potent vanilloid molecules have been synthesised by pharmaceutical companies and other researchers in the field. Most of these compounds have a physiological effect as an ethical drug, in addition to being potent warming molecules. Therefore I don't expect them to be accepted as flavour or fragrance molecules anytime soon.

In the area of taste, reduction of salt is the Holy Grail. If researchers can find an organic molecule to replace salt, then this will move the molecular performance from a compound imparting preferred consumer preference to a compound providing health benefit. This would mean big business for the large salt users on the world, primarily manufacturers of soups, sources, snack foods and other savoury foodstuffs. This would, in turn, mean significant business for a company manufacturing flavours.

Until that day comes, its back to the day job.

References

1. *The Neurobiology of Taste and Smell* (2000) T.E. Finger, W.L. Silver and D. Restrepo (eds), 2nd edn, Wiley-Liss Inc., New York, ISBN 0–471–25721–4.
2. *Chemical Senses*, Vol. 2, Irritation (1990) B.G. Green, J.R. Mason and M.R. Kare (eds), Marcel Dekker Inc., New York.
3. Breslin, P.A.S. (2000) Human gustation. In T.E. Finger, W.L. Silver and D. Restrepo (eds), *The Neurobiology of Taste and Smell*, 2nd edn, Wiley-Liss, Inc., New York, pp. 423–461.
4. Bryant, B. and Moore, P.A. (1995) Factors affecting the sensitivity of the lingual trigeminal nerve to acids. *Am. J. Physiol.*, **268**, R58–R65.
5. Bryant, B. and Silver, W.L. (2000) Chemesthesis. In T.E. Finger, W.L. Silver and D. Restrepo (eds), *The Neurobiology of Taste and Smell*, 2nd edn, Wiley-Liss, Inc., New York, pp. 73–100.
6. Parker, G.H. (1912) The relations of smell, taste and the common chemical sense in vertibrates. *J. Acad. Nat. Sci. Phila.*, **14**, 221–234.

7. For reviews of the alkamides see Greger, H. (1984) Alkamides: structural relationships, distribution and biological activity. *Planta Medica*, **50**, 366–375; Greger, H. (1988) Comparative phytochemistry of the alkamides. In J. Lam, H. Breteler, T. Arnason and L. Hansen (eds), *Chemistry and Biology of Naturally-Occurring Acetylenes and Related Compounds (NOARC)*, Amsterdam, Elsevier, pp. 159–178 and references therein.
8. Nakatani, N. and Nagashima, M. (1992) Pungent alkamides from *Spilanthes acmella* L. var. *oleracea* Clarke. *Biosci. Biotech. Biochem.*, **56** (5), 759–762 and references therein.
9. Ramsewak, R.S. *et al.* (1999) Bioactive N-isobutylamides from the flower buds of *Spilanthes acmella. Phytochem.*, **51**, 729–732 and references therein.
10. Nagashima, M. and Nakatani, N. (1992) LC-MS analysis and structure determination of pungent alkamides from *Spilanthes acmella* L. Flowers. *Lebensm.-Wiss. U.-Technol.*, **25**, 417–421 and references therein.
11. Molina-Torres, J., Garcia-Chavez, A. and Ramirez-Chavez, E. (1999) Antimicrobial properties of alkamides present in flowering plants traditionally used in Mesoamerica: affinin and capsaicin. *J. Ethnopharmacol.*, **64**, 241–248 and references therein.
12. Navarrete, A. and Enrique, H. (1996) Anthelmintic properties of alpha-sanshool from *Xanthoxylum liebmannianum. Planta Medica*, **62**, 250–251.
13. Kashiwada, Y. *et al.* (1997) Amides of the fruit of *Zanthoxylum* spp. *Phytochem.*, **44**, 1125–1127.
14. Adesina, S.K. and Akinwusi, D.D. (1986) New constituents of *Zanthoxylum tessmanii* (Engl.) ayafor root. *J. High Res. Chromat.*, **9**, 412–414.
15. Adesina, S.K. (1987) Further new constituents of *Zanthoxylum leprieurii. Fitoterapia*, **58**, 123–126.
16. Adesina, S.K. (1986) Further novel constituents of *Zanthoxylum zanthoxyloides* root and pericarp. *J. Natural Products*, **49**, 715–716.
17. Tsai, I.-L., Lin, W.-Y., Huang, M.-W., Chen, T.-L. and Chen, I.-S. (2001) N-Isobutylamides and butyrolactone from the fruits of *Zanthoxylum intergrifoliolum. Helv. Chim. Acta*, **84**, 830–833 and references therein.
18. Xiong, Q., Shi, D., Yamamoto, H. and Mizuno, M. (1997) Alkamides from pericarps of *Zanthoxylum bungeanum. Phytochem.*, **46**, 1123–1126.
19. Bryant, B.P. and Mezine, I. (1999) Alkamides that produce tingling paresthesia activate tactile and thermal trigeminal neurons. *Brain Res.*, **842**, 452–460.
20. Bryant, B.P. (1999) Trigeminal sensory stimuli and animal repellents from plants. Int. Pat. Appl., WO 9921425 A1 19980923.
21. Tirillini, B. and Stoppini, A.M. (1994) Volatile constituents of the fruit secretory glands of *Zanthoxylum bungeanum* maxim. *J. Essent. Oil Res.*, **6**, 249–252.
22. Gamboa-Leon, R. and Chilton, W.S. (2000) Isobutylamide numbing agents of toothache grass, *Ctenium aromaticum. Biochem. Systematics and Ecology*, **28**, 1019–1021.
23. Binns, S.E., Livesey, J.F., Arnason, J.T. and Baum, B.R. (2002) Phytochemical variation in *Echinacea* form roots and flowerheads of wild and cultivated populations. *J. Agric. Food Chem.*, **50**, 3673–3687.
24. Saadali, B., Boriky, D., Blaghen, M., Vanhaelen, M. and Talbi. M. (2001) Alkamides from *Artemisia dracunculus. Phytochem.*, **58**, 1083–1086.
25. Adesina, S.K., Adebayo, A.S., Adesina, S.K.O. and Groning, R. (2002) GC/MS investigations of the minor constituents of *Piper guineense* stem. *Pharmazie*, **51**, 622–627.
26. DaCosta, N.C., Dewis, M.L. and John, T.V. (2002) IFF, Unpublished work.
27. Ikeda, Y., Ukai, J., Ikeda, N. and Yamamoto, H. (1987) Stereoselective synthesis of disubstituted 1,3-diene from aldehyde using organotitanium reagent. *Tetrahedron*, **43**, 731–741 and references therein.
28. Sloley, B.D., Urichuk, L.J., Tywin, C., Coutts, R.T., Pang, T. and Shan, J.J. (2001) Comparison of chemical components and antioxident capacity of different *Echinacea* species. *J. Pharmacy and Pharmacol.*, **53**, 849–857.

29. Wolfson, P. (2002) Herbal composition for improving oral hygiene, for providing local anesthesia, for use as an oral sensate, flavor enhancer and potentiator, and methods of using same. US Pat. Appl., 2002/0122778 A1 20020905.
30. Shimada, T. and Gomi, T. (1995) Spilanthol-rich essential oils for manufacturing toothpastes or other oral compositions. Japan Kokai Tokkyo Koho, JP 07090294 A2 19950404.
31. Sugano, H., Yoshida, F., Watanabe, Y. and Tokumoto, N. (1987) Dentifrices containing organic acids and flavors. Kokai Tokkyo Koho, JP 62198611 A2 19870902.
32. Moroe, M. and Hosoda, F. (1985) Bath preparations containing spilanthol. Japan Kokai Tokkyo Koho, JP 60215610 A2 19851029.
33. Christensen, S.B. and Kharazmi, A. (2001) Antimalarial natural products. In Trinagli and Corrado (eds), *Bioactive Compounds from Natural Sources*, Chapter 10. Taylor & Francis Ltd, London, UK, pp. 379–431.
34. Dewis, M.L., Huber, M.E. *et al.* IFF patent pending.
35. Watson, H.R. (1977) Flavor Characteristics of synthetic cooling compound. In C.M. Apt (ed.), *Flavor: It's Chemical Behavioral & Commercial Aspects*, Westview Press, Boulder, pp. 31–50.
36. Watson, H.R., Hems, R., Roswell, D.G. and Spring, D.J. (1978) New compounds with the menthol cooling effect. *J. Soc. Cosmet. Chem.*, **29**, 185–200.
37. Eccles, R. (1994) Menthol and related cooling compounds. *J. Pharm. Pharmacol.*, **46**, 618–630.
38. *Common Fragrance and Flavor Materials* (1991) K. Bauer, D. Garbe and H. Surburg (eds), 2nd rev. edn, VCH Publishers, New York, ISBN 0–89573–919–4.
39. Pelzer, R., Surburg, H. and Hopp, R. (1994) Preparation of compounds, particularly menthyl carbonate derivatives, having a cooling sensation. DE 4226043 A1 19940201.
40. Greenberg, M.J. (1992) Preparation of methone ketals for use as flavor enhancers. Eur. Pat. Appl. EP 485170 A1 19920513.
41. Jacobs, P. and Johncock, W. (1999) Some like it cool. Menthol derivatives as effective cosmetic cooling agents. Results of a worldwide study. *Parfuemerie und Kosmetic*, **80**, 28–31.
42. Jabloner, H., Dunbar, B.I. and Hopfinger, A.J. (1980) A molecular approach to flavor synthesis. I. Menthol eaters of varying size and polarity. *J. Polymer Sci., Polymer Chemistry Edition*, **18**, 2933–2940.
43. Shimizu, T. *et al.* (1996) Preparation of dicarboxylic acid monoesters. Japan Kokai Tokkyo Koho, JP 08198812 A2 19960806; Shimizu, T. *et al.* (1995) Synthesis of dicarboxylic monoesters from cyclic anhydrides under pressure. *Synlett.*, **6**, 650–652.
44. Mane, J.M. and Ponge, J.-L. (1998) Coolant compositions. US 5,725,865 and 5,843,466.
45. Grainger, B. *et al.* (2002) Oral sensory perception-affecting compositions containing dimethyl sulfoxide, complexes thereof and salts thereof. US 6365215 B1 20020402.
46. Amano, A. *et al.* (2003) Method for producing 3-L-menthoxypropane-1,2-diol, US 6,515,188 B2; Amano, A. (2002) Method for producing 3-L-menthoxypropane-1,2-diol. Eur. Pat. Appl., EP 1 253 130 A2.
47. Nakatsu, T., Green, C.B., Reitz, G.A. and Kang, R.K.L. (1998) Fragrance compositions containing 4-(1-menthoxymethyl)-2-phenyl-1,3-dioxolane or its derivatives. US 5753609 A 19980519; Nakatsu, T., Green, C.B., Reitz, G.A. and Kang, R.K.L. (1996) 4-(1-Menthoxymethyl)-2-phenyl-1,3-dioxolane or its derivatives and flavour compositions containing them. US 5545424 A 19960813.
48. Watson, H.R., Roswell, D.G. and Spring, D.J. (1979) N-Substituted paramenthane Carboxamides. US 4,150,052; Watson, H.R., Roswell, D.G. and Spring, D.J. (1980) N-Substituted paramenthane Carboxamides. US 4,226,988; Watson, H.R. (1974) Substituted p-menthane compositions having a physiological cooling effect. Brit. Amended, GB 1315625 19741126.
49. Barcelon, S.A. *et al.* (1999) Enhanced flavouring compositions containing N-ethyl-p-menthane-3-carboxamide and methods of making and using same. *Int. Pat. Appl.*, WO 9907235 A1 19990218; Luo, S.J. (1997) Breath-freashening edible compositions comprising menthol and an N-substituted-p-menthane carboxamide. US 5698181 A 19971216.

50. Roswell, D.G., Spring, D.J. and Hems, R. (1981) Acyclic carboxamides having a physiological cooling effect. US 4,296,255.
51. Boden, R.M. and Ramirez, C. (2002) Method for making carboxamides from nitriles and sulfates. US 6303817 B1 20011016.
52. Lebedev, M.Y. and Eman, M.B. (2002) Process for obtaining N-monosubstituted alkyl amides. US 6482983 B1 20021119; Lebedev, M.Y. and Erman, M.B. (2002) Lower primary alkanols and their esters in a Ritter-type reaction with nitriles. An efficient method for obtaining N-primary alkyl amides. *Tetrahedron Letters*, **43**, 1397–1399.
53. Watkins, S.D., Hills, M.J. and Birch, R.A. (2002) Use of menthyl-2-carboxylate as an insect repellent. US 6,451,844 B1 20020917.
54. Wei, E.T. and Seid, D.A. (1983) AG-3-5: a chemical producing sensations of cold. *J. Pharm. Pharmacol.*, **35**, 110–112.
55. Ottinger, H., Bareth, A. and Hofmann, T. (2001) Characterization of natural 'cooling' compounds formed from glucose and L-proline in dark malt by application of taste dilution analysis. *J. Agric. Food Chem.*, **49**, 1336–1344; Ottinger, H., Soldo, T. and Hoffmann, T. (2001) Systematic studies on structure and physiological activity of cyclic alpha-keto enamines, a novel class of 'cooling compounds'. *J. Agric. Food Chem.*, **49**, 5383–5390.
56. Furman, D.K. (1993) Coolant compositions. Int. Pat. Appl. WO 93/23005.
57. Tsuchiya, Y. *et al.* (2003) Hair cosmetics containing cool feeling agents and guanidine surfactants. Japan Kokai Tokkyo Koho, JP 2003104846 A2 2030409.
58. Lee, K.-C.M. (2002) Oral compositions containing coolants and sweeteners having improved consumer aesthetics. US 6391886 B1 20020521.
59. Anderson, G.T. (2000) Methods for alleviating discomfort of a hot flash and compositions therefor. Int. Pat. Appl., WO 0045815 A1 20000810.
60. Raman, A. and Lin, Z. (2000) Piperine and piperine analogues for treatment of skin depigmentation disorders. Int. Pat. Appl., WO 0002544 A2 20000120.
61. Constant, H.L., Cordell, G.A. and West, D.P. (1996) Nonivamide, a constituent of capsicum oleoresin. *J. Natural Products*, **59**, 425–426.
62. Brand, L., Berman, E, Schwen, R., Loomans, M., Janusz, J., Bohne, R. *et al.* (1987) NE-19550: a novel, orally active anti-inflammatory analgesic. *Drugs Under Experimental and Clinical Research*, **13**, 259–265.
63. Johnson, S.S. and Greenberg, M.J. (2003) Hydrophobic sweetner-containing chewing gum having prolonged sensory benefits. US Pat. Appl., US 2003072842 a1 20030417.
64. Jean, D. and Cariel, L. (1998) Use of gingerols for preparing deodorant and antiseptic compositions. French Pat. Appl., FR 2758087 A1 19980710.
65. Ishizaki, T., Green, C.B., Nakatsu, T. and Lupo, A. (2001) A (1'R, 2'S, 5'R)-3-L-menthoxyalkan-1-ol cooling sensate. Eur. Pat. Appl., EP 1122233 A1 20010808.
66. Kumamoto *et al.* (2002) Warming compositions. US Pat. Appl., 2002/0142015 A1 20021003.
67. Nakatsu, T., Mazelko, P.J., Lupo, A.T., Green, C.B., Manley, C.H., Spence, D.J. and Ohta, H. (2001) A composition causing different skin sensations. Eur. Pat. Appl., EP 1121927 A2 20010808.
68. Kazutoshi, S. and Kenichiro, S. (2003) Compositions for improving cooling sensation effects. US 6,544,535 B2 20030408.
69. Greenberg, M.J. and Johnson, S.S. (2002) A chewing gum having prolonged sensory benefits. Int. Pat. Appl., WO 02091849 A2 20021121.
70. Shallenberger, R.S. (1998) Sweetness theory and its application in the food industry. *Food Technol.*, **52**(7), 72–76.
71. Merkel, C.M. and Lindley, M.G. (2002) Method of improving sweetness delivery of sucralose. US 6,461,658.
72. Ponakala *et al.* (2002) N-[N-(3,3-dimethylbutyl)-L-alpha-aspartyl]-L-phenylalanine 1-methyl ester as a sweetener in chewing gum. US Pat. Appl., US 2002/0164397 A1 07112002.

73. Arnoldi, A., Bassoli, A., Borgonovo, G., Drew, M.G.B., Merlini, L. and Morini, G. (1998) Sweet isovanillyl derivatives: synthesis and structure–taste relationships of conformationally restricted analogues. *J. Agric. Food Chem.*, **46**, 4002–4010.
74. Dick, W.E. and Hodge, J.E. (1979) Substituted benzodioxan sweetening compound. US Pat. 4,146,650.
75. Kurihara, Y. and Nirasawa, S. (1994) Sweet antisweet and sweetness-inducing substances. *Trends in Food Sci. & Tech.*, **5**, 37–42.
76. Hofmann, T., Ottinger, H., Frank, O., Soldo, T. and Blank, I. (2003) Preparation of pyridinium-betaine compounds as taste enhancers. Eur. Pat. Appl., EP 1291342 A1 03122003.
77. Ley, J.P. *et al.* (2002) Use of hydroxyflavanones for masking bitter taste. US Pat. Appl., US 2002/0188019 A1 12122002 and Eur. Pat. Appl., EP 1258200 A2 04292002.
78. Amino, Y. *et al.* (2003) Aspartyl dipeptide ester derivatives and sweeteners. US Pat. 6,548,096 B1 04152003.
79. Bonorden, W.R., Giordano, D.A. and Lee, B.L. (2003) Salt flavor enhancing compositions, food products including such compositions, and methods for preparing such products. US Pat. 6,541,050 B1 20030401.
80. Glendinning, J.I., Chaudhari, N. and Kinnamon, S.C. (2000) Taste transduction and molecular biology. In T.E. Finger, W.L. Silver and D. Restrepo (eds), The Neurobiology of Taste and Smell, 2nd edn, Wiley-Liss, Inc., New York, pp. 315–351.
81. Song, H. *et al.* (2001) Determination of Umami substances and primary odorant of Gushi chicken. *Shipin Yu Fajiao Gongye*, **27**(9), 55–58.
82. Watanabe, K. *et al.* (2002) Taste-active components in a Vietnamese fish sauce. *Fisheries Science*, **68**, 913–920; Koriyama, T. *et al.* (2002) Fatty acid compositions of oil species affect the 5 basic taste perceptions. *J. Food Science*, **67**, 868–873; Cha, Y.-J. *et al.* (1999) Identification of aroma-active compounds in Korean salt-fermented fishes by aroma extract dilution analysis. 1. Aroma-active components in salt-fermented anchovy on the market. *Han'guk Sikp'um Yongyang Kwahak Hoechi*, **28**, 312–318; Kuramitsu, R. *et al.* (1996) Studies on flavored peptides. Part XII. Tastes produced by peptides containing ionic groups and by related compounds. *Biosci., Biotechnol. Biochem.*, **60**, 1637–1642; Hosokawa, M. *et al.* (1990) Flavor compositions of Katsuobushi. I. Volatile flavor components in dorsal, abdominal and red meat parts of dried skipjeck (katsuobushi). *Nippon Shokuhin Kogyo Gakkaishi*, **37**, 814–818.
83. Hofmann, T., Soldo, T. and Blank, I. (2003) (+)-(S)-Alapyridaine – a general taste enhancer? *Chem. Senses*, **28**, 371–379; Villard, R., Robert, F., Blank, I., Bernardinelli, G., Soldo, T. and Hofmann, T. (2003) Racemic and enantiopure synthesis and physicochemical characterization of the novel taste enhancer N-(1-carboxyethyl)-6-(hydroxymethyl)pyridinium-3-ol inner salt. *J. Agric. Food Chem.*, **51**, 4040–4045.

Chapter 10
Stability of Aroma Chemicals

Chris Winkel

10.1 Introduction

'All flavours could be made from only 200 different molecules if only those 200 molecules were stable.' This statement made by Professor W. Grosch at the Weurman Flavour Symposium of 1998 in Garching, Germany, describes well the challenges flavour and fragrance manufacturers face today. Consumers want to buy food that has an optimal taste upon consumption; but due to chemical and physical effects food is changing over time. And consumers want their detergent powder to smell the same just after opening the package and when the package is finished a few weeks later.

Everybody prefers freshly baked bread compared to stale bread or a fresh cup of coffee compared to one that is kept warm for an hour. On the other hand most people prefer freshly squeezed orange juice compared to orange juice from a pack but everybody accepts that commercial orange juice has a different taste than the fresh juice and the market for commercial juice is incomparably larger than the fresh juice market. In other words, food manufacturers not only deal with the technical restrictions, they also have to deal with consumer perception.

Fragrance ingredients suppliers deal with the same issues as flavour ingredients suppliers, although their molecules need to be able to withstand more severe conditions. People expect a perfume or an after-shave to be stable for five years or so. Moreover, the pH of products that contain fragrances varies between 1 and 12.

Although we realize that a consumer will judge the total food or perfume sensation, this chapter will focus only on the (added) aroma ingredients in a food, a personal product or a cleaning product. So we will not look at the loss of water as one of the main reasons for bread getting stale but we will look at the staling of coffee as positive flavour notes disappear and negative notes/off-flavours prevail.

We will start this chapter with the reasons for the instability of flavour and fragrance (ingredients). Subsequently we will deal with methods to prevent flavour and fragrance degradation like precursors, encapsulation and analogues. At the end, we will discuss two publications more extensively to show how flavour companies investigate flavour instability.

10.2 Flavour stability

Flavours are complex mixtures of natural and/or synthetic ingredients. Those ingredients can either be single ingredients, like benzyl acetate, or mixtures, like a vanilla bean extract. The flavour industry is producing these flavours to improve the flavour of (processed) food.

Nowadays, consumers like to smell the taste of the food as if it was prepared in the conventional way but they do not want to spend the time in the kitchen to obtain such a result. The food manufacturer needs to be able to supply a product that has a long shelf life and has the same palatability over this time period. And since the flavour of the product is one of the main determining factors of the consumer's satisfaction, flavour manufacturers need to know what happens with the natural flavour of the food, as well as what happens with the added flavour as such and when incorporated in the product.

As already described by Grab [1], one has to look after:

- the stability of the flavour itself;
- its stability in food;
- flavour changes during processing;
- flavour changes during storage.

Moreover, one has to realize that there are physical parameters like:

- evaporation of the volatile components;
- crystallization of non-soluble material (only in liquid flavours);
- phase separation (for example, in emulsions);
- solution (for example, in fat-containing food);
- absorption and absorption effects (with a complicated food matrix).

and chemical parameters like:

- reactions with food components;
- reactions of flavour components (for example, through degradation, rearrangement and/or oxidation).

The stability of the (added) flavour itself is mainly determined by chemical parameters. With liquid flavours, the solvent is very important. Alcohols like ethanol and propylene glycol are commonly used. However, one has to realize that alcohols can react with aldehydes and ketones to form acetals and ketals (see Chapter 2). Usually, these acetals/ketals have different odour properties than the original keto-compound. Triacetin (glycerol tri-acetate) is another solvent, and because this is an ester, it may trans-esterify with other esters, formate esters most easily, or with alcohols with the release of acetic acid.

2-Acetyl pyrroline (1) is responsible for the characteristic note of several roasted products. However, its intrinsic stability is so low that flavour companies cannot use this ingredient in commercial flavours; the molecule is gone before the food has reached the consumer.

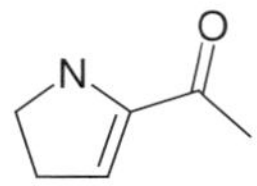

2-Acetylpyrroline
(**1**)

Flavour ingredients that are known to react relatively easily are the thiols (also described by Simon Jameson in Chapter 6). Thiols are oxidized to disulfides and can loose their characteristic smell. Aldehydes and amines can react to imines and it is well known that proteins present in a food/flavour can capture aldehydes (Scheme 10.1).

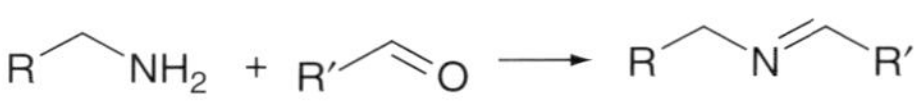

Scheme 10.1 Imine formation.

In coffee it has been shown that the melanoidins, the macromolecular dark brown pigments in coffee, play a crucial role in the staling of coffee [2]. The common belief was that coffee staling occurred by the oxidation of volatile thiols, mainly furfuryl thiol, to form disulfides with each other or with cystein residues of proteinaceous material in the coffee. Hofmann and Schieberle showed that reactive pyrazynium ions react in a catalytic redox cycle with the thiols to form dihydropyrazine-furfuryl thiol adducts (Scheme 10.2).

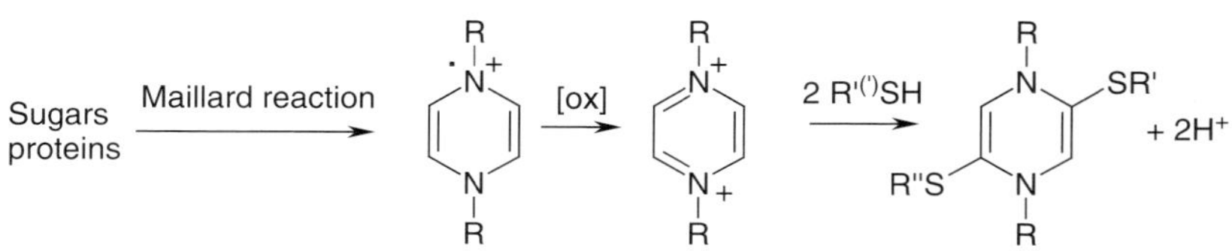

Scheme 10.2 Simplified reaction scheme to explain removal of volatile thiols in coffee.

A special type of flavour ingredients are the process flavours. Process flavours are a separate group in the EU Food laws and they are generated when ingredients normally occurring in food, like fats, fatty acids, sugars, amino acids, peptides or proteins are reacted together at higher temperatures. Especially, the reaction between sugars and amino acids is of crucial importance to foods. This so-called Maillard reaction [3] is responsible for the taste and colour of, especially, baked goods like the crust of bread or the crust of a piece of meat. The Maillard reaction is a very complex reaction and many hundreds of different molecules are formed. It is therefore not difficult to understand that a process flavour changes continuously over time. And, as a consequence it is almost impossible to keep a process flavour stable in solution. Normally, process flavours are indeed spray-dried until use. In the second part of this chapter, a case study on the stability of sulfur-containing ingredients is described.

The stability of the flavour in the food is an even more complex issue. Proteins can react with aldehydes, as mentioned before. But, in general, the physical condition of the food needs to get the most attention. A low pH, as in most beverages, will help acid-catalysed reactions to occur like hydrolysis of esters in an acid and an alcohol or acid-catalysed rearrangements like citral into *p*-cymene (Scheme 10.3).

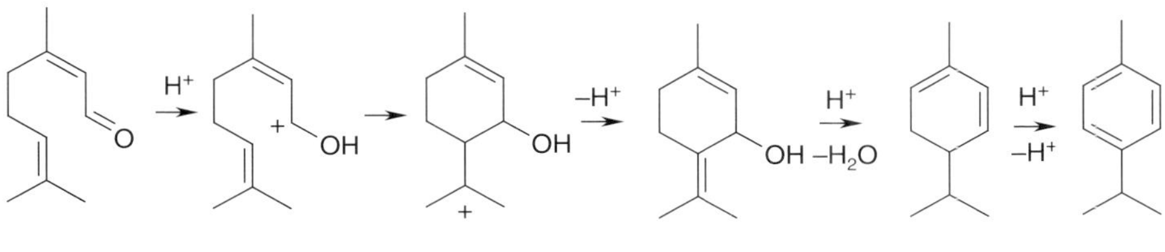

Scheme 10.3 Acid-catalysed degradation of citral.

A high oxygen concentration will make fat-containing products rancid. If one wants to make use of anti-oxidants one has to realize that both vitamins C and E are anti-oxidants but the first one only dissolves in water and the latter dissolves only in fat. From the above, it is clear that processing is key for the flavour of the product. Sensorially, one can hardly believe that strawberry jam is made from fresh strawberries; their tastes are so different [4].

A new trend in food processing is aseptic flavour dosing; the sterile flavour is only added after the food processing steps to reduce its degradation [5]. A strawberry jam with the flavour of fresh strawberries will be quite a change.

The flavour stability in the food will also be determined by the packaging. A classic example here is the so-called sun-struck flavour, 3-methyl-2-butene-1-thiol (3-MBT or prenyl mercaptan) in beer [6]. In the proposed pathway for 3-MBT formation hop-derived isohumulones are decomposed to 3-methyl-2-butenyl radicals due to sunlight exposure. Sulfur-containing amino acids and proteins decompose to SH radicals through riboflavin-photosensitized reactions. These two radical types then combine and form 3-MBT (see Scheme 10.4).

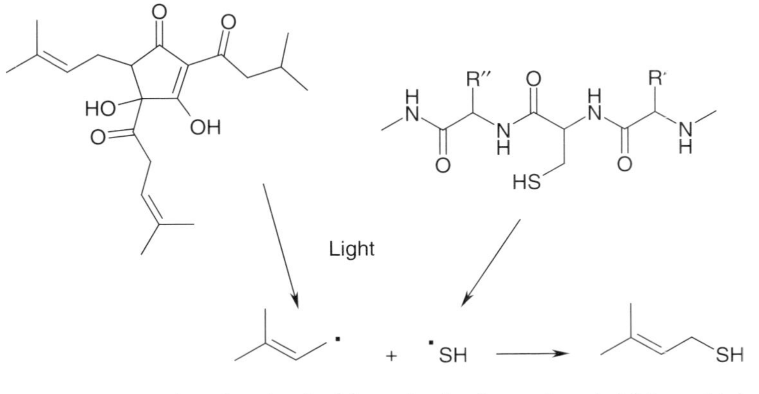

Scheme 10.4 Light-induced radical formation leading to 3-methyl-2-butenthiol.

Glass bottles are usually impermeable for oxygen. Plastic bottles have the advantage of being light and unbreakable; however, the oxygen permeability is usually much higher.

10.3 Flavour precursors

As described above, flavour ingredients are instable, and a food technologist needs to have tricks to make sure that the best flavour is obtained at the moment the consumer eats the product. One way to do this is by physically separating the flavour (ingredient) from its environment. This is called encapsulation and is described in the next paragraph. This paragraph describes chemical means to protect flavour ingredients. In the flavour industry we normally speak about flavour precursors; in fragrances we normally speak about pro-perfumes (see also Chapter 13). The technical approach in flavours and fragrances is the same; find molecules that, after a physical trigger, release the desired volatile molecule. Triggers can be pH and temperature but also more sophisticated triggers like enzymes, oxygen, light or even pressure change could be used theoretically.

To obtain such molecules, we can either look at how nature does this or come up with purely synthetic approaches. In foods, the latter is not an easy way as food laws usually require flavours to contain molecules with a smell, and flavour precursors usually do not have that. Moreover, new molecules need extensive toxicity testing. Still, flavour companies find it worthwhile to invest in this type of research.

Common flavour precursors in nature are alliin, as a precursor for allicin and other disulfides in garlic [7], and vanillin-O-glucoside, in green vanilla beans [8]. Both precursor and volatile are depicted in Scheme 10.5. Both precursors are quite ideal as they have no smell themselves, and nature has developed efficient release mechanisms to generate aroma. In garlic there is the enzyme allinase, and in vanilla beans there is the enzyme glucosidase, although this is still under scientific discussion [8].

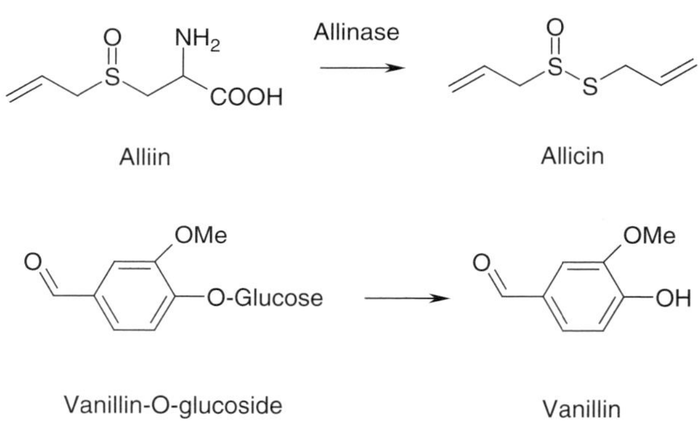

Scheme 10.5 Natural precursor in garlic and vanilla beans.

Givaudan has investigated a series of nature-identical flavour precursor systems [9]. As already mentioned for vanillin, nature has chosen more often glycosides as flavour precursor, and Givaudan has tested their commercial applicability. Sugars have been used as aldehyde precursor but it turned out that tartrate acetals of aldehydes had better precursor properties. (See (2), (3) for a structure). Also, the Amadori rearrangement product is said to be a useful precursor. This first, relatively stable condensation product of a sugar and an amino acid in the Maillard reaction can be used to boost the Maillard reaction. Especially in microwave-heated foods, in which the temperature is relatively low and the water content relatively high, the Maillard reaction does not run very spontaneously. If you help the Maillard reaction by adding an intermediate/precursor, the food will get a better taste.

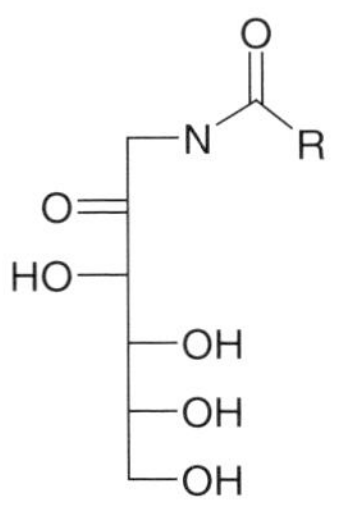

Amadori rearrangement product

(**2**)

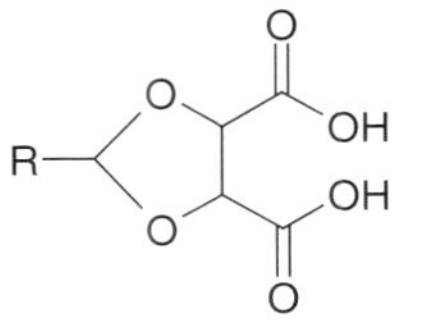

Tartrate-protected aldehyde

(**3**)

Thiocarbonates are artificial ingredients that have been claimed as thiol precursors (see (4) for a structure). Unilever [10] and Quest [11] have used different esters for different applications. Unilever used tertiary esters to release thiols in instant soups whereas Quest claimed primary esters to help a thiol survive a sterilization process.

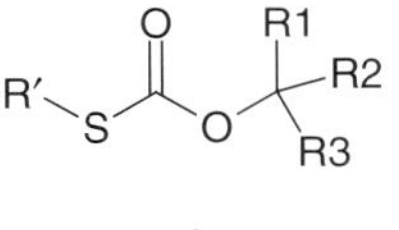

(**4**)

10.4 Encapsulation

As described earlier, flavours tend to degrade over time. If one would be able to protect the flavour against its hostile environment, the degradation could be stopped or certainly slowed down. Encapsulation is a technique in which a physical barrier between the flavour and its environment is made. Before consumption, the encapsulation material should have released the flavour, of course.

Several good books [12, 13] and articles [14, 15] have been published in this area. A simplified drawing of two of the most commonly used encapsulation types is given in Figure 10.1.

There are five reasons for the encapsulation of flavours:

(1) to prevent oxidation (by air oxygen);
(2) to prevent hydrolysis (by water);
(3) to prevent reaction of flavour ingredients with each other;

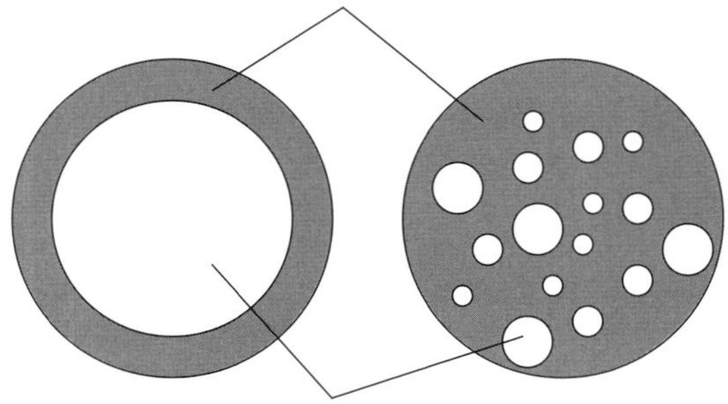

Figure 10.1 Schematic diagrams of two representative types of flavour encapsulates.

(4) to prevent evaporation;
(5) to change (difficult to handle) liquids into (free-flowing) powders.

Encapsulation traditionally has been used to protect flavours but current research is focusing on delivery systems. In the past one wanted to store a powder until it was dissolved; nowadays the encapsulates help the food technologist to deliver the flavour at the right time, at the right quantity, at the right rate. One often speaks about slow release mechanisms. Chewing gum is a nice example in which, usually, different encapsulation techniques are used to guarantee a more or less constant taste over a longer period.

Encapsulation is currently an active field of research. Quest has filed a patent [16] in which they claim that the addition of smaller sugars significantly helps to protect the encapsulated flavour against oxidation. Shelf life of more than two years can be obtained.

A nice story to tell is that customers smelling the new encapsulated flavours were disappointed because they did not smell anything. Only when it was told that the flavour encapsulation in the powder was so effective that no flavour material could escape from the matrix, the customer became enthusiastic. The powder showed normal behaviour when used in application.

The production of encapsulates can be done in many ways but the main routes are spray-drying and extrusion. In spray-drying, first a pre-mix is made in which 'solvent', flavour, encapsulating agent and, sometimes, some additives are mixed and homogenized. Subsequently the solvent, usually water, is removed by injecting the liquid through a nozzle in a hot expansion vessel under reduced pressure. For extrusion, the encapsulating material must be able to melt at high temperature to get effective mixing with the flavour.

The encapsulating material is usually polysaccharide-based as the carrier material must be (at least partly) water-soluble. Most commonly used materials are hydrophobically modified starches, although gums are also used.

It is good to realize that (liquid) emulsions can also be used to stabilize a molecule. When the instable molecule is in the other phase, than where the cause for its instability is, one can increase the shelf life of the product. A good example is an emulsion of a citrus oil in a lemonade.

10.5 Analogues

The statement of Professor Grosch at the beginning of this chapter can also be seen as a call to come up with 200 stable molecules that cover the whole flavour palette. So far, nobody has even come close to such a collection. But people are looking at structural units in a molecule that give the same flavour impression. It is well known that nitriles usually have similar smells as the analogous aldehydes. And since nitriles are much more stable than aldehydes, they could be good replacements. Due to the fact that the smells are not identical and many nitriles are artificial, nitriles have not got general application in flavours.

2-Formyl-thiophene smells very similar to benzaldehyde, so one could argue that an S-atom can replace a *cis* double bond. Indeed, Boelens *et al.* [17] indicate that *cis*-3-hexenol and ethyl 2-hydroxy-ethyl sulfide smell similarly.

Analogues seem an attractive route to search for more stable molecules, and this actually happens much more heavily in fragrance research. However, the complexity of the human smell sensation is so complex that it is hardly impossible to change anything in a molecule without the nose smelling it. Quite a few studies have been published on structure–function relationships of aroma chemicals but the predictability was only valid within a small group of molecules of similar structure. And to take an example from fragrance molecules: no programme or experienced perfumer would have ever predicted that nitro-musks and macrocyclic musks smell similarly [30]. In (5), (6), (7) it is clearly seen how different those structures are. Either by still hitting the same smell receptors or, just by luck, generating the same signal to the brain these molecules have similar smells.

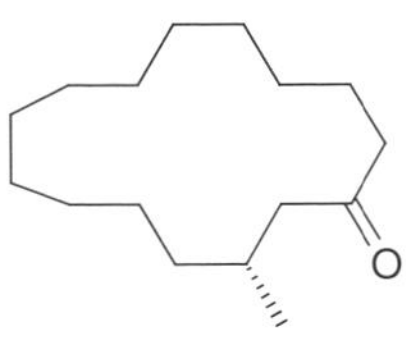

Muscone

(**5**)

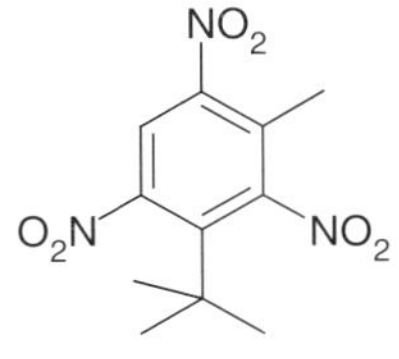

Musk baur®

(**6**)

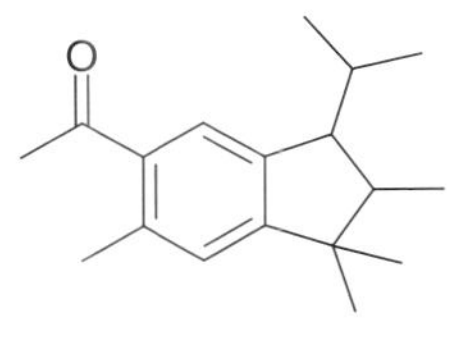

Traseolide®

(**7**)

10.6 Case study 1: citral and vanillin stability in milk-based products [18]

Gubler has published a nice piece of work that shows how complex a food system is and, therefore, how difficult it is to predict the behaviour of a flavour (ingredient). He shows that two aldehydes, citral and vanillin behave differently in a milk base. In (8), (9) their structures are drawn.

Citral is the character impact component in fresh lemon juice. When lemon juice is used in sodas or dairy products, it only survives for a few days. Gubler has shown that

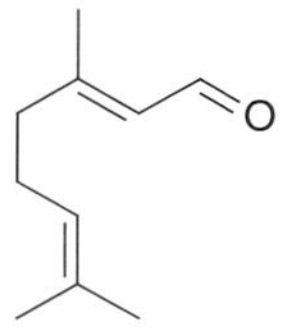

Citral
(**8**)

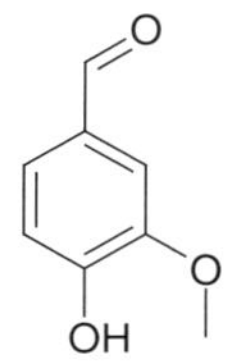

Vanillin
(**9**)

indeed pH is a major factor in citral degradation in dairy products but that other factors play a role as well. Casein, a milk protein, caused an even quicker citral degradation, probably due to the reaction of the aldehyde group in citral with the amino groups in the casein. Lactose, to a limited extent, and butter fat, to a significant extent, increase the stability of citral. Citral will probably dissolve in the milk fat droplets and is less accessible to acids. Why lactose is protective is unclear; may be because of reversible acetal formation.

Another remarkable result was obtained when synthetic citral and citral, in a cold-pressed oil, were compared. The synthetic citral was much more instable; 50% was degraded in a 5% casein in water solution at 45°C in 0.5 days whereas 50% of the natural citral in the oil was degraded after 3 days. The oil effectively protects the citral to be degraded.

As an example in real food Gubler looked at ice-cream. They were surprised again by the fact that now the citral in the 5% casein/water solution, containing 8% fat, was the most stable system. Apparently, fat, water and casein are 'crystallized' in three different phases and the citral is dissolved and protected by the fat.

Comparing citral and vanillin in the same ice-cream mix showed that vanillin was clearly more stable than citral. However, when other proteins were used differences started to occur. Citral became more stable both in the model systems and in the ice-cream mix when the casein was replaced by soy protein. However, vanillin became much more instable in the presence of soy protein. Vanillin is just as unstable in a soy isolate ice-cream mix as citral in a casein ice-cream mix. Gubler does not give an explanation for these phenomena.

10.7 Case study 2: stability of thiols in an aqueous process flavouring [19]

Process flavourings, which are generally prepared by thermal treatment of an aqueous solution of reducing sugars and amino acids, show considerable flavour instability in aqueous solution. To overcome this instability, process flavours are generally spray-dried. The resulting powder has sufficient flavour stability, if temperature and humidity do not get too high. The stability in water is so limited however, that a great deal of effort in planning is required, to prevent significant flavour loss, before the product is spray-dried. In particular, thiols, which are essential for a good meaty character of savoury process flavours, are particularly unstable. This urged us to look in more detail at flavour instability of a model thiol containing process flavour in aqueous solution.

Thermal treatment of a solution of ribose and cysteine results in the formation of several specific character impact components with a meaty overall impression [20, 21, 22]. The use of cysteine in combination with a pentose is responsible for the formation of very potent sulfur-containing meat flavour compounds, such as 2-methyl-3-furanthiol (MFT), 3-mercapto-2-pentanone (3MP), 2-mercapto-3-butanone (MB) and 2-furfurylthiol (FFT) [20, 21]. Some of these compounds were similar to those identified in beef [23] and were recently quantified in several types of heated meat [24](see Table 10.1). In the study described here, the ribose + cysteine-based process flavour reported by Hofmann was used [2, 20] as a model for meaty process flavours. As thiols are known for their reactivity/instability such as oxidation and involvement in nucleophilic or radical reactions, we assumed that the degradation of these compounds is a likely cause of the limited shelf life or flavour instability of savoury process flavours.

Table 10.1 Stability at 50°C of five character impact components of the aqueous model process flavour (entry 1) and reconstituted models (entries 2–5). Influence of ribose and cysteine on the decrease in concentration (% per day)

Compound	Decrease in concentration (% per day)				
	1[a] model process flavouring	2[b] no addition	3[b] +250 mM ribose	4[b] +150 mM cysteine	5[b] +150 mM cysteine +250 mM ribose
2-Mercapto-3-butanone (MB)	<20	>90	>90	<30	<30
2-Methyl-3-furanthiol (MFT)	<60	>90	>90	<80	<50
2-Furfurylthiol (FFT)	n.d.[c]	>90	>90	<50	<50
4-Hydroxy-2,5-dimethyl-3 (2*H*)-furanone (HDF)	<10	<10	<10	>90	<70
3-Hydroxy-4,5-dimethyl-2(5*H*)-furanone (sotolone)	n.d.	<10	<10	<30	<10

[a] According to the procedure of Hofmann [21] a phosphate buffered (0.5 M, pH 5.0) solution of D-ribose (100 mM) and L-cysteine (33 mM) was heated in an autoclave from room temperature to 130°C in 10 min and was kept at 130°C for 20 min. Character impact components were formed in the order of concentrations of around 5 μM.
[b] 50 μM of each character impact component present in aqueous phosphate buffered solution (0.5 M; pH 5.0).
[c] Not determined.

The interaction of thiol and disulfide flavour compounds with food components has been studied by Mottram [25], who indicated that these compounds undergo redox reactions with proteins. Hofmann [26] and Guth [27] reported model studies on the oxidative stability of odour-active thiols and disulfides, respectively. Eisenreich [28, 29] has reported studies on antioxidative activities of volatile sulfur-containing heterocyclic compounds confirming their reactivity in free-radical oxidation reactions.

The study described in this paper was designed to determine the (in)stability of the most important character impact components in an aqueous solution of the model process flavour described by Hofmann [20, 21], and to investigate why some of the character impact components are unstable.

The stability in an accelerated storage test at 50°C of some important flavour compounds in the aqueous model process flavouring is shown in Table 10.1 (entry 1) [21, 22]. MFT was found to be the most instable component. The furanones 4-hydroxy-2,5-dimethyl-3(2*H*)-furanone (HDF) and 3-hydroxy-4,5-dimethyl-2(5*H*)-furanone (sotolone) were relatively stable. Furthermore, an interesting observation was the fact that there was almost no difference between storage of the aqueous model process flavouring under air or argon atmosphere (data not presented here), probably due to an anti-oxidative effect of the matrix. The anti-oxidative properties of the matrix components cysteine and products formed in the Maillard reaction are well known. Entry 2 shows the results obtained from a storage experiment of a reconstituted mixture of flavour compounds in aqueous solution, using concentrations that are approximately ten fold higher than in the original process flavour. The concentrations were increased to allow more facile analysis of the components of interest.

We anticipated that the presence of residual cysteine and/or ribose could have an effect on the stability of some of the character impact components of the model process flavour. The effect of additional cysteine and ribose on the stability of the flavour impact components was therefore investigated in the reconstituted mixture (entries 3–5). Residual cysteine and ribose concentrations were determined in a freshly prepared model process flavour, and were found to be 15 mM, and 25 mM, respectively.

Entries 3 and 4 of Table 10.1 show the results obtained when the reconstituted mixture was exposed to accelerated storage conditions with addition of ribose and cysteine, respectively. Comparison of entry 2 with entry 3 shows that ribose addition has no effect on stability. A stabilizing effect of the addition of cysteine was clearly present (entry 4). Comparison of entry 4 with entry 5 shows that, when both ribose and cysteine are present (entry 5), an even larger stabilizing effect on MFT and the furanones can be observed, than with cysteine alone.

Because of the stabilizing effect of cysteine on the model process flavouring, the stability of the reconstituted aqueous mixture with character impact components in concentrations of 50 μM was investigated at various cysteine concentrations. Figure 10.2 shows the stabilities of MB, FFT and MFT at various cysteine concentrations. Interestingly, the decrease in concentration of MFT versus cysteine concentration graph has a clear minimum, which is different from what is observed for MB and FFT. The stability of MFT shows an optimum at a cysteine concentration of around 50 mM.

Based on these results, we suggest that the observed stabilization of MFT by additional ribose (see Table 10.1; entries 4 and 5) was caused by the decrease in the cysteine concentration due to the formation of the thiazolidinecarboxylic acid derived from cysteine and ribose. The actual concentration of cysteine after addition of ribose (250 mM) will be lower than 150 mM, as a result of the formation of this thiazolidinecarboxylic acid.

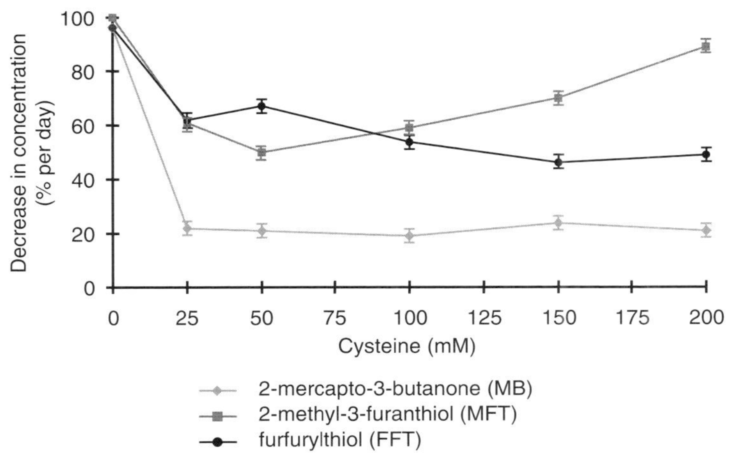

Figure 10.2 Stability of a reconstituted model process flavouring: influence of cysteine at pH 5.0 (0.5 M phosphate).

The results shown in Figure 10.2 suggest that the mechanism of degradation of MFT in the presence of cysteine is slightly different from that of MB and FFT. Apart from an oxidative degradation pathway via (mixed) disulfide formation, another pathway leading to non-volatile or polar degradation products by reaction of MFT with a.o. cysteine may be possible, which would account for the decreased stability of MFT at higher cysteine concentrations.

Evidence for the exceptional reactivity of MFT came from the storage of a mixture of only the three thiols, MB, FFT and MFT, in phosphate-buffered solution (0.5 M; pH 5.0) in the absence of cysteine. The amount of MB and FFT at the beginning was equal to the total amount of both thiols and (mixed) disulfides at the end of storage. This was absolutely not the case for MFT. After storage, a large amount of MFT was not detected, neither as thiol nor as (mixed) disulfide, when GC was used as analytical tool. When the same experiment was performed with 2,5-dimethylfuranthiol (DMFT), instead of MFT, the total amount of DMFT at the beginning was equal to the total amount of both DMFT and (mixed) disulfides at the end of storage (data not presented).

We think this can be explained by the greater ability of MFT to undergo oligomerization/polymerization. We know that MFT easily undergoes polymerization, when stored neat at ambient temperature, as some of the commercial samples already contain considerable amounts of polymeric material. Therefore, we hereby propose a hypothetical mechanism for the polymerization of MFT, as shown in Scheme 10.6, which can explain these observations. Protonation at the 2-position leads to an electrophilic species, which would easily react with a nucleophile. As the thiol group of MFT is a good nucleophile, protonation would catalyse dimerization/polymerization. Apart from the reaction of MFT with itself, other thiols, when present in a large enough concentration, could react as well, e.g. cysteine. This mechanism could explain the different effects of the cysteine concentration on the stability of MFT in comparison to other thiols such as MB and FFT (Figure 10.2). In the case of DMFT, polymerization at the 5-position would be more difficult due to steric hindrance caused by the 5-methyl group.

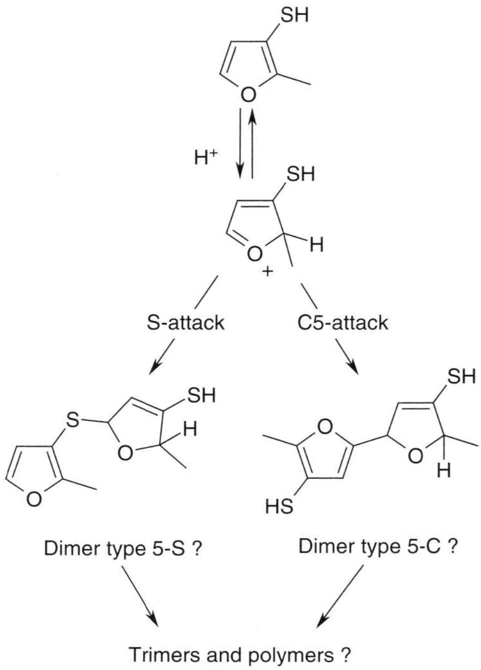

Scheme 10.6 Proposed mechanism for polymerization of 2-methyl-3-furanthiol.

If the left-hand part of the mechanism is of practical importance, coupling with other thiols would be expected as well. Since in the reconstituted mixture experiments no loss of the other thiols was observed, we must conclude that this part of the proposed mechanism is not significant.

To confirm our hypothesis of the degradation mechanism of MFT in aqueous solution, a deuterium exchange experiment with MFT was performed. Using ^{1}H-NMR, some proton–deuterium exchange was observed at the 4-position, but to a much larger extent at the 5-position (Figure 10.3) when MFT was kept in CH_3OD solution containing DCl at room temperature. This is in agreement with the preference of electrophilic substitution at the position next to the hetero-atom of aromatic heterocycles. However, except for the formation of the corresponding disulfide, no evidence for the formation of dimers was found, as was expected according to the mechanism in Figure 10.2. An explanation

4-position: H/D-exchange of 10%

5-position: H/D-exchange of 85%

(thiol/disulfide 5-6/1)

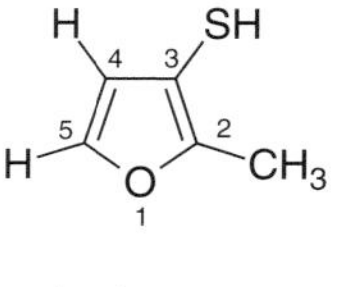

Figure 10.3 Deuterium exchange experiment with 2-methyl-3-furanthiol in CH_3OD/DCl for 64 days at room temperature.

for this could be that in the model NMR experiment a methanolic solution had to be used instead of an aqueous phosphate buffered (0.5 M; pH 5) solution.

To obtain deuterium exchange data for the 2-position of 3-furanthiols, the synthesis of 3-furanthiol was performed according to Scheme 10.7 [21]. Deuterium exchange experiments with this compound could provide evidence for the proposed mechanism of the degradation of MFT (**4**). Some difficulties were encountered during the isolation of 3-furanthiol. The purity of the isolated compound was not satisfactory (^{1}H-NMR), possibly caused by the inherent instability of 3-furanthiol.

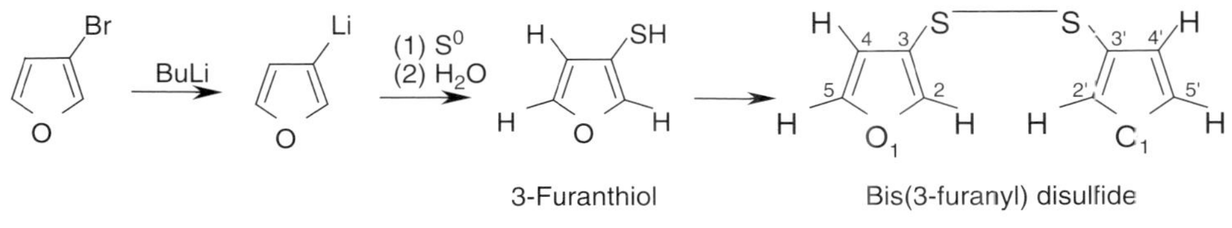

Scheme 10.7 Synthesis of bis(3-furanyl) disulfide.

However, bis(3-furanyl) disulfide was isolated in a pure form and was used for the deuterium exchange experiment. The high exchange rate at the 2- and 2′-positions was remarkable, when compared to that of the other positions (Figure 10.4). The preference for exchange at the 2- and 2′-positions is the result of the directing and activating effect of the S-atom linked to C-3 and C-3′. These results suggest that 2-protonation is facile in MFT, and is possibly a crucial step in the degradation of MFT, as shown in Scheme 10.6. Addition of nucleophiles to the 5-position of protonated MFT will most likely proceed via an ionic mechanism. Nevertheless, other mechanisms (e.g. a radical mechanism) could play a role in the degradation of MFT.

Bis(3-furanyl) disulfide: deuterium exchange in CH_3OD/DCl (8 days, RT):

2,2′-position: H/D-exchange of 71%

4,4′-position: H/D-exchange of 17%

5,5′-position: H/D-exchange of 10%

Figure 10.4 Deuterium exchange experiment with bis(3-furanyl) disulfide.

The data presented indicate that in meat process flavours, based on cysteine and ribose, 2-methyl-3-furanthiol is the one of least stable character impact components. The instability is not due to disulfide formation, but appears to result from electrophilic coupling reactions, resulting in products that cannot be detected by GC. H–D exchange showed that H-5 exchanges much faster than any other proton in MFT, suggesting that the reactivity of the 5-position is responsible for the rapid decomposition of MFT. This is in agreement with the lower stability of MFT in comparison with 2,5-dimethyl-3-furanthiol. Residual cysteine stabilizes MFT and other thiols.

10.8 Stability and fragrance applications

In the previous sections I have discussed the stability of flavour ingredients. However, most of the comments made can also be applied to fragrance applications. Of course, there are some differences as well and they will be shortly commented here.

In foods the pH varies between 2.5 for beverages, like ice-tea and cola, and 6–7 in naturally occurring foods. In fragrances, the pH range is much broader as acid cleaners have a pH as low as 1, and a bleach-containing detergent can have a pH as high as 12. A list is shown in Table 10.2. Therefore, the stability requirements for fragrance molecules are more severe. Citral is also here a nice example. Citrus smell is also desired in air fresheners and WC-cleaners. In WC-cleaners citral is not stable. Fragrance ingredients suppliers are still looking for stable citral replacers. Currently the nitrile analogue of citral is used but this molecule lacks the freshness of the natural citral and smells more 'synthetic'. Other molecules have been suggested (Figure 10.4) but none of them is the ideal replacer.

Table 10.2 pH range of fragrance-containing products

Application	pH
Acid cleaner	1
Cream rinse	3.5
Toilet cleaner	4
Deodorant stick	4.3
Fabric softener	4.5
Deodorant pump spray	6.6
Eau de Toilette	7.1
Hand cream	7.5
Liquid detergent	7.8
Shampoo	8.3
Talc	9.7
Bar soap	10.2
All purpose cleaner	10.4
Detergent powder	10.9
Compact detergent powder	11
Liquid bleach	11

This brings us to the second difference between flavours and fragrances: flavour molecules must be stable for a year as a maximum, fragrance molecules must be stable for at least five years. So also here, fragrance molecules need to survive more severe conditions.

Precursors, as they are usually called in flavours, or pro-perfumes, as they are usually called in fragrances, sometimes can have the same release trigger. A (nature-identical) acid-instable citral-precursor (suppose it would exist) can be used in ice-tea and in an acid cleaner. In pro-perfumes light is also considered as a release trigger as the perfume in a fabrics detergent is fully exposed to sunlight when the cloth is worn. In flavours, this would not work since food is never very long exposed to light (suppose again that a nature-identical precursor could be found).

The third difference is the production scale of flavour and fragrance molecules. A yearly production of over a tonne is quite a lot for a flavour molecule, and for fragrance molecules several tens of tonnes are usually produced. This has quite an influence on the production capabilities.

The fourth difference is the way a molecule is considered safe. In Europe a flavour molecule must be NI and in the US a flavour molecule needs to be GRAS (generally regarded as safe). For fragrance molecules the main requirement is that they do not cause allergic reactions. Moreover, the biodegradability needs to be investigated.

So, although the fragrance molecules must obey very many different requirements, fragrance scientists have a much greater flexibility in the choice of the molecules they make, compared with the flavour scientist. Any molecule that has the right smell, is stable in aqueous ethanol solution, is non-allergic and is biodegradable, can be used. In flavours, the European requirement for natural occurrence limits the possibilities of the flavour chemist in his choice of molecules. A good example here is the group of molecules that smell like musk [30]. Originally, natural musk molecules like muscone (see (**5**), (**6**), (**7**)) were used but these molecules were extracted from animalic sources and therefore were very expensive. At the end of the ninteenth century nitro-musks were found and used for commercial purposes but they turned out to be allergenic. In the 1950 a new class of musk compounds was introduced, the polycyclic musks (see (**5**), (**6**), (**7**)), and they quickly replaced the nitro musks. However, the polycyclic musks were not very biodegradable and the fragrance industry is now focused on making the natural, macrocyclic musks on a commercial scale synthetically.

10.9 Conclusion

The aim of this chapter was to show the huge amount of requirements flavour and fragrance molecules must have before they can be used in commercial preparations. For flavour and fragrance scientists, this is a very interesting and challenging time as these molecules are not easily found. As new European food laws focus on safety as the main requirement for a flavour molecule, the development work for new flavour and fragrance molecules will become more similar.

References

1. Grab, W. (1994) Trends in flavour research. In H. Maarse and D.G. van der Heij (eds), *The Stability of Industrial Flavours*, Elsevier Science BV, pp. 381–385.
2. Hofmann, T. and Schieberle, P. (2002) Chemical interactions between odor-active thiols and melanoidins involved in the aroma staling of coffee beverages. *J. Agric. Food Chem.*, **50**, 319–326.
3. Mottram, D.S. (1998) Flavour formation in foods. *Food Chem.*, **62**, 415–424.
4. Gimenez, J. *et al.* (2001) A study on the colour and sensory attributes of high-hydrostatic-pressure jams as compared with traditional jams. *J. Sci. Food Agric.*, **81**, 1228–1234.

5. Honer, C. (1994) Aseptic dosing. *Dairy Field*, **177**, 92.
6. Blockmans, C. *et al.* (1981) Photodegradation and formation of carbonyl and sulphur compounds in beer. Proceedings, European Brewery Convention, 347–357.
7. Block, E. (1992) The organosulfur chemistry of the genus Allium and its importance to the organic chemistry of sulfur. *Angew. Chemie Int. Ed. Engl.* **31**, 1135.
8. Dignum, M.J.W. *et al.* (2002) Vanilla curing under laboratory conditions. *Food Chem.*, **79**, 165–171.
9. Anderson, D.A. (1993) A matter of taste. *Chem. Technol.* **23**, 45–49.
10. van der Heijden, A. and Schutte, L. (1976) Foodstuff containing a diester of monothiocarbonic acid. US Pat. 3,978,240.
11. Fitz, W. *et al.* (2002) Thiocarbonates as flavour precursors. EU Pat. 1 170 295.
12. Vilstrup, P. (ed.) (2001) Microencapsulation of food ingredients. Leatherhead Publishing, ISBN 1 904007 08 2.
13. Risch, S.J. and Reineccius, G.A. (eds) (1995) *Encapsulation and Controlled Release of Food Ingredients*, American Chemical Society, Washington DC.
14. Bakan, J.A. (1973) Microencapsulation of foods and related products. *Food Technol.*, **27**, 34–44.
15. Dziezak, J.D. (1988) Microencapsulation and encapsulated ingredients. *Food Technol.*, **4**, 136–151.
16. Burger, J. *et al.* (2002) A moisture and oxygen stable composition and a process for obtaining said composition. EP 1 214 892.
17. Boelens, M. and Heydel, J. (1973) Chemische konstitution und geruch. *Chemiker Zeitung*, **97**, 6–14.
18. Gubler, B.A. (1981) Trends in flavour research. In Schreier, P. (ed.), *Stability of Flavouring Substances in Food Models Related to Milk*, de Gruyter, Berlin, **35**, pp. 717–728.
19. van Seeventer, P.B. *et al.* (2001) Stability of thiols in an aqueous process flavouring. *J. Agric. Food Chem.*, **49**, 4292–4295.
20. Hofmann, T. (1995) Charakterisierung intensiver geruchstoffe in kohlenhydrat/cystein-modelreaktionen und klärung von bildungswegen. Thesis, Technische Universität München.
21. Hofmann, T. and Schieberle, P. (1995) Evaluation of the key odorants in a thermally treated solution of ribose and cysteine by aroma extract dilution techniques. *J. Agric. Food Chem.*, **43**, 2187–2194.
22. Schieberle, P. and Hofmann, T. (1996) Untersuchungen zum einfluß von herstellungsparametern auf den aromabeitrag intensiver aromastoffe in cystein/kohlenhydrat – reaktionsmischungen. *Lebensmittelchemie*, **50**, 105–108.
23. Gasser, U. and Grosch, W. (1988) Identification of volatile flavour compounds with high aroma values from cooked beef. *Z. Lebebnsm. Unters. Forsch.*, **186**, 489–494.
24. Kerscher, R. and Grosch, W. (1998) Quantification of 2-methyl-3-furanthiol, 3-mercapto-2-pentanone, and 2-mercapto-3-pentanone in heated meat. *J. Agric. Food Chem.*, **46**, 1954–1958.
25. Mottram, D.S., Szauman-Szumski, C. and Dodson, A. (1996) Interaction of thiol and disulphide flavour compounds with food components. *J. Agric. Food Chem.*, **44**, 2349–2351.
26. Hofmann, T., Schieberle, P. and Grosch, W. (1996) Model studies on the oxidative stability of odor active thiols occurring in food flavours. *J. Agric. Food Chem.*, **44**, 251–255.
27. Guth, H., Hofmann, T., Schieberle, P. and Grosch, W. (1995) Model reactions on the stability of disulphides in heated foods. *J. Agric. Food Chem.*, **43**, 2199–2203.
28. Eisenreich, J.P. and Shibamoto, T. (1994) Antioxidative activity of volatile heterocyclic compounds. *J. Agric. Food Chem.*, **42**, 1060–1063.
29. Eisenreich, J.P., Wong, J.W. and Shibamoto, T. (1995) Antioxidative activities of furan- and thiophenethiols measured in lipid peroxidation systems and by tyrosyl radical scavenging assay. *J. Agric. Food Chem.*, **43**, 647–650.
30. Pybus, H.P. and Sell, C.S. (eds) (1999) *The Chemistry of Fragrances*, Royal Society of Chemistry, Cambridge.

Chapter 11
Rational Odorant Design

Luca Turin

11.1 Introduction

From its beginnings in the mid-nineteenth century, synthetic organic chemistry has discovered vast numbers of odorant molecules. The advantages of synthesis in cost and purity, as opposed to extraction, were apparent from the first. Houbigant's *Fougère Royale* (1881), thought to be the first fragrance containing a synthetic material, set a pattern for most great fragrances to follow; a skilful mixture of the natural (oakmoss) and the synthetic (coumarin) respectively, giving complexity and power per unit cost. (*Fougère Royale* can be smelled at the Osmothèque, the perfume museum attached to the ISIPCA school, in Versailles.) Synthetic coumarin added nothing new to the perfumer's palette, since the natural material was readily available and very pure already, though not cheap. The discovery of the ionones [1], by contrast, went further by replacing fabulously expensive naturals (violet flower absolute) by cheap aroma chemicals. A few years earlier, nitro musks [2] had gone further still and added an entirely new type of smell (the powdery musk) to the perfumer's palette. From then until today, the cheap and the new have been coming out in a continuous stream from the few, and getting fewer, companies involved in odorant discovery. The perfumer's palette now consists of approximately 4000 raw materials, approximately half of which are synthetic. How were they found? Five types of discovery processes account for all but a handful of materials.

(1) **Grind and find** Evolution's "four billion year R&D programme" has developed thousands of molecules chiefly intended to be attractive to animals. The fact that we share their tastes (particularly those of pollinating insects) to some extent has been a blessing to the fragrance industry. Analysis of the components of rose oil, for example, has yielded hundreds of interesting compounds (the damascones [3], in addition to its main components, phenylethyl alcohol, eugenol, geraniol and citral).
(2) **Imitation and development** Chemists are not bound (or helped) by the rules of biological synthesis, elegant though the latter are. Once a natural molecule is identified by painstaking analysis, synthetic organic chemists can design dozens, sometimes hundreds of variations, hoping to find some that are either interesting and cheap (e.g. seco-ionones [4]), or irresistible and expensive (ambrox [5]).
(3) **Serendipity** Synthetic organic chemistry is an art as well as a science, and the surprises are sometimes more interesting than the intended result. Famously, Baur's nitro musks were originally intended as explosives. But chance only favors the

prepared mind, and someone has to notice the serendipitous result. Materials like Osyrol® [6] and Karanal® [7] are the result of chance discoveries by fragrance chemists, followed by recognition and careful analysis.

(4) **Brute force** The availability of combinatorial chemistry methods from the pharmaceutical world [8] opens up the possibility of making novel odorants by mass synthesis. The usefulness of this method is limited by the necessity to assay each by a nose, and the difficulty in making sure that the main component, not some impurity, is responsible for the smell. There are a number of cases where trace impurities have proven to be the key odorant material, not the "named" product, and this has led to the discovery of important compounds. This approach is in its early days.

And lastly,

(5) **Rational design** In order to design odorants rationally, one needs to have some understanding of how smell character is written into the molecule. Odor character belongs to biology, i.e. it is a property of the molecule *as perceived* by the nose. Unlike the four previous approaches which relied on chemistry alone, this approach will only be as good as our understanding of human olfaction. Unfortunately, the latter is still sketchy, and as a result very little has been done in this area. Since this is the subject of this article, a brief review of olfactory reception is in order.

11.2 Theories of olfaction

For obvious reasons, in its early days, understanding of olfaction could not progress faster than that of molecular structure. Early theories of olfaction were mere speculation, and often assumed fanciful physical mechanisms such as rays emitted by odorant molecules, etc. In the late 1920s, one is surprised to find G.M. Dyson still patiently explaining to an audience of fragrance chemists the notion that there is no action at a distance involved in smell [9]. In 1946, an article by Linus Pauling [10] first hypothesized that shape was responsible for molecular interactions, and that the recognition of odorants was just one example among many. The idea was taken up by Moncrieff [11] and Amoore [12] two years later and its modified and refined form survives to this day. What has changed in the intervening years is that we have gained a large database of odors and structures, and a vastly better understanding of the ways in which a ligand can interact with a receptor. What has not changed, however, is our ignorance of the exact structure of the receptor, which makes proper modeling of SORs virtually impossible.

Nevertheless, until recently, a poll among specialists as to what gave a molecule its odor would have elicited a uniform answer – "shape" [13, 14]. By molecular shape is meant the location in space of the atoms of the molecule. By convention, with some sacrifice of physical realism, this usually means the lowest-energy position of the atoms, as determined by a suitable quantum chemistry computational package. The effect of molecular flexibility will be discussed below.

Until 1991, when the receptors were finally discovered [15], studies of the relationship between molecular shape and odor were made without reference to the biological sensor. The discovery that the olfactory receptors were seven-transmembrane helix proteins (7-TM) finally brought this problem into focus. When probed by a biological system, shape now translates into the sum total of all the repulsive and attractive interactions that a molecule feels when bound to a receptor: exchange repulsion plus hydrogen bond donors and acceptors, lone pairs, etc. The discovery of the receptors was widely expected to confirm the prevailing "shape" theory and to reduce odorant design to a branch of rational drug design. However, several unexpected factors intervened.

The first was the revival [16] of a hitherto discredited contender among olfaction theories – vibration. Dyson [17] and, later, Wright [18] had argued that there was a better correlation between the odor of a molecule and its vibrational spectrum than between odor and shape. After proving remarkably hardy in the face of opposition, this idea abandoned the field in the late 1970 when (a) it became clear that some enantiomer pairs have different odors (and identical vibrations when measured by a conventional spectroscope) and (b) the proponents of vibrational theories failed to come up with a mechanism by which vibrations might be sensed by a biological system. Turin [16] showed that both problems could be addressed and solved, at least in principle, and put this long-forgotten contender back into the arena.

The second was, in retrospect, that part of the attractiveness of "shape" theories (there are two variations, to be discussed below) lay chiefly in the absence of competition. The effect of this scientific "one-party system" was to discourage a critical look at shape theory itself. The revival of vibration theory, regardless of its intrinsic merits, has led to a critical re-examination of shape theory in the light of (a) new experimental evidence on receptor selectivity (b) a large number of discrepant facts which had accumulated over time and (c) the growing realization that since nobody seemed to be designing odorants rationally by shape, the theory could not be *that* good. Turin and Yoshii [19] set out the pros and cons of shape and vibration theories in a recent review. Their arguments, and others which have recently come to light, will now be summarized.

11.2.1 For shape

(1) **Plausibility** Most-known molecular recognition mechanisms (including, it now appears, a classic exception – general anaesthesia [20]) require a fit between receptor and agonist, enzyme and substrate, or antibody and antigen. In receptor mechanisms, this fit is not, as it is for enzymes, a prelude to chemistry, but the very event that (sometimes) leads to receptor activation.

(2) **The nature of the receptors** The fact that olfactory receptors are 7-TM proteins linked to a G-protein cascade suggests that a shape-based mechanism similar to, say, that of other 7-TM types such as adrenergic receptors and rhodopsin [21] is a good place to start.

11.2.2 Against shape

(1) **No predictive ability** Note that I refer here to predictions of odorant *character*, not intensity. As pointed out in a recent review [19], intensity has everything to do with molecular shape, and some good correlations have been found. Odor character is a different story; despite 150 years of fragrance chemistry and a vast database of odorants, structure–character relations are still in "a sorry state" [14]. Reviews [13]

of SORs are essentially catalogues of exceptions. Accordingly, discovery of odorants is still "empirical", i.e. proceeds by trial and error. Claims of odorant discovery by "rational" design based on shape are rarely backed up by disclosure of the complete set of odorants tried before hitting the correct one.

(2) **Isosteric molecules smell different** There are many examples in the literature of molecules of very similar shapes having very different smells. The most striking is undoubtedly the series of molecules investigated by Wannagat and coworkers [22] in which carbon was replaced by Si, Ge and Sn. Despite possessing very similar geometries, these molecules smell different, and not only between C and Si, but also between Si and Ge or Sn where the size change is a great deal smaller.

(3) **"Strong" shape looks unlikely** In "strong" shape theories, every odorant fits one receptor with very high affinity and the others with little or not at all. This one-odorant, one-receptor idea is incompatible with the fact that there are many more (tens of thousands) known odorants, *each of which has a distinctive odor* than receptors (347 according to the latest count [23], some of which may in fact reside in other tissues). Further, whenever they have been tested, and with only few exceptions so far, odorant receptors appear to respond to a broad spectrum of odorants [24].

(4) **"Weak" shape appears untestable** "Weak" shape comes in two versions. The first is "odotope" theory, wherein parts of the molecule are felt by the receptors and the odor, like the proverbial elephant described by blind observers, is the sum of the parts. The other is a low-affinity theory in which odorants are "swallowed whole" by the receptors, but, on average, offer a poor fit. Once again, the pattern of receptor activation determines odor. But odotopes are "hidden variables" and hard to infer, and receptor structure remains unknown at the kind of resolution needed to make sense of structure–odor relations. Further uncertainty is added by the conformational flexibility of the vast majority of odorant molecules, which makes odotopes literally moving targets. It is hard to think of a conclusive experiment to test these theories. This does not mean that they are false, but weakens, in principle, the case that can be made in their favor.

(5) **The chiral receptor problem** Weak and strong shape theories suffer from a problem due to the chiral nature of both the odotopes and the proteins which detect them. Consider an odorant bearing four (generally chiral) odotopes. It will be perceived by four (generally chiral) receptors, and the receptor activation pattern will determine odor character. Now consider the enantiomer of the odorant; each of the four odotopes will (generally) give a poor fit to the original receptors, and as a result, the enantiomer will be "recognized" by a *different* set of receptors.

One would therefore generally expect enantiomers to have completely *different* smells. This is emphatically not the case. In a compilation of 277 known enantiomer pairs [25] 59% are found to smell similar (one descriptor in common), 5% "identical" (all descriptors in common), 17% different (no descriptors in common) and the remainder 19% are unknown. Even if all the unknowns turned out to have completely different odors, however, *the majority of enantiomer pairs would still smell similar.*

Contrast this with the different picture from conventional pharmacology [26]. There, the difference in activity between eutomer [27] (the correct enantiomer) and distomer (the incorrect one) is frequently marked. An example is *R*- and *S*-propranolol which differ in potency on the *beta*-adrenergic receptor by a factor of 130. This case resembles the odorant enantiomer pairs which differ in intensity, though the problem for odorants is raised to the *N*th power if odorants activate *N* different receptors. But frequently, a pattern with no parallel in smell is found; for example, the enantiomers quinine and quinidine have different actions altogether, respectively, antimalarial and anti-arrythmic, while (+) and (−) dobutamine have opposing actions (agonist vs antagonist) on the same receptor. This brings up the following related problem.

(6) **No odorant antagonists have been found** A fundamental feature of shape-based pharmacology is the existence of antagonists. The mechanism by which receptor antagonists work is thought to be that they bind more tightly to the "off" than to the "on" state of the receptor [28]. Frequently, the difference in structure between agonists and antagonists is small, amounting to as little as the replacement of one hydrogen by a fluorine atom. To my knowledge, no such molecule has ever been found in the case of odorants. If it existed, it would not be hard to notice. Imagine a molecule derived, from, say, santalol, which had a weak (sandalwood) smell of its own or no smell, but which prevented you from smelling santalol alone for some time thereafter, or when mixed with it. The fact that no such thing has been found in any odor group indicates either that (a) there is a problem with the shape idea, (b) for some unknown reason the on–off states of smell receptors differ from those of others or (c) we haven't looked hard enough.

(7) **We smell functional groups** This puzzling feature of human olfaction has so far defied explanation. The most striking instance, and the best-known, is the -SH (thiol, mercaptan), which imparts to any molecule, *regardless of its shape*, a character appropriately called "sulfuraceous". Less familiar, but equally easy to detect by trained observers, are the smells of nitriles, oximes, isonitriles, isothiocyanates, etc. [29]. What makes this remarkable from a molecular recognition point of view is that these functional groups constitute the smallest odotopes, capable of making only one, at most two hydrogen bonds. How this translates into unfailingly accurate detection is hard to understand. For example, the fact that -OH (alcohols) never smell like -SH (thiols) at *any* concentration is hard to reconcile with known molecular recognition mechanisms. Again, the comparison with conventional pharmacology is telling. There, substitutions designed to modify binding to a given receptor are legion. For example, thiobarbiturates, where the purine carbonyl is replaced by a thiocarbonyl [30], and fluorinated benzodiazepines behave similarly to the parent compound. A chemist colleague humorously summed up this molecular recognition puzzle by saying "It's easy, sulfur is yellow and oxygen red."

11.2.3 For vibration

(1) **Functional group recognition** As any IR spectroscopist knows, functional group recognition from their distinctive stretch frequencies is straightforward [31]. The most remarkable instance of this is once again the -SH group which has a "unique" stretch frequency signature around 2550 wavenumbers. A prediction from the vibration theory is that any other group possessing the same vibration frequency should smell sulfuraceous. This is the case; only the B–H stretch vibration of boranes

falls in the same range, and boranes alone, among all compounds known to date, smell sulfuraceous, as was noticed as early as 1912 by Stock [32], the first to synthesize most of them. There is, of course, very little in common between BH and SH in terms of chemistry. Further, a vibrational theory accounts for *similarities* in smell between molecules. The well-known replacement of nitriles for aldehydes, for example, is understandable by the proximity of their stretch vibrational frequencies.

(2) **Isosteric molecules accounted for** In the case of the C-Si-Ge-Sn series, for example, the big differences in smell, despite similarity in shape, are easily accounted for, because almost every molecular vibration will be affected by the large changes in mass.

(3) **Isotopes smell different (maybe)** The ultimate test of a vibrational theory is a pair of molecules differing markedly in vibrations and not at all in shape. This theoretical goal is in fact impossible to attain. Zero-point molecular motion will always differ if two molecules have different vibrations, which means that the average "shape" will differ slightly. Nevertheless, one can come pretty close using isotope substitutions. The substitution of deuterium (mass 2) for hydrogen (mass 1) has the largest relative effect on vibrations involving H atoms, typically the various modes involving CH bonds. The experiment requires that the isotope substitution takes place on (a) non-exchangeable protons; otherwise D will exchange with H rapidly in the nose, (b) hydrogens not involved in H bonds, because these will be slightly affected by substitution. Remarkably, there is good evidence that insects and fishes can tell isotopes apart. Recently, a test on human subjects using fully deuterated benzaldehyde showed that subjects could tell it apart from the undeuterated form. Though tantalizing, these experiments suffer from the difficulty of making sure that small amounts of impurities are not affecting the smell. A GC-smelling test of deuterated acetophenone revealed a (small) difference with normal acetophenone [16]. The ideal would be a combination of GC-smelling and large observer set with randomized trials. Such an experiment is planned for the near future.

(4) **Enantiomers accounted for (differently)** In a vibrational theory, the diversity of receptors is due to them having to accommodate a vast number of unforeseen ligands. Nevertheless, drawing a parallel from color vision, one may expect these receptors to fall in a small number of classes spanning the vibrational spectrum, each class approx. 400 wavenumbers in width. Turin [16] has argued that the number of classes is probably less than 10. For enantiomers to smell identical, they would have to stimulate each receptor *class* to the same average extent. In the limit of an infinite number of receptors, therefore, all enantiomers would smell identical because they would always, in each class, find a receptor to which they bind equally tightly. But humans have 347 olfactory receptor types. Neglecting the possibility that some may not be in the nose at all, assume that each spectrum region is equally represented. That gives ca. 30 receptors *per class* to bind efficiently with the tens of thousands of known odorants. This line of reasoning leads to a surprising conclusion; the smell differences between enantiomers are a manifestation of the imperfection of the system. Turin [16] has suggested a mechanistic explanation of the difference in smell between the enantiomers of carvone. Given that a majority of enantiomers smell similar, the system is doing remarkably well. This idea nevertheless implies the existence of limits on odor prediction (see below).

(5) **The chiral limit** An advantage of vibrational theory is that it makes no recourse to "hidden variables" like odotopes. Molecules with similar vibrational spectra *as*

perceived by the nose should smell similar. Vibrational frequencies can be calculated, and an educated guess can be made of mode intensities perceived by an IETS spectroscope. This approach has been used to predict the smells of existing, well-documented odorants, with some success [33]. More recently, a much-expanded version has been used by Flexitral to predict novel odorants successfully.

The discussion in "Enantiomers accounted for (differently)", however, raises an interesting problem. Enantiomers must still bind to receptors, and from that standpoint, enantiomers are as different from each other as any two molecules can be. Therefore the probability, given a particular molecule with a particular odor, that a structurally *unrelated* one with identical vibrations will smell the same cannot be higher than for enantiomers, i.e. 5% or so, a fraction we call the "chiral limit". This problem will crop up whenever one is trying to improve on a known odorant and modify its chemical properties without altering its smell. It is probably best in that case to look for structural modifications that change shape as little as possible as well as preserving vibrations.

11.2.4 Against vibration

(1) **Mechanism novel and unproven** Turin [16] proposed that the receptors detect vibrations by a solid-state mechanism involving inelastic electron tunnelling. While the different parts of the mechanism involve only plausible chemistry and physics, it is fair to say that there is to date no direct evidence in favor of (or, for that matter against) this idea. The olfactory receptors are sufficiently different from other 7-TM receptors to make it at least possible that they have diverged to take advantage of this remarkable possibility. Experiments designed to test this idea directly are of course possible, but none have so far been devised.

(2) **Inability to account for odorant intensity** A feature of a vibrational theory is that to a first approximation, all molecules will have vibrational spectra of comparable intensity. To be sure, larger molecules will have more modes, and groups bearing intense charges may give larger peaks, but intensity differences of no more than a factor of, say, three can be accounted for in this fashion. Needless to say, this will not account for the range of odorant intensity which spans at least seven orders of magnitude. As Turin and Yoshii [19] have discussed, vibrational theory is a theory of odor *character* only. This means that odorant intensity is not part of odor character, and can only be due to the more or less tight binding of the odorant to the receptors.

11.3 Rational design by shape

If one narrowly defines "rational" to mean quantitative, then very little rational odorant *character* design by shape has been done to date. A recent comprehensive review of fragrance chemistry [14] illustrates odor classes with diagrams of 3D olfactophores (structural features necessary to a given odor type) obtained using Catalyst software. While emphasizing the use of "modern molecular similarity techniques", they nevertheless conclude that the finding of a new odorant is "based almost exclusively on a broad chemical knowledge, experience, fantasy and instinct". While these tools are arguably rational, they are certainly not quantitative. In general, it is hard to assess the success rate of shape-based synthetic programs for odorant discovery, for two reasons. First, successes are more likely to have been reported than failures. Second, even when the research is performed by one of the major manufacturers, it seems that researchers do not compare the "successful" shape to their extensive (proprietary) databases of molecular structures to see whether counterexamples can be found.

The evidence is therefore fragmentary. Many "rules" for the design of odorants, e.g. the triaxial rule for ambers [5], and steric rules for sandalwoods [34] have been described over the years, but they seem to be of limited validity insofar as counterexamples can always be found [13]. Similarly, structural changes which frequently leave odor "unchanged", such as the isobutenyl–phenyl replacement [34] have been described, but it is not clear how general they are. Turin [33] has argued that some can be accounted for vibrationally. A celebrated study by Sestanj [4] was among the first to report the successful (qualitatively) rational design of violet odorants (seco-ionones) by excision of half of the cyclohexene ring. A similar qualitative approach has been followed by Rossiter [35] in the design of muguet aromachemicals. This was followed more recently by a comprehensive study of violet odorants [36], in which, shape similarity was used to achieve successful predictions of novel violet molecules. The data in the latter study has been re-examined by vibration theory [33], and a reasonable fit of vibration theory to the published odor data was found, suggesting that the correlation between structure and odor may in this case go via vibrations.

By contrast, the relationship between odorant *intensity* and shape is well documented [13]. Most early studies took a classical QSAR approach, and used statistical techniques to fit the intensity of the odorants to a variety of molecular descriptors. More recently, greater attention has been paid to biologically plausible receptor-binding mechanisms. In particular, though it clearly does not constitute the whole answer [19], the notion first proposed by Turin [16] that binding to a zinc ion in the receptor is an important factor in odorant intensity is gaining ground [37, 38]. Although the 3D structure of the receptor is still unknown, recent studies have given more emphasis, than in the past, to the involvement of frontier orbitals Recent studies of musks [39], for example, appear to support the idea that odorant intensity depends on the position and prominence of certain electronic features. In view of the remarkable computational power now available on desktop computers, this area is clearly ripe for extensive study.

11.3.1 Rational design by vibration

Few things are more exciting (and, it must be said, scary) to a scientist than a chance to put theory to the test in practical use, all the more so since it is unusual for a theory to be reduced to practice before being validated scientifically. The creation of Flexitral, Inc. by private investors early in 2002 afforded me this opportunity. What follows is a progress report on our vibrational approach to rational odorant design, as candid as is compatible with the necessity to protect intellectual property. A description of an early attempt at calculating odor character has been given [33]. The research effort at Flexitral greatly extended and refined this approach, with equal emphasis on both character and intensity prediction.

11.4 Replacement molecules

Regardless of their merits and the soundness of the public-health strategy on which they are based, regulatory requirements are a fact of life for the fragrance industry. Recently, a number of staple aroma chemicals, among which are citral, isoeugenol, Lyral®, etc., have appeared on a list that advises restricting their use [40]. This is not a tragedy for perfumery; perfumers, like artists in general, are very adept at getting around regulations (think of piano concertos written for the left hand). Nevertheless, there is a pressing need for replacement molecules, particularly to be used in existing (legacy) fragrances.

11.4.1 Acitral®

Flexitral began with a conversation with flavorists in September 2001 that made us aware of the need for an acid-stable lemon. Citral (**1**), the main character-impact molecule of lemon oil, is unstable in acid media (e.g. lemonade) and cyclizes quite rapidly (weeks) to unpleasant-smelling compounds. The origin of this instability resides in the tail double bond of citral, and any replacement should seek to replace that feature with another that preserves odor as far as is possible. In the search for possible replacements, we came across the interesting possibility of replacing a double bond with a cyclopropane ring. Givaudan's Javanol® (**2**), a very powerful and superb-smelling sandalwood material, is derived from (**3**) by cyclopropanation. Bajgrowicz *et al.* [41] attribute the similarity in smell to the cyclopropane ring being "bioisosteric" to a double bond, though the exact meaning of the word is not defined. To be sure, there are interesting chemical similarities between cyclopropane rings and double bonds, not least the ability to conjugate to neighbouring carbonyls, etc. These similarities have led cyclopropane rings to be informally dubbed as "fat olefins".

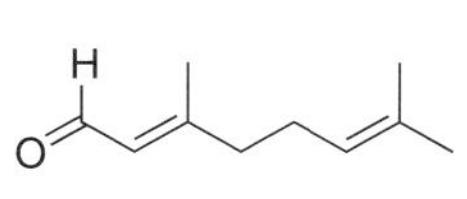

(**1**)

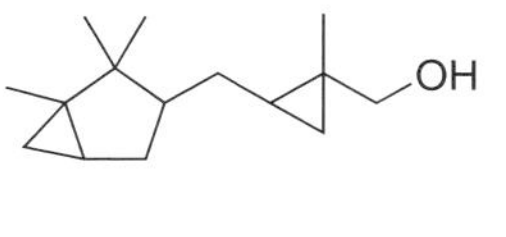

(**2**)

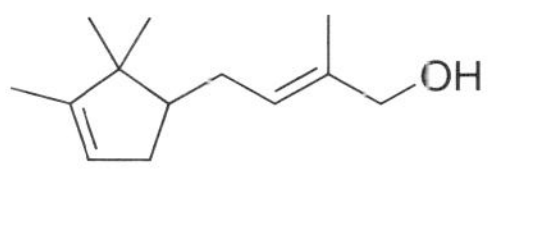

(**3**)

It seemed at least possible that the preservation of smell character might be due to vibrational, rather than steric or electronic features. Calculations showed that cyclopropanation of the tail double bond would preserve spectra. The *cis* and *trans* isomers of 5-(2,2-dimethylcyclopropyl)-3-methyl-2-pentenal (**4**) were synthesized from citral by cyclopropanation of the tail double bond and the reaction product assayed by GC olfactometry. The isomer mixture smelled lemony, though with a drier, slightly more aldehydic, powdery character than citral. The *trans* isomer (from geranial) appears to have a greener, more intense note than the *cis* isomer (from neral). In 2 M acid solutions, compared to citral which breaks down in minutes, acitral decays slowly (30% loss after 24 h) and gracefully (odorless or pleasant-smelling degradation products). Acitral's acute oral toxicity is low (LD >2000 mg/kg) and, subject to further regulatory approval, it is presently on its way to commercial use.

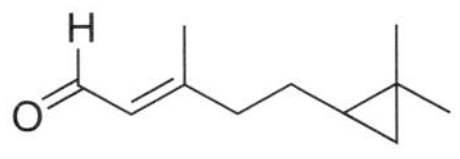

(**4**)

11.4.2 Lioral®

Two prominent materials on the EU list of potential allergens are the lily-of-the-valley odorants Lilial® (5) and Lyral® (6). Despite the abundant supply of related odorants, there appears to be a need for novel, powerful muguets with a soft odor profile. Here again, our strategy was to find vibration-conserving replacements. The resemblance in smell in between benzene and thiophene rings in various odorants has been noted many times. Inspired by Boelens' insightful reviews [42], we decided to test the possibility of replacing thiophene for benzene in cyclamenaldehyde (7) [43]. As has been pointed out many times, sulfur-containing odorants present unique problems with regard to control of impurities, because these can overwhelm the odor profile when present even in tiny amounts. Thiophene poses fewer problems than most, however, because of its remarkable chemical stability.

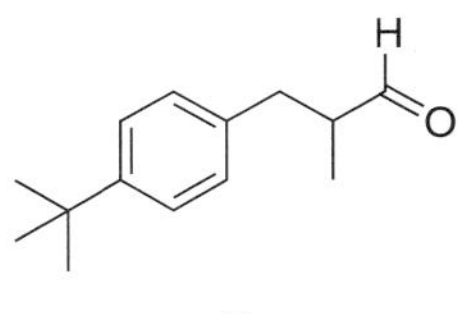

(**5**)

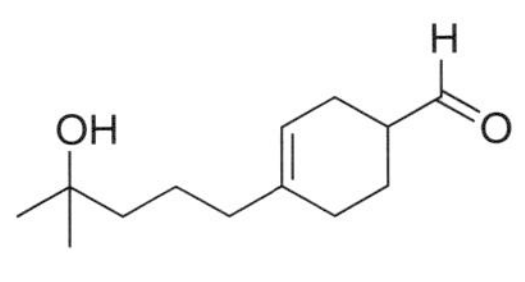

(**6**)

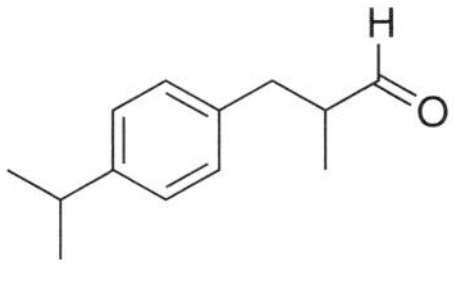

(**7**)

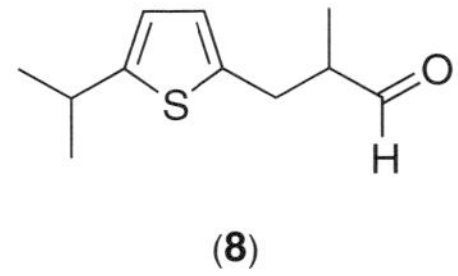

(**8**)

When faced with replacing a six-membered ring with a five-membered one, the question comes up of where to put the substituents. Fortunately, in this case, there was good agreement between ease of synthesis and computed odor profiles. A computational search through the substituted derivatives revealed that 3-(5-isopropylthiophen2-yl)-2-methylpropional (8) was the likeliest candidate and also the easiest to synthesize, given the reactivity of positions 2 and 5 on the thiophene ring. The resulting molecule was named Lioral® (from the Hebrew *li-or*, "light-giver"). It has a classic lily-of-the-valley profile, with a slight cuminic note and is devoid of aggressive "plastic" connotations. In potency it is comparable to Florhydral®.

11.5 Prospects for the future

A good theory is, among other things, a labor-saving device. In this respect, vibrational theory is impressive. Using it, Flexitral has developed three novel raw materials in its first six months of operation. In each case, hundreds of molecules were explored computationally but the number of syntheses required to achieve the desired odor profile was less than five. Since Jasphene was developed, we have, on behalf of clients, gone on to address other requirements of the industry and have come up, in a similar fashion, with five other products now in various stages of development and certification. It would seem that either we have been blessed with supernatural luck or there is some correspondence between our calculations and odor character.

The human element remains essential, because computational brute force, though vastly cheaper and faster than chemical brute force, is still an inefficient strategy. We improve our calculation tools continuously to broaden the range of structures explored, speed up computation, and achieve greater accuracy in odor character and intensity prediction. But we also try to make them fun to use, because most ideas come from playing. Tasks for the near future include pruning molecular search trees according to ease of synthesis and incorporating other properties such as substantivity and chemical stability in the criteria by which candidate molecules are searched.

Acknowledgements

I thank Jacquelin S. Grant, CEO, Flexitral, Inc. for her valuable advice and constant support.

References

1. Tiemann, F. and Kruger, P. (1893) *Berichte*, **26**, 2675.
2. Baur, A. (1889) German Pat. 47 559 May 21.
3. Demole, E., Enggist, P., Saeuberli, U., Stoll, M. and Kovats, E. (1970) Structure and synthesis of damascenone [2,6,6-trimethyl-1-(trans-crotonoyl)-1,3-cyclohexadiene], odorous constituent of Bulgarian rose oil (*Rosa damascena*). *Helv. Chim. Acta*, **53**, 541–551.
4. Sestanj, K. (1962) Synthesis of an acyclic analog of ionone. *Croatica Chemica Acta*, **34**, 211–217.
5. Ohloff, G. (1971) Relationship between odor sensation and stereochemistry of decalin ring compounds. *Gustation and Olfaction*, **54**, 7633–7703; Ohloff, G. (ed.) (1971) Relationship between odor sensation and stereochemistry of decalin ring compounds. In *Gustation and Olfaction, Int. Symp.*, Academic, London, England, 178–183.
6. Jones, B.N., Ansari, H.R., Jaggers, B.G. and Janes, J.F. (1973) Odorous 7-hydroxy-2-methoxy-2,6-dimethyloctane. *Chem. Abstr.*, **79**, 41918e.

7. Newman, C.P., Rossiter, K.J. and Sell, C.S. (1989) Preparation of dioxane derivatives as perfume components. *Chem. Abstr.*, **110**, 13401.
8. Nicolaou, K.C. *et al.* (eds) (2002) *Handbook of Combinatorial Chemistry: Drugs, Catalysts, Materials* (2-Vol. Set). Wiley-VCH.
9. Dyson, G.M. (1928) Some aspects of the vibration theory of odor. *Perfumery and Essent. Oil Record*, **19**, 456–459.
10. Pauling, L. (1946) Analogies between antibodies and simpler molecules. *Chemical & Engineering News*, **24**, 1064–1065.
11. Moncrieff, R.W. (1946) *Chemical Senses*. John Wiley, New York.
12. Amoore, J.E. (1970) *Molecular Basis of Odor*. Thomas, Springfield IL.
13. Rossiter, K.J. (1996) Structure–odor relationships. *Chem. Rev.*, **96**, 3201–3240.
14. Frater, G., Bajgrowicz, J.A. and Kraft, P. (1998) Fragrance chemistry. *Tetrahedron*, **54**, 7633–7703.
15. Buck, L. and Axel, R. (1991) A novel multigene family may encode odorant receptors: a molecular basis for odor recognition. *Cell*, Apr 5, **65**(1), 175–187.
16. Turin, L. (1996) A spectroscopic mechanism for primary olfactory reception. *Chem. Senses*, **21**(6), 773–791.
17. Dyson, G.M. (1938) The scientific basis of odor. *Chemistry* & *Industry* (London, United Kingdom) 647–651.
18. Wright, R.H. (1982) *The Sense of Smell*. CRC Press, Inc., Boca Raton, Florida, USA.
19. Turin, L. and Yoshii, F. (2003) Structure–odor relations: a modern perspective. In R.L. Doty (ed.), *Handbook of Olfaction and Gustation*, 2nd edn, Marcel Dekker, New York.
20. Franks, N.P. and Lieb, W.R. (1994) Molecular and cellular mechanisms of general anaesthesia. *Nature*. Feb 17, **367**(6464), 607–614.
21. Meng, E.C. and Bourne, H.R. (2001) Receptor activation: what does the rhodopsin structure tell us? *Trends Pharmacol. Sci.*, **22**, 587–593.
22. Wannagat, U., Damrath, V., Huch, V., Veith, M. and Harder, U. (1993) Sila-Riechstoffe und Riechstofffisostere XII. Geruchsvergleiche homologer organoelementverbindungen der vierten hauptgruppe (C, Si, Ge, Sn). *J. Organometallic Chem.*, **443**, 153–165.
23. Zozulya, S., Echeverri, F. and Nguyen, T. (2001) The human olfactory receptor repertoire. *Genome Biol.*, **2**(6), RESEARCH0018. Epub 2001 Jun 01.
24. Malnic, B., Hirono, J., Sato, T. and Buck, L.B. (1999) Combinatorial receptor codes for odors. *Cell*, **96**, 713–723.
25. http://www.leffingwell.com/chirality/chirality.htm.
26. Kennedy, I.J. and Jane, D.E. (2002) Stereochemistry in drug design. In *Textbook of Drug Design and Discovery*, 3rd edn, pp. 53–85.
27. Ariens, E.J. (1986) Chirality in bioactive agents and its pitfalls. *Trends Pharmacol. Sci.*, **7**, 200–205.
28. Colquhoun, D. (1998) Binding, gating, affinity and efficacy: the interpretation of structure–activity relationships for agonists and of the effects of mutating receptors. *Brit. J. Pharmacol.*, **125**, 923–947.
29. Klopping, H.L. (1971) Olfactory theories and the odors of small molecules. *J. Agri. Food Chem.*, **19**, 999–1004.
30. Goodman, L.S. *et al.* (2001) *The Pharmacological Basis of Therapeutics*. McGraw-Hill.
31. Socrates, G. (2001) *Infrared and Raman Characteristic Group Frequencies: Tables and Charts*, 3rd edn. John Wiley & Sons, New York.
32. Stock, A. and Massenez, C. (1913) Boron hydrides, *Ber.*, **45**, 3539–3568.
33. Turin, L. (2002) A method for the calculation of odor character from molecular structure. *J. Theor. Biol.*, **216**, 367–385.

34. Buchbauer, G., Hillisch, A., Mraz, K. and Wolschann, P. (1994) Conformational parameters of the sandalwood-odor activity: conformational calculations on sandalwood odor. Part X. *Helv. Chim. Acta*, 77, 2286–2296.
35. Rossiter, K.J. (1997) The design and synthesis of novel muguet materials. *Special Publication – Royal Society of Chemistry*, **214**(Flavours and Fragrances), 21–35.
36. Kraft, P. (1999) Design and synthesis of novel violet odorants with bicyclo[6.4.0]dodecene and bicyclo[5.4.0] undecene skeletons. *Synthesis*, **4**, 695–703.
37. Rakow, N.A. and Suslick, K.S. (2000) A colorimetric sensor array for odour visualization. *Nature*, Aug 17, **406**(6797), 710–713.
38. Wang, J., Luthey-Schulten, Z.A., Suslick, K.S. (2003) Is the olfactory receptor a metalloprotein? *Proc. Natl. Acad. Sci. USA*, **100**, 3035–3039.
39. Ishikawa, Y. and Kishi, K. (2000) Molecular orbital approach to possible discrimination of musk odor intensity. *Int. J. Quantum Chem.*, **79**, 109–119.
40. EU document SCCNFP/0017/98, Final December 1999 http://europa.eu.int/comm/food/fs/sc/sccp/out98_en.pdf.
41. Bajgrowicz, J.A., Frank, I., Frater, G. and Hennig, M. (1998) Synthesis and structure elucidation of a new potent sandalwood-oil substitute. *Helv. Chim. Acta*, **81**, 1349–1358.
42. Boelens, H. (1975) Relationships between the chemical structure of compounds and their olfactive properties. *Cosmetics and Perfumery*, **89**, 1–7.
43. Knorr, A. and Weissenborn, A. (1929) US Pat. 1,844,013.

Chapter 12
Applications I: Flavors

David Baines and Jack Knights

12.1 Introduction

12.1.1 The early days of flavour analysis

At the start of the 1970s there were just under 1500 flavour chemicals identified in food products. By the year 2000 this number had risen to over 7000, with over 1000 identified in coffee alone. Flavour chemicals were found to be ubiquitous, occurring widely across food groups, and individual foods were found to contain hundreds of flavour chemicals. In each food, only a few make a significant contribution to flavour, this being determined by their concentration in the food and by their flavour threshold. The flavour of food products, therefore, very much depends on the balance of the flavour chemicals present but in some cases the real character is determined by the presence of key aroma compounds, often referred to as character impact compounds, without which it would be impossible to reproduce the flavour. It would be impossible, for example, to create nature-identical vanilla flavours without vanillin or nature-identical orange flavours without *n*-decanal.

The growth in the identification of flavour chemicals over the past five decades was a direct result of the development of new analytical techniques. The combination of gas chromatography and mass spectrometry (GC-MS) has played a pivotal role since the mid-1950s. In these early days, it was thought that these techniques would spell the end of flavour creation by exposing the ingredients used in all flavourings. In the event, the opposite happened by the discovery of a vast array of new flavour compounds of novel utility. It was not until the mid-1970s, when the data being generated by mass spectrometry was computerised, that these emerging technologies could be fully exploited. Prior to the use of computers, the analysis of mass spectral data was painstaking and laborious. Fragmentation and molecular ions produced by the mass spectrometer were measured as abundance signals on UV-sensitive paper, and the mass of each ion determined by counting signal peaks by eye. The molecular ions derived from nitrogen, oxygen and carbon dioxide from traces of air in the ion chamber were used as markers. Ion peak heights were measured by ruler and the mass spectra drawn by hand. The potential flavour compounds were identified either by comparison with tables, which bore no resemblance to today's vast mass spectral libraries, or more often by interpreting the spectra to deduce the most likely chemical structure, followed by proof through chemical synthesis.

Computerised mass spectrometry changed this overnight and significantly reduced the time required to analyse mass spectra, but more importantly uncovered information generated by the mass spectrometer that had hitherto been hidden in the complexity of the data. The technology heralded an era of discovery, as nature was peeled open and the secrets of flavour chemistry revealed. The major flavour companies invested heavily in fundamental research, using the new technologies of gas chromatography and computerised mass spectrometry, in order to be the first to discover and, sometimes, patent new flavour chemicals. Those occurring in foods that are harvested and eaten without further processing, such as fruits and vegetables, were identified first. It took considerably longer for flavour chemists to come to terms with the vast number of chemicals formed in the Maillard and similar reactions, as food is cooked or as flavours are formed through fermentation and other traditional food processes. As the technology evolved and became more sensitive, and as extraction methods such as headspace analysis became more sophisticated, flavour chemists could dig ever deeper and identify many of the more elusive flavour chemicals with very low thresholds.

Eventually, the law of diminishing returns dictated that the investment needed to discover new and even more elusive flavour compounds was no longer worthwhile, and the era of discovery came to an end. Legislation also played its part in determining the research strategies of flavour companies, as they were required to submit their lists of flavour chemicals to the European Commission for compilation into a positive list of flavouring substances. The final list is not published yet, but over 2800 flavour compounds have been submitted. Any new ones which are discovered and which are found to be useful in flavourings will have to be submitted to the Commission with appropriate toxicological data in order to be added to the published list. Clearly, the commercial return on research investment aimed at discovering new flavour chemicals in foods is now so low that it has largely been abandoned.

Unfortunately, there still remain many flavour chemicals to be discovered because flavour scientists have concentrated their efforts on the study of individual foods such as beef, tea, coffee, cheese, onion, etc., which have commercial significance for flavour companies. But, of course, many foods are cooked together as stews, casseroles, soups, sauces, bouillon's, pies, baked goods – the list is endless. The flavour precursors derived from each raw material may interact to produce new flavour chemicals which contribute to flavour and give many dishes their distinctive character.

12.1.2 The role of the flavourist

If the challenge of analysis or 'taking nature to pieces' has largely passed, the job of the flavourist, the 'piecing together of nature' to re-create flavour in all its guises, not only continues but has become more involved and complex. Formulating flavours using volatile organic chemicals is a skill that takes many years to acquire. Flavourists have to memorise, recognise and familiarise themselves with an enormous number of flavour chemicals, understand their flavour profiles at different concentrations, and their chemical and sensory interactions with other flavour chemicals and flavour carriers.

It not only requires experience and understanding to produce commercially successful flavours but also needs that extra spark of creativity. To overcome the complexity of the millions of permutations possible in formulating a flavour and create what the customer or the market expects in the shortest possible time scale the flavourist must rely on experience, judgement, flair and intuition to be successful.

The flavourist also needs to understand the food into which the flavour will be applied and the processing and storage conditions to which it will be subjected to select the carriers and release agents needed to protect the flavour and make it available and enjoyable at the point of consumption. To apply the enormous range of flavour chemicals with such wide variations in thresholds, polarity, volatility and stability, the flavourist must have a sound understanding of solvents and of methods of converting liquid products into free-flowing powders and encapsulates.

12.2 Liquid flavourings

12.2.1 Water-soluble liquid flavourings

Permitted solvents for use in the formulation of water-soluble liquid flavours are listed in Table 12.1.

Probably, the most widely used solvent for flavourings is propylene glycol, which satisfies the technical and commercial requirements of many flavour systems, being a good solvent for most flavour compounds and being relatively inexpensive. It forms acetals and ketals with carbonyl compounds, which can be an advantage in some cases, conferring extra protection against oxidation, but in other cases the acetals or ketals may be insoluble in propylene glycol, and two layers may form in the flavour. Hydrolysis of acetals and ketals takes place in an aqueous food product or in the mouth during consumption, and water can be used in the flavour formulation to inhibit their formation. The propylene glycol level in the final food should not exceed 3 g/kg in the EU to meet the requirements of the amended legislation 95/2/EC [1]. This gives ample scope for the flavouring of most food products.

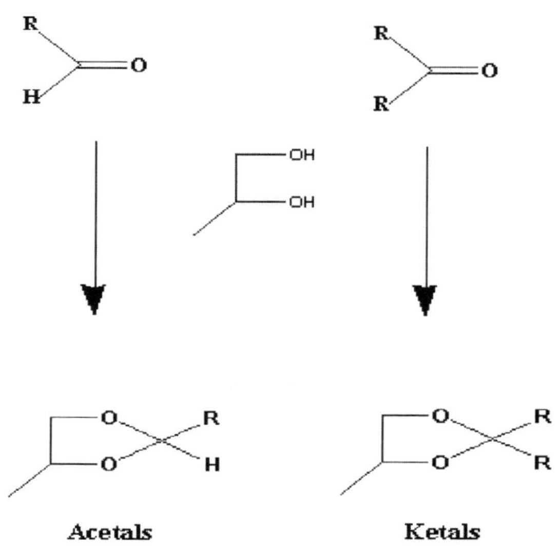

Scheme 12.1 Acetal and ketal formation.

Table 12.1 Permitted solvents for water-soluble flavour formulations

Permitted solvents	Advantages	Disadvantages
Water	Cheap, readily available	Poor solvent for most flavouring ingredients
Propylene glycol (Propan-1,2-diol) (E1520)	Fairly cheap, readily available, widely acceptable, low volatility	Poor solvent for citrus oils, limited acceptability in some countries
Glycerol (Propan-1,2,3-triol) (E422)	Cheap, readily available, low volatility	Poor solvent for most flavouring ingredients, normal grades not vegetarian, Kosher or Halal
Ethanol	Fairly cheap, widely accepted, good solvent for flavour ingredients, bactericidal at high concentrations	Highly volatile, flammable, not Halal
Isopropyl alcohol (Propan-2-ol)	Fairly cheap, good solvent for flavour ingredients, bactericidal at lower concentrations, Halal	Highly volatile, flammable, EU status doubtful, not acceptable in some countries as a food solvent, e.g. USA
Diacetin (Glyceryl diacetate) (E1517)	Fairly cheap, moderate solvent for flavour materials, does not react with aldehydes, low volatility and flammability	Fairly easily hydrolysed to acetic acid, normal grades not vegetarian, Kosher or Halal, not widely acceptable, e.g. USA
Triacetin (Glyceryl triacetate) (E1518)	Fairly cheap, good solvent for flavour materials, does not react with aldehydes, low volatility and flammability	Basically oil-soluble (10% soluble in water), normal grades not vegetarian, Kosher or Halal

Ethanol is the second solvent of choice, especially in countries where excise duty is not a requirement. Religious restrictions are a growing barrier to its use, and like propylene glycol, it forms acetals and ketals which are inhibited by the presence of water. Unlike propylene glycol, it is natural and is an excellent bactericide, inhibiting the growth of microorganisms in natural products. The other aliphatic alcohol used as a flavour solvent, isopropyl alcohol, has a flash point near to room temperature and its vapours represent an explosion hazard. It is a good solvent for essential oils but possesses an ethereal, acetone type of aroma, which is unpleasant, and not a good medium to present a flavour to a potential customer. It was extensively used in countries where excise duty on ethyl alcohol was a problem but doubts over its toxicity and its other negative attributes are limiting its use, especially in the EU.

Glycerol or glycerine is a natural product but unfortunately has poor solvent properties for many flavour compounds. It can be used in combination with ethanol and has a sweet taste about 0.6 times that of cane sugar and is non-cariogenic. It is a useful solvent for vanillin and vanilla extracts, and because of its heat stability it finds use in flavours for baked goods. It can also be used to create consistent pastes for Maillard-based process flavours and to provide 'body' in alcoholic beverages.

Diacetin (glyceryl diacetate) and triacetin (glyceryl triacetate) find use when there is a restriction on the use of propylene glycol or where it is vital to avoid acetal formation. Both are miscible with ethanol and with oils. Triacetin possesses a mild sweet taste at levels less than 500 ppm, but above this it is bitter and may also have an acetic acid character, depending on the extent of decomposition. Diacetin, by comparison, has no perceptible flavour at normal use levels. These two solvents find greatest use in chewing gum flavours where they act as plasticisers, unlike propylene glycol, which has an undesirable effect on chewing gum texture. The levels of these two glyceryl acetates should not exceed 3 g/kg in finished food products [1].

12.2.2 Solvents for special uses

Citrus oils contain varying proportions of terpene hydrocarbons which are insoluble in water, creating problems for their application in soft drinks and other clear aqueous-based solutions. Hydrocarbons separate in the drink forming an unsightly ring at the top of the bottle and an oil layer on the surface. To create a water-soluble orange flavour, for example, based on the essential oil, the terpene hydrocarbons, especially limonene which is present at between 90 and 95% of the oil, must be removed to leave the more soluble and flavoursome oxygenated monoterpenes and sesquiterpenes. Deterpenation of essential oils can be achieved through distillation, column chromatography, supercritical carbon dioxide extraction or by the simple use of solvents. The solvent method (washing) is usually the most successful because it retains the flavour chemicals responsible for the fresh juicy character of citrus oils. A simple solvent system used for this purpose is:

Ethanol or isopropanol	60–70%
Water	40–30%

The solvent blend and essential oil are thoroughly mixed and in some cases chilled, and the mixture separates. The top layer containing the terpene hydrocarbons is removed to yield a bottom layer consisting of a solution of the deterpenated essential oil.

Fruit juices are often supplemented with natural or nature-identical topnote blends that are often insoluble in the juice. Carrier solvents are used to aid dispersion and dissolve the topnote blend in the juice. A useful mixture is:

Propylene glycol	70%
Ethanol or isopropanol	20%
Water	10%

The juice is mixed with water, and propylene glycol is added. The topnote flavour is dissolved in the alcohol and added to the aqueous juice blend with gentle agitation.

12.2.3 Oil-soluble liquid flavourings

Permitted solvents for use in the formulation of oil-soluble liquid flavours are listed in Table 12.2.

Hydrogenated oils and fats come in many forms but probably the most utilised fractionated solvents are those derived from coconut oil. Commercially they are blends of C6–C10 triglycerides and are a reasonable solvents for many lipophilic flavour applications.

Triethyl citrate is practically insoluble in water and possesses a mild winey, fruity plum-like odour and is commonly used for carrying berry, butterscotch, cherry, honey and other flavours for application to baked goods [2]. The taste is bitter at concentrations above 500 ppm and this restricts its use in many food applications. Legislation limits its application to less than 3 g/kg but at this level it would be far too bitter for most food products.

Benzyl alcohol has a rather sharp, chemical taste with a slightly sweet, faintly floral aroma. It is used as a flavouring substance in many fruit- and floral-type flavours. As

Table 12.2 Permitted solvents for oil-soluble flavour formulations

Permitted solvents	Advantages	Disadvantages
Natural hydrogenated oils and fats	Natural, normal ingredients of foodstuffs	Poor solubility for some ingredients. Stabilised by antioxidants. May contain *trans* fatty acids. Concerns about allergic responses
Triethyl citrate (E1505)	High stability, good solvent properties	Expensive, bitter taste
Triacetin (E1518)	Good solvent properties, high boiling, low flammability, water-soluble at 10%	Expensive, poor stability – hydrolysed to acetic acid in aqueous media
Benzyl alcohol (E1519)	High stability, good solvent properties, high boiling, low flammability	Expensive and has inherent taste
Sorbitan esters (E491–495)	Good solvent properties, high boiling, low flammability, good water-in-oil emulsifiers	Very expensive, poor legislative acceptability in EU
Polyoxyethylene sorbitan esters (E432–436)	Good solvent properties, high boiling, low flammability, good oil-in-water emulsifiers	Very expensive, poor legislative acceptability in EU

a carrier solvent, its utility has been curtailed in the EU by the 2002 amendment to the Miscellaneous Additives Directive. Its application level in non-alcoholic-flavoured drinks is restricted to 50 mg/kg, in liquors and aromatised wine products to 100 mg/kg and in confectionery and fine bakery wares to 250 mg/kg.

Sorbitan esters and polyoxyethylene sorbitan esters (polysorbates) are emulsifiers and surfactants with varying oil-in-water and water-in-oil properties and are available as stearates, laurates, oleates and palmitates. Sorbitan esters are lipophilic non-ionic emulsifiers and the polysorbates are non-ionic hydrophilic surfactants. The sorbitan esters are used to hold aqueous solutions in suspension in fatty materials, and find use in ice cream, fat spreads and desserts. They are also used as antifoaming agents for boiled sweets and preserves and to inhibit the development of 'bloom' in chocolate. The polysorbates are used for forming oil-in-water emulsions in products such as dressings, sauces and margarines. They are also used to assist the dispersion of coffee whiteners and improve the volume and texture of cakes. Both sorbitan esters and polysorbates are excellent solvents for flavourings and are sometimes used as the vehicle to deliver flavour to food products that utilise emulsifiers.

12.2.4 Emulsion liquid flavourings

Emulsion liquid flavourings are used in the preparation of cloudy, flavoured soft drinks to ensure that essential oils remain evenly distributed throughout the drink over the shelf-life period. The essential oils have a lower density than the water in the drink, and will separate to the top of the container without the presence of a 'weighting' agent, so called because they hold an essential oil in suspension by increasing its density. Weighting agents in use are:

- Sucrose acetate isobutyrate (SAIB) (E444);
- Glyceryl esters of wood rosins (rosin esters) (E445);
- Deodorised gum resins – pine, dammar and benzoin;
- Deodorised waxes – beeswax and carnauba;
- Brominated vegetable oil (BVO) – not permitted in the EU.

The weighting agents act in conjunction with water-phase additives to further stabilise the emulsion such as gum arabic, cellulose esters and permitted emulsifiers.

12.3 Powder flavours

Flavour manufacturers make use of a number of processes for the conversion of flavourings from liquid to solid form. There are many reasons for expending time and cost to achieve this:

- Improved handling;
- Stabilisation of volatile organic compounds to minimise losses;
- Encapsulation of flavour;

- Controlled flavour release;
- Extended product shelf life and improved storage options;
- Flavours protected from oxidation;
- Create barrier between two or more incompatible ingredients;
- Create barrier between the flavour and the ingredients in the food product.

The main technologies for producing powdered flavours used by the industry are:

- Plating;
- Spray drying;
- Spray cooling;
- Coacervation;
- Melt extrusion;
- Molecular encapsulation.

The new technology of yeast encapsulation, which involves the bioencapsulation of flavours in the protective outer shell of yeast cells, is also covered in Section 12.3.4.

12.3.1 Plating

Plating is the oldest and simplest process for converting a liquid flavouring into a solid and was originally developed to convert oleoresins, liquid spice extracts and oils such as onion oil into convenient powdered products for application to meat seasonings and directly to food products. The liquid to be plated is added gradually to a suitable support material being activated in a mixer, and a number of mixer designs can be used with varying degrees of efficiency. The screw auger mixer is the least efficient design for plating liquids. This mixer is an inverted cone, and the material being plated is lifted by a vertical rotating screw from the apex of the cone and returned to the apex by gravity. Segregation of larger particles can take place around the periphery of the mixer and agglomerates can form, and this reduces the evenness of the liquid coating and the consistency of the finished product. The 'Z' blade mixer and the ubiquitous ribbon blender, the most common of all dry powder blenders, are horizontal rotating screw designs, and both can be used successfully to plate liquids and both have a long history of use in the production of standardised spices. The most effective design for spreading a variety of liquids onto a particulate support is the high-shear ploughshare mixer [3]. This mixer has a horizontal mixing drum with plough-shaped shovels rotating on a central shaft to produce a three-dimensional movement of the powder. Choppers can be fitted on the inside surface of the mixing drum to disperse any agglomerates being formed as the liquid is being added but this has the disadvantage that it can cause attrition to friable support materials. High-speed impeller mixers can also be used and are described by Patterson [4] as an effective method for producing dispersed standardised spice oleoresin powders. The criteria for selecting a mixer is listed by King [5] and depends on a number of factors such as: efficiency of mixing, mixing time, volume throughput, discharge mechanism, friability of products being mixed, dust and ease of cleaning.

Carrier materials in common use for plating liquids are salt, rusk, maltodextrins, dextrose and lactose, and these give flavour loadings ranging between 2 and 7%.

The main problem with plated products is the rapid loss of volatiles during storage. Salt-plated products are the worst affected because the flavour is dispersed over the surface of a crystalline structure. To alleviate this to some extent, a very fine grade of salt is used. With other carriers, some absorption takes place. Rusk is a baked product used extensively in sausage manufacturing to absorb both fat and water, and this capability can be put to good effect to absorb and retain flavours in plated dry powder products. Flow agents such as calcium silicate, tricalcium phosphate and silicon dioxide are used to control the final flow properties of plated products.

Properties of plated flavours and spices
The main advantages of plated flavours and spices are:

(1) Low production costs;
(2) Simple process;
(3) Easier handling of problem ingredients such as chilli and pepper oleoresins.

The major disadvantages are:

(1) Rapid loss of volatiles;
(2) Low flavour loads;
(3) Poor protection of flavour/ease of oxidation.

12.3.2 Spray drying

Spray drying is the most widely used method for the production of powdered flavours and is an essential technology for the flavourist to understand and use to reproduce in powder form, flavours created as liquid formulations. In the conversion of a liquid flavour into a powder using spray drying, consideration must be given to a number of factors including the use and concentration of flavour solvents, flavour volatile losses during drying, the active loading of flavour, flavour encapsulation and flavour release from the finished product. In addition, the physical parameters of the powder being produced such as particle size distribution, bulk density, dispersibility, hygroscopicity, flowability and appearance must also be considered. A good spray-dried encapsulated flavour will not have a strong aroma, will flow freely, have a good shelf life and will disperse evenly and release flavour into the target food matrix.

Spray drying is the atomisation of a liquid feedstock into a spray of droplets which contact hot air in a drying chamber and are converted into a powder. It can be considered as four sequential processes: emulsification, atomisation, drying and separation. Figure 12.1 shows a schematic diagram of the process.

12.3.2.1 Emulsification

Flavours based on blends of topnote chemicals and other ingredients such as essential oils are hydrophobic, and the formation of a successful emulsion is key to the production of an evenly dispersed powdered flavour and the retention of volatile flavour chemicals [6].

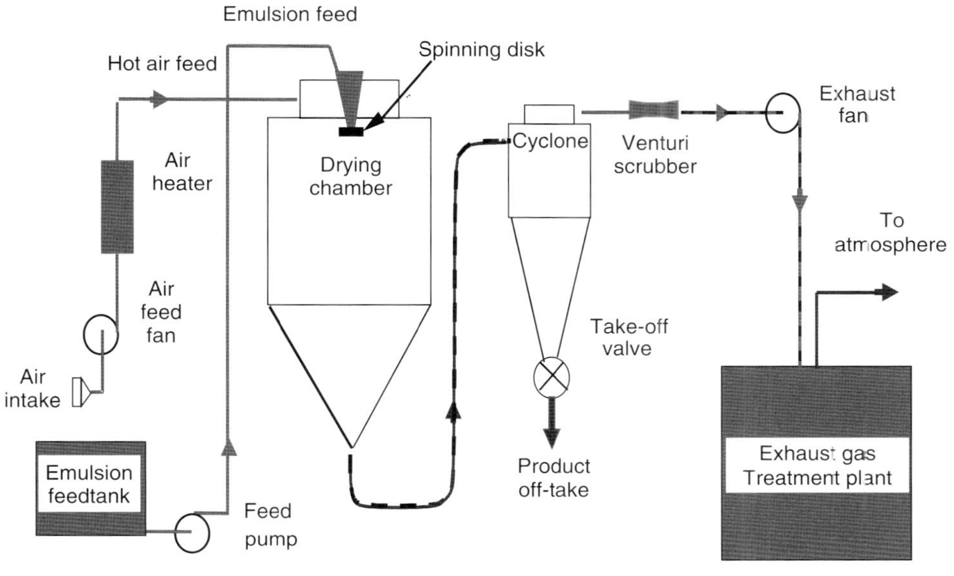

Figure 12.1 Spray drying – schematic diagram.

Obtaining a stable, low-viscosity emulsion with a high solids content is key to the success of the spray-drying process. The more stable the emulsion the greater the flavour retention in the finished powder [7]. The emulsion formed must remain intact until atomisation takes place, and this is achieved by homogenisation and through the use of carrier materials that assist emulsion stability. Gums, starches and proteins can be selected to assist the generation of stable emulsions as well as serving as carriers and encapsulants for the finished flavour. The solids level of the feedstock is usually maximised at a concentration and viscosity that will atomise successfully to make the most efficient use of the drying equipment. Maximising throughput is an important consideration in the management of a flavour production plant. A high feedstock solids is also a key factor influencing the retention of volatile flavour chemicals [8] during drying, by reducing the time necessary to form a semi-permeable membrane on the surface of the emerging particle, thus locking in the flavour chemicals. This semi-permeable membrane remains permeable to water vapour allowing evaporation to continue whist retaining all but the smallest molecular size flavour chemicals [9, 10].

12.3.2.2 Atomisation

Atomisation is achieved in the flavour industry by two techniques: spinning disc atomisation which utilises centrifugal energy and nozzle atomisation which relies on pressure. Spinning disc atomisation is used by most flavour companies because of its flexibility in handling viscous and thixotropic systems and its ease of operation. In this process, the feedstock emulsion is pumped to the atomiser, and feedstock droplets are formed as a mist as they emit from radial holes in the rotating disc. Evaporation of water takes place as the droplets contact the hot air being drawn through the drying chamber. The particle size of flavoured powders produced by this technique is determined by the speed of rotation of the centrifugal wheel and the solids content of the feedstock. One of the main limitations of spinning disc dryers is the difficulty of handling products which contain significant amounts of sugars, particularly sucrose, fructose and glucose. These sugars produce semiplastic particles which adhere to the sides of the drier and build up as glasses. It is virtually impossible, for example, to spray dry concentrated fruit juices on a spinning disc dryer, and nozzle methods are used for these materials.

With the nozzle method, the feedstock is pumped through nozzles into the drying chamber, and the droplet size and the final particle size are determined by the pressure exerted and the nozzle diameter, known as the nozzle orifice pressure drop. The higher the pressure and smaller the orifice, the smaller the particle. The advantage of nozzle drying is that it has better control over particle size and produces a narrower distribution of particle sizes in the finished powder. This has a beneficial effect on the appearance, flowability and dispersibility of the product. Centrifugal atomisation, on the other hand, produces a broad spectrum of particle sizes but from the flavour point of view the broad range of particle distribution produced by centrifugal atomisation is satisfactory for most flavour applications. The variation in particle sizes from the two atomisation techniques is shown on Figure 12.2.

From a practical point of view particle shape is as important as particle size, and drying conditions which yield spherical particles will improve the flowability of the finished powder. Spherical particles also pack together better and this increases bulk density making the flavour more compact, less dusty, and easier to bag off and transport. The spherical particles formed in spray drying are not completely round but are usually partially collapsed taking on the appearance of deflated footballs. This particle shape can be seen in the electron micrograph image in Figure 12.3. Particles which are not spherical are produced when in-feed viscosity is too high, atomiser rotation is too slow or inlet

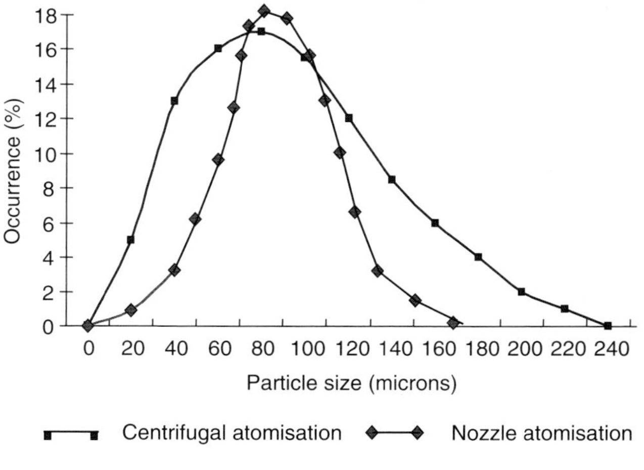

Figure 12.2 Particle size distribution.

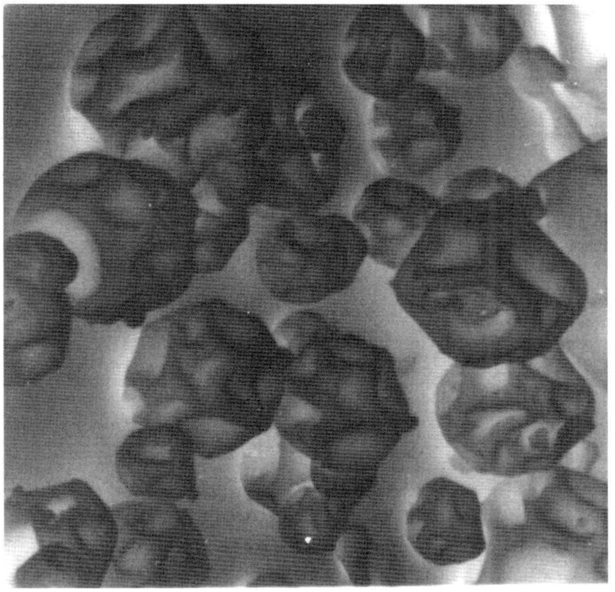

Figure 12.3 Electron micrograph image of spray-dried pepper oil in maltodextrin and gum arabic.

air flow is not high enough. Particle shape does not have a great bearing on flavour retention because the oil-phase flavour material is dispersed throughout the spray-dried particle as fine droplets encased in a honeycomb-like structure. This honeycomb effect can be seen in the electron micrograph images in Figures 12.4 and 12.5. This limits the amount of flavour which can be carried to around a maximum loading of 35%, in contrast to other microencapsulation techniques employed by the flavour industry, which produce spherical particles in which a continuous core region of flavouring is encased in a continuous shell. Coacervation is one example of this technology and is discussed in Section 12.3.5.

12.3.2.3 Drying

Air inlet temperature is an important factor in flavour volatile retention. Keeping the inlet temperature as low as possible and minimising the residence time of the actives in the dryer are both beneficial in spray drying flavours. Inlet temperatures ranging from 160 to 210°C are usual [11, 12] although temperatures as high as 245°C have been successfully used. At higher temperatures the phenomenon of 'ballooning' [13] can occur in which the particle approaches or exceeds the boiling point of water, fills with steam and expands. This significantly alters the structure of the particle matrix reducing its capability to retain flavour molecules. To avoid this, an air inlet temperature is selected which favours the formation of a diffusion-resistant semi-permeable membrane over the surface of the droplet during the transition between it being an aqueous emulsion and a powder. Water first evaporates from the surface of the droplet to create a denser area and a transient

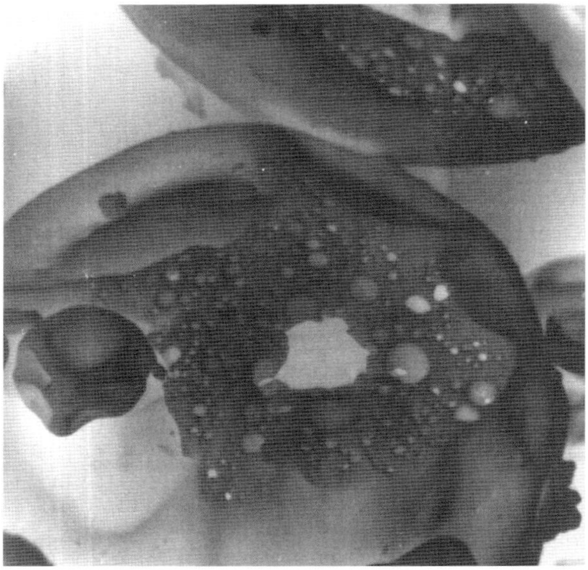

Figure 12.4 Electron micrograph image of a damaged spray-dried particle showing the internal honeycomb structure of encapsulated pepper oil in a matrix of maltodextrin and gum arabic.

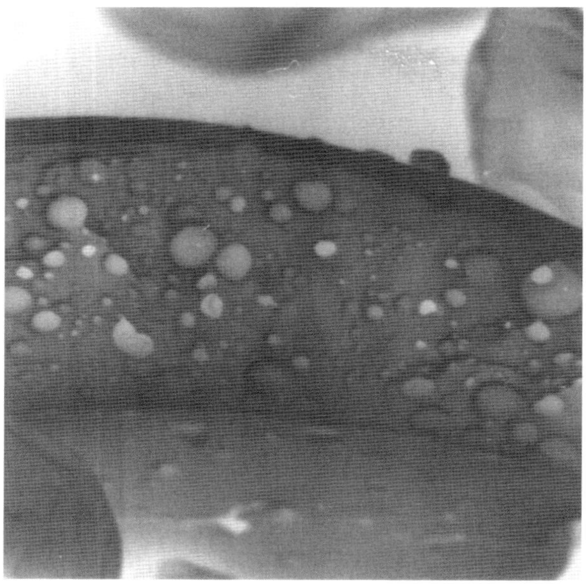

Figure 12.5 Section of a spray-dried particle containing pepper oil in maltodextrin and gum arabic showing the honeycomb structure of oil globules in the particle shell wall and the shell membrane.

membrane which continues to allow the passage of water vapour with a minimum loss of volatile flavour chemicals. The formation of the membrane is dependent on the in-feed solids level and the nature of the carriers being used, and it acts to deter the evaporation of flavour molecules by size exclusion. The larger the flavour molecule, the slower the rate of diffusion through the membrane [14]. The shell membrane which forms over the surface of a spray-dried particle from the transient semi-permeable membrane can been seen in Figure 12.5. The vapour pressure of the flavour molecule also has a bearing on volatile losses but is considered to be secondary to molecular weight. Thus, low molecular weight flavour chemicals with high vapour pressures are lost more readily in spray drying than high molecular weight flavour chemicals with lower vapour pressures. This is an obvious statement to anyone experienced in the technology of drying flavours, and a number of methods are used to reduce the loss of low molecular weight flavour compounds during drying.

(1) *Flavour adjustment* Flavour formulations used in spray drying are often adjusted by increasing proportionately the levels of flavour volatiles lost during spray drying, such that, after drying, the formulation of the liquid flavour is reproduced in the powdered product. Gas chromatography can be useful in determining the percentage increase of each volatile needed to compensate for the losses. A tentative relationship exists between the retention time of the flavour compound on a non-polar column and the percentage retention of the flavour compound during spray drying, and this can be expressed in graphical form and used to make the necessary formulation changes. The calculation outlined in Figure 12.6 can be used for each flavour chemical in a given formulation to provide a reasonable estimate of the flavour formulation needed to reproduce the liquid flavour in powder format.
(2) *Solvent selection* In preparing a flavour for spray drying a minimum amount of flavour solvent is employed to improve the loading of flavour chemicals in the final flavour and to reduce costs. Reineccius *et al.* [15] have shown that solvent selection has a bearing on flavour volatile retention. Fat-soluble flavour solvents such as triethyl citrate, triacetin and vegetable oils are better at retaining flavour chemicals than water-soluble solvents such as propylene glycol. This is because the flavour chemicals are dissolved in the hydrophobic solvent droplets in the in-feed emulsion, thus reducing their vapour pressure and increasing their resistance to diffusion. In a water-soluble solvent the flavour chemicals are dispersed as droplets in the aqueous environment exposing them to loss. Greatest losses are observed when ethanol is used as a flavour solvent in spray drying. In addition to it being a water-soluble flavour solvent, it also promotes ballooning of the drying droplet in the spray dryer and this leads to distorted particle structure and further losses.
(3) *Delayed addition* Simply, addition of the most volatile flavour compounds to the flavour emulsion immediately prior to transfer to the dryer can reduce losses. Compounds such as dimethyl sulfide, acetaldehyde, etc. can be lost to some extent during emulsification, and delayed addition can reduce this.

Dryer outlet temperatures are much lower than input temperatures and are determined by the rate at which the feedstock is pumped through the atomiser. The normal range is between 65 and 90°C. During drying the evaporation of water from the droplet creates

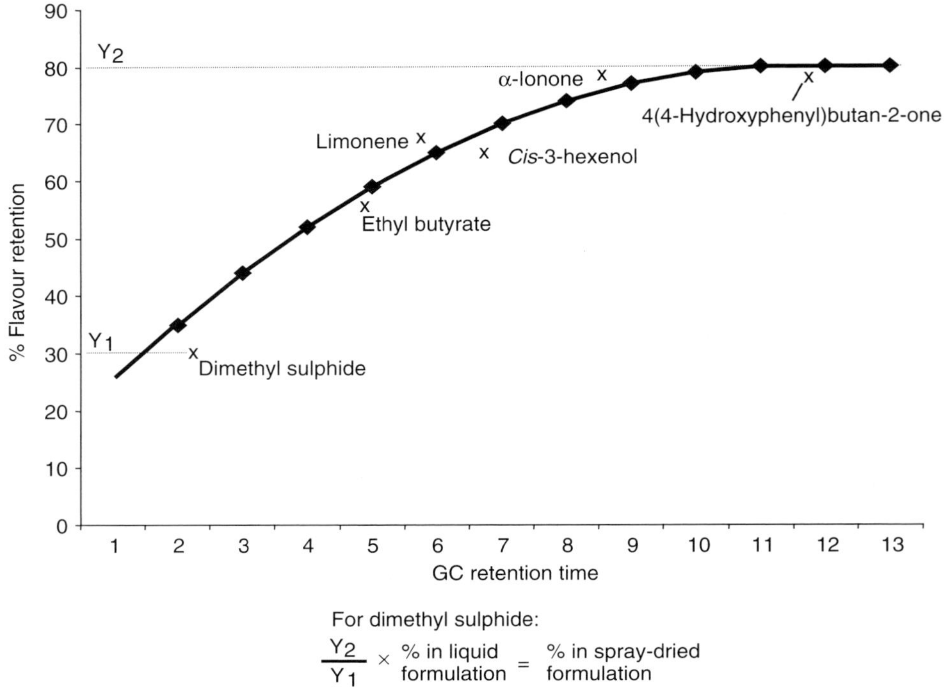

Figure 12.6 Compensation for spray-drying losses using gas chromatography.

cooling with the net result that the droplet temperature and the temperature of the finished powder rarely exceed that of the outlet temperature and are typically around 40–50°C.

12.3.2.4 Separation

Once the spray-dried powder has formed it is discharged from the bottom of the chamber and transferred to a cyclone where it is separated from the air stream and cooled before being bagged off. If the flavour-active loading in the feedstock is too high, drying may be incomplete and the powder may stick to the surface of the dryer. Flavour loadings of up to a maximum of 25% are normal with the optimum level usually between 10 and 20%, depending on the properties of the materials being dried and the requirements of the finished food product.

Flavour chemicals lost during drying end up in the exhaust gasses and these, especially some of the high-intensity savoury meat chemicals, can create a serious environmental nuisance. Exhaust air should be treated and can be incinerated, passed through a caustic scrubber, through carbon filters or through an active bed containing, for example, peat, to remove the volatile flavour chemicals.

12.3.2.5 Carriers and encapsulating agents

Selection of a suitable carrier system is an important factor in the development of a successful spray-dried product. A number of factors should be considered:

- How much emulsification is necessary for a stable feedstock?
- What level of encapsulation is required?
- Does the flavour need protection against oxidation?
- What properties are required in the finished product?

12.3.2.5.1 Maltodextrins

Maltodextrins are derived from edible starch such as corn starch by chemical or enzymatic hydrolysis and are referred to by their DE (dextrose equivalent) value which indicates the degree of hydrolysis. The higher the DE value the greater the extent of hydrolysis. Chemically, maltodextrins are glucose polymers of high molecular weight, possess very little sweetness and have an upper DE value of around 20. Maltodextrins are bland in flavour, inexpensive and very soluble. Selecting the correct DE maltodextrin is important for protection against oxidation – the higher the DE value the more the protection conferred. The main disadvantages of maltodextrins are that they lack emulsifying capacity, are not good at retaining volatiles and the higher DE maltodextrins can cause caking problems.

Other starches derived from potato and tapioca are also used in spray drying and these are designed to be better at retaining and encapsulating flavours than the maltodextrins but are usually more expensive.

12.3.2.5.2 Chemically modified starches

The lipophilic properties of starch polymers are enhanced by partial hydrolysis, followed by the addition of 1-octenyl succinic anhydride to the starch backbone. This confers on the hydrolysed starch molecule surfactant properties, converting it into an excellent emulsifier and a very useful spray-drying carrier. These modified starches are also very good at retaining volatile flavour chemicals during spray drying. However, because they are not natural, they are being used less and less.

12.3.2.5.3 Gums

Gum arabic obtained from the *Acacia* tree is extensively used in spray drying. It has excellent encapsulating properties, is a good emulsifier, has a bland flavour and retains volatiles very well during drying. Its main advantage over other gums is its water solubility. As it is a natural product and its properties can vary considerably, tight purchasing specifications are required to obtain material with good spray-drying properties.

12.3.2.5.4 Proteins

Gelatin and sodium caseinate are the two proteins in the greatest use in spray drying but do not find broad usage. They have excellent emulsifying properties but are unstable at low pH and create very viscous solutions, and this limits the solids content of the spray-drying feedstock. As they are both animal-derived this has limited their use in recent years.

The above carriers and encapsulating agents are used by the industry in various combinations to take advantage of their different properties. Combinations of maltodextrin and gum arabic have excellent encapsulating and carrier properties for topnote flavours. Gum arabic provides emulsification and encapsulation properties, and maltodextrin provides particle stability, bulk density and free-flow carrier properties, and these systems are being used more and more with the decline in the use of modified starches. In high-fat systems, such as some meat process flavours, the use of sodium caseinate as a carrier can be beneficial in combination with maltodextrin and gum arabic to improve the emulsification of a high-fat feedstock. Two reviews on spray drying and spray drying of flavours are recommended for further reading [9, 16].

Conventional spray-drying processes, using carrier systems such as maltodextrin and gum arabic, produce water-soluble powder flavours. For certain applications, an encapsulated water-insoluble powder is preferable, and such systems can be produced by spray cooling, an adaptation of spray drying, and by encapsulation in yeast cells followed by conventional drying to produce powder flavours that are totally insoluble in water. These technologies are covered in Sections 12.3.3 and 12.3.4.

12.3.3 Spray cooling

Spray cooling, alternatively known as spray congealing, is a variation of spray drying in which ambient or chilled air is used to solidify a molten droplet [17]. In spray drying, liquid droplets formed by atomisation are exposed to hot air, producing evaporation from the droplet, which initiates cooling of the particle matrix and the formation of a powdered product. In spray cooling, the feedstock is heated to a sufficient temperature that will allow it to flow easily through a heated atomiser into a cooled chamber, where it solidifies. The temperature and the rate of solidification depend on the nature of the carrier materials used. Blends of fats/vegetable oils with a range of melting points are the main carrier ingredients used for flavourings, and can be combined with other ingredients such as waxes, hydrocolloids, gums and hard mono- and di-glycerides with melting points between 45 and 65°C. The mono- and di-glycerides, being good emulsifiers, have been demonstrated to facilitate the dispersion of the encapsulated flavour into the finished food matrix [18]. For carrier materials with a solidification temperature above 45°C particle formation can be achieved with ambient temperature air. With a cooled chamber, finished products with a melting point range of 32–45°C can be produced and this process is often referred to as spray chilling.

Both spinning disc and nozzle spray dryers are used for the process, depending on the particle properties required for the finished product. High-speed spinning disc atomisation in a conical base cooling chamber will produce fine particle products, whereas nozzle atomisation with the nozzle pointing upwards into the chamber, fountain style, yields coarse particle products. Other designs are possible, and many companies using this technology have bespoke purpose-designed systems.

In application, spray-cooled flavours have the advantage of controlled flavour release by tailoring the product to suit different product applications by altering the melting point. At ambient temperatures or above, spray-cooled products are insoluble in an aqueous food environment but during cooking the shell material melts liberating the flavour into the food product. This has distinct advantages in baked products where the flavour compounds can interfere with yeast activity. By selecting a carrier system with a melting point above that where the yeast is active, flavour release is controlled at a temperature where it will not interfere with fermentation. It also ensures that flavour volatiles are not lost during the fermentation stage. Other applications where the technology has benefits are in dry-soup mixes and high-fat foods such as margarines.

12.3.4 Yeast encapsulation

This technology was developed and patented in by AD2 Ltd [19] and the British Textile Technology Group [20–22], and the patents were assigned to Bestfoods, formerly CPC International. Patent rights are now held by Micap plc Ltd who have developed the technology for use in drug delivery, insecticide and herbicide delivery, cosmetics, textiles and food ingredients, including flavourings.

The process is relatively straightforward and involves mixing an aqueous suspension of yeast cells with a hydrophobic flavour oil, which diffuses through the yeast cell wall without rupturing it, displacing the intracellular material in the process. The flavour resides in the cell as oil droplets, in the hydrophobic moieties of the cell structure, and flavour loadings as high as 70% can be achieved, with 30–50% being common. Once the flavour has been incorporated into the yeast cells, water is removed and the resultant slurry can be spray dried without any additional carriers other than the yeast cells themselves. A schematic diagram of the process is shown in Figure 12.7.

The yeast used, *Saccharomyces cerevisiae*, need not be viable and is grown on corn syrup to eliminate any yeasty taste which may interfere with the encapsulated flavour. The flavour is well protected as the yeast cell walls are very strong, being composed of a β-1,6 glucan and β-1,3-glucan weaving to form a tough ball protecting the inside of the yeast cell, covered by a surface mannoprotein layer. Yeast cells also produce chitin

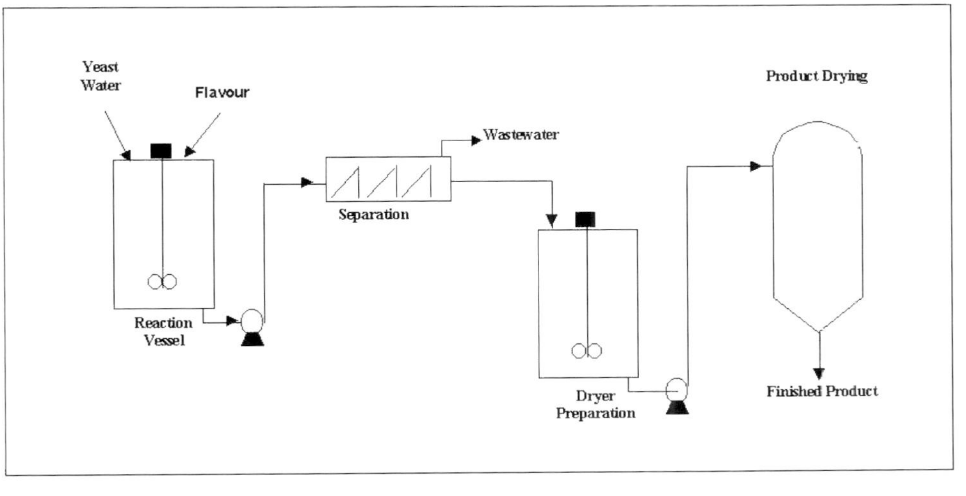

Figure 12.7 Schematic diagram of the yeast encapsulation process.

to patch up any holes in the weave, giving the cell wall an extremely strong mechanically rigid structure.

Spray drying of the yeast slurry produces spherical agglomerates of yeast cells, which are non-hygroscopic, flow well and are totally insoluble in water. Figures 12.8 and 12.9 show electron micrograph images of spray-dried yeast cells. Each particle is a globular cluster of yeast cells forming a regular spherical shape, conferring good flowability and bulk density on the powdered product.

The spray-dried yeast agglomerates also have good shelf-life properties. Figure 12.10 shows the stability of bio-encapsulated peppermint oil over several months.

One of the most interesting properties of yeast-encapsulated products is associated with the release of flavour from the yeast cells. Flavour is not released when the particles are added to water or to a water-based food product, but is released immediately from the yeast cells in the mouth during mastication, producing an immediate flavour impact on contact with the tongue. The actual mechanism of this process is not fully understood but is likely to involve cell emptying of flavour onto lipophilic sites on the tongue, possibly assisted by enzymatic action from saliva. The natural cell walls of yeast also confer a number of other useful properties on encapsulated flavours such as protection from oxygen, UV and heat, and flavours encapsulated in this way can survive most cooking processes including frying, boiling, roasting, baking and extrusion. Application areas therefore include heat-processed products such as baked goods, extruded snacks, soups, sauces and hot beverages, and products requiring specific slow-release properties such as chewing gum.

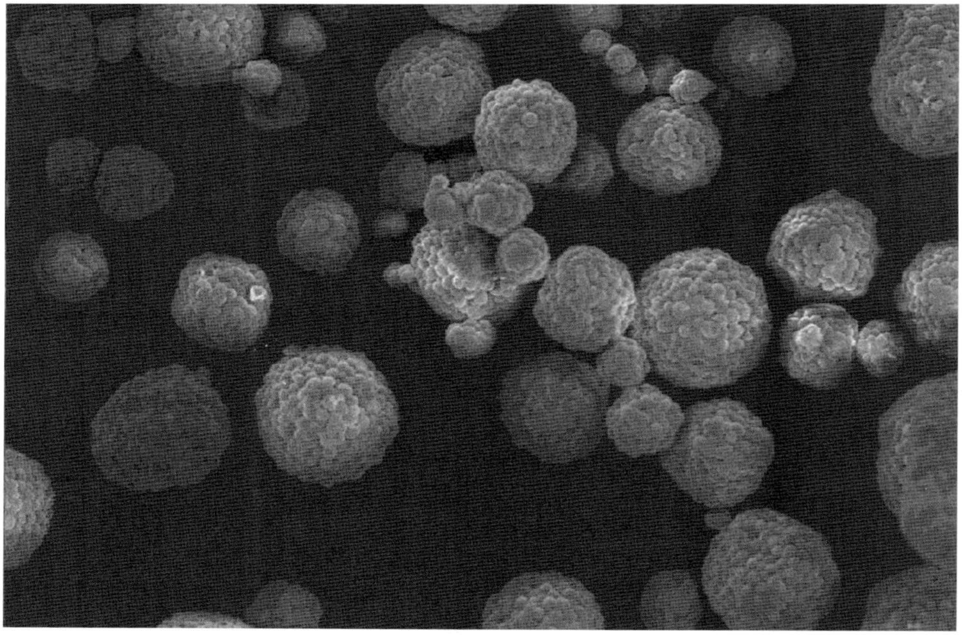

Figure 12.8 Spray-dried yeast cells (reproduced with kind permission of Micap plc, Haydock, WA12 0JQ, UK).

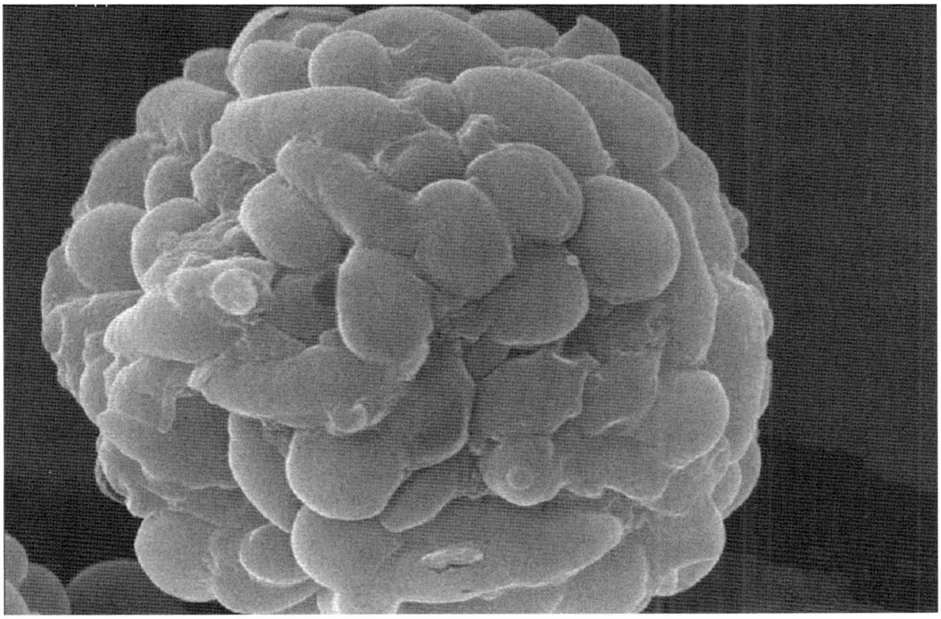

Figure 12.9 Enlargement of a spray-dried yeast particle (reproduced with kind permission of Micap plc, Haydock, WA12 0JQ, UK).

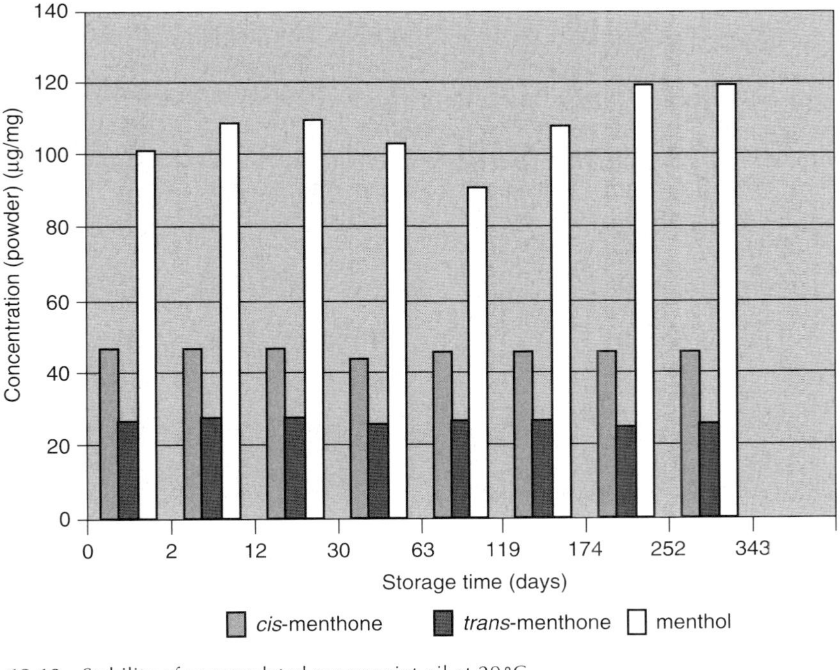

Figure 12.10 Stability of encapsulated peppermint oil at 20 °C.

12.3.5 Coacervation

The original idea that it might be possible to use this technique to encapsulate flavourings and fragrances came as a spin-off from the National Cash Register Company, USA, patent describing a means of producing a two-component ink system for carbonless copying developed in the 1950s. In this patent is described the production of hardened microcapsules which are applied to copy paper such that when they are ruptured by the application of pressure by writing on the top paper they react with a second coating to produce a coloured reaction product.

It was thought that this technique might be used for the protection of flavour and fragrance ingredients that would only be released under specified conditions [23, 24]. The suggested conditions were either rupture as in the original patent or predictable solubility in water.

12.3.5.1 Method of production

When the pH of a protein solution is reduced to the protein's isoelectric point, the protein is precipitated. If this is done in the presence of a suitable hydrocolloid, the precipitate consists of hydrocolloid-protected protein particles known as a coacervate. If the precipitation is done in a suspension of oil droplets the precipitation is in the form of a thin continuous coating on the droplets.

The production of such coacervates is very simple (see Figure 12.11). An emulsion of oil droplets in a solution of a soluble protein such as gelatin, and a water-soluble hydrocolloid such as a carboxymethylcellulose (CMC) at room temperature and pH of about 7 is gently stirred with a mechanical stirrer. The temperature is slowly reduced

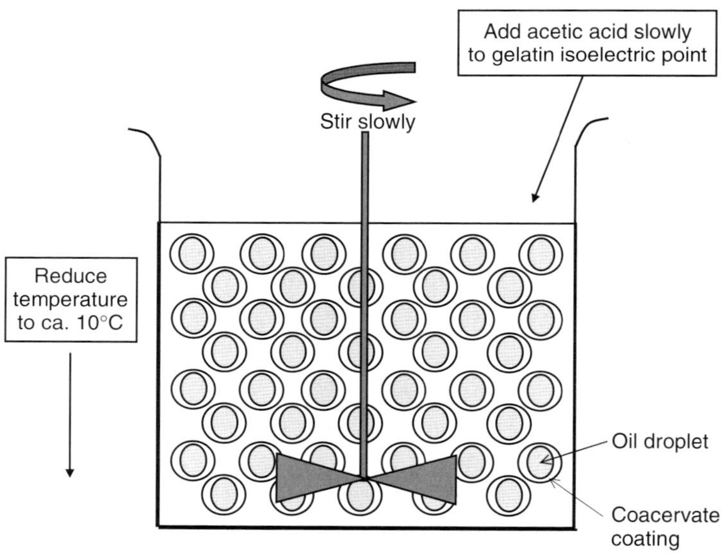

Figure 12.11 Schematic production of coacervated oil droplets.

to about 10 °C and acetic acid slowly added until the isoelectric point of the protein is reached. This produces the coacervate on the outside of the oil droplets as a continuous coating.

The stirring is stopped and the encapsulated oil droplets filtered off and air-dried at about 50 °C. They are quite stable at this temperature. The coating is soft and pliable, and somewhat permeable. It is also possible to prepare similar microcapsules by precipitation of a complex of positively charged protein and negatively charged hydrocolloid on the oil droplets.

If the capsules are required to be ruptured by abrasion or pressure they must now be hardened. This is performed by cross-linking the gelatin with a suitable aldehyde. For flavour use, this is normally glutaraldehyde, which is added as an aqueous solution. This results in capsules which, when dried, are quite brittle and easily ruptured by pressure.

Much effort has been used to try to produce soft capsules, which rupture by dissolution in water at a specified temperature, for use in such food applications as powdered milk-based desserts, table jellies or soup mixes which are prepared by heating in water. These have generally been unsuccessful due to the cross-linking properties of the internal flavour after several months' storage. This either makes the capsules insoluble or results in a chemical rupturing of the coating, resulting in the loss of the capsule contents.

12.3.5.2 Properties of coacervated flavourings

The main advantages of coacervated flavourings are:

(1) The internal phase is totally protected against evaporation and oxidation as the coating of dried capsules is virtually continuous and impervious to oxygen.
(2) Since the production process is very gentle there is almost no flavour degradation during manufacture.
(3) Compared with spray-dried flavourings the products are much stronger, as the capsules normally contain 80–85% of active ingredients.

The major disadvantages are:

(1) Only oil-soluble flavour concentrates can be encapsulated, making it very difficult to produce authentic fruit flavours (other than lemon oil).
(2) Many flavour materials, particularly onion and garlic oils, degrade the capsule coating leading to insolubility or instability.

A further development of the coacervation technique has been developed and patented by Givaudan (Figure 12.12) [25]. This consists of producing 'blank' capsules containing vegetable oil and then loading them with flavour by direct adsorption through the coating. This allows the adsorption of water-soluble flavour compounds such as diacetyl. The maximum flavour loading of the capsules using this technique is 30–50%.

Many experimental food products have been market-tested with some success. Microcapsules, particularly those produced by the adsorption technique have found application in heated foods, such as bakery products. Also an obvious use of flavour-type products

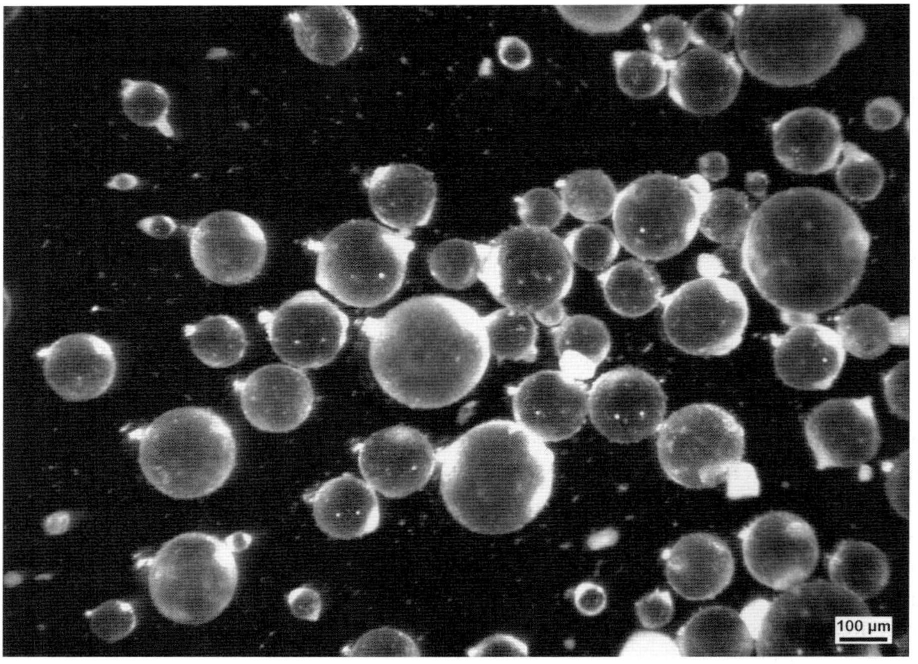

Figure 12.12 Photomicrograph of coacervate capsules (reproduced by kind permission of Givaudan Ltd).

is the 'scratch and sniff' application, where hardened capsules are printed onto paper or card and release their flavour contents on rupture. Less obvious is the use of coacervated microcapsules in cigarettes. This has been proposed for menthol both in the tobacco and also in the filter. In the former case the flavour is released on heating as the cigarette is smoked, whereas in the latter it is released by squeezing the filter. An extreme case of flavoured tobacco is seen in 'kretek' cigarettes smoked in Indonesia. These contain a mixture of tobacco and kibbled clove buds and are generally flavoured with fruit flavourings such as strawberry. Using coacervated flavourings has vastly increased the shelf life of these products.

12.3.6 Melt extrusion

Melt extrusion technology converts a liquid flavouring into a durable free flowing amorphous glassy particulate with a shelf life as long as two years or more. In extruded products the flavour is totally encapsulated and, depending on the formulation of the carrier, is stable at elevated temperatures, will dissolve slowly in aqueous environments and release flavour in a controlled manner in the food product. Achieving an amorphous state for the encapsulating particle rather than a crystalline structure is important for the stabilisation of the flavour which would not be held securely by a crystalline mass, and production techniques and carrier systems have been developed to achieve this.

12.3.6.1 Method of production

The method is relatively simple and involves mixing and melting the carrier materials followed by addition of the flavour blend just prior to extrusion. Usually an emulsifier is incorporated into the blend at a level of 1–2.5% plus a small amount of water to form a stable emulsion with the correct viscosity. The amount of water added is important because it has an effect on the glass transition temperature and the robustness of the finished product and levels from around 5–10% are usual. The moisture content of carrier materials has to be taken into account in calculating the optimum water content of the feed emulsion. The molten formulation is extruded as filaments at temperatures ranging from 85 to 150°C using a single- or twin-screw extruder, or purpose-designed equipment into a cool environment or a cold solution of a flavour solvent such as isopropyl alcohol. The advantage of using a solvent is that any free flavour is washed from the filament surfaces during the process to yield a relatively odour-free particulate with excellent resistance to oxidation during storage. In the case of a purely maltodextrin-based product with a moisture content of 4–7%, the flavour can be plated onto the maltodextrin and the mixture extruded in the dry state but this method gives a low flavour loading of 2–5% in the finished particle. For oxygen-sensitive flavours the process can be adapted to operate in an atmosphere of an inert gas such as nitrogen or carbon dioxide. The finished product is dried and sized using conventional processes such as chopping and grinding, followed by size classification to remove fines or oversized particles, and this produces a consistent product with excellent visual appeal, especially if a colour is used in the formulation. Flavour loadings of 5–10% are typical, and products can be designed to control flavour release in a range of food environments.

12.3.6.2 Carriers

The selection of carrier materials is based on the desired properties of the end product such as solubility, temperature stability, pH resistance, particle hardness and the desired release properties of the finished flavour in the food system. The process is very flexible and allows the use of a large number and variety of carrier materials. The main groups are:

- *Carbohydrates* Mono- and di-saccharides such as glucose, lactose, fructose, sucrose and maltose, and polyols such as sorbitol, mannitol, maltitol, glycerol and propylene glycol. These are used at levels of between 10 and 60% of the formulation and are selected for their melt and flow properties as well as glass transition state, plasticising ability and flavour-release properties.

  Maltodextrins with a DE value between 5 and 20 are used at levels of 25% to as high as 90%. The lower DE maltodextrins give good protection against oxidation of flavour compounds.

  Modified starches such as CMC and hydroxypropyl cellulose (HPC) are used. HPC has a marked effect on flavour retention in extruded encapsulates used in products subjected to elevated temperatures, especially in confectionery products such as boiled sweets and hard candies [26].
- *Food acids* Citric, tartaric, malic and lactic acids can be used from 5 to 50% of the formulation, and will provide improved acidity and tartness in food products.

- *Gums* Gum arabic is used to increase encapsulation strength and flavour-release parameters.
- *Fats* Vegetable oils and hydrogenated fats improve the lipophilic properties of the extruded particles.
- *Proteins* Gelatin, zein and gluten have gelling, cross-linking, flavour-binding and film-forming properties.
- *Food emulsifiers* Monoglycerides of fatty acids and sorbitan fatty-acid esters.
- *Other ingredients* Food colourings, antioxidants, preservatives, silicates, etc. can be employed in the process.

The formulation possibilities are endless and a number of patents are currently operative governing the use of the technology [26–31].

12.3.6.3 Properties of melt extruded flavourings

The main advantages of melt extruded flavourings are:

(1) Excellent shelf life in the order of two years or more with low volatile losses over that period.
(2) Flavour protected against oxidation due to the highly dense structure of the extruded particles – much more stable in this respect than spray-dried flavours.
(3) Can be used in high-temperature food processes and remain intact releasing flavour at the point of consumption.

The major disadvantages are:

(1) Low flavour loading of up to 10%.
(2) High-temperature process which is not suitable for some flavour types.
(3) Slow solubility in aqueous food systems.

Extruded flavours are commercially used in dry beverages such as tea bags where protection against flavour loss and oxidation are important. Prevention of 'bag spots' triggered by moisture softening of flavours which stain the tea bag can be significantly reduced with extruded flavours [32]. Extruded flavours also find use in confectionery products such as boiled sweets and candies where stability to high-temperature processing is essential.

12.3.7 Molecular encapsulation

Individual flavour chemicals can be encapsulated by forming inclusion complexes with cyclodextrins. These are cyclic molecules composed of glucose units with the ability to encapsulate other smaller molecules within a hydrophobic central core. They were first isolated from digests of potato starch in 1891 but it took until the 1970s before each of the naturally occurring cyclodextrins had been characterised. Three types were identified, the α-, β- and γ-forms, composed of 6, 7 and 8 glucose units, respectively, with the β-form predominating when they are manufactured by enzymatic processes from starch. Cyclodextrins have a crown-like structure and the β-form is shown in Figure 12.13.

The molecule is orientated such that the primary hydroxyl groups sit on one outside rim of the crown and the secondary hydroxyl groups sit on the other, forming a hydrophilic outer surface. The internal cavity has both hydrogen atoms and glycosidic oxygen bridges pointing inwards, conferring hydrophobic properties on the central core. This hydrophobic cavity makes it suitable to accept organic hydrophobic molecules [33], such as most flavour chemicals, ranging from 80 to 250 molecular weight. Complexation and inclusion of flavour chemicals is simply achieved by stirring in water followed by precipitation and drying or by mixing the flavour chemical with an aqueous cyclodextrin paste followed by drying [34]. The solubility of β-cyclodextrin in water is 5% at 25°C and 40% at 40°C, and the complexed form is insoluble in water, and hence precipitation takes place when inclusion has been achieved.

Cyclodextrin flavour complexes are very stable [35] and the encapsulated flavour chemicals are protected against oxidation and evaporation, and the products are shelf-stable for several years [36]. The technology is particularly useful for the encapsulation of

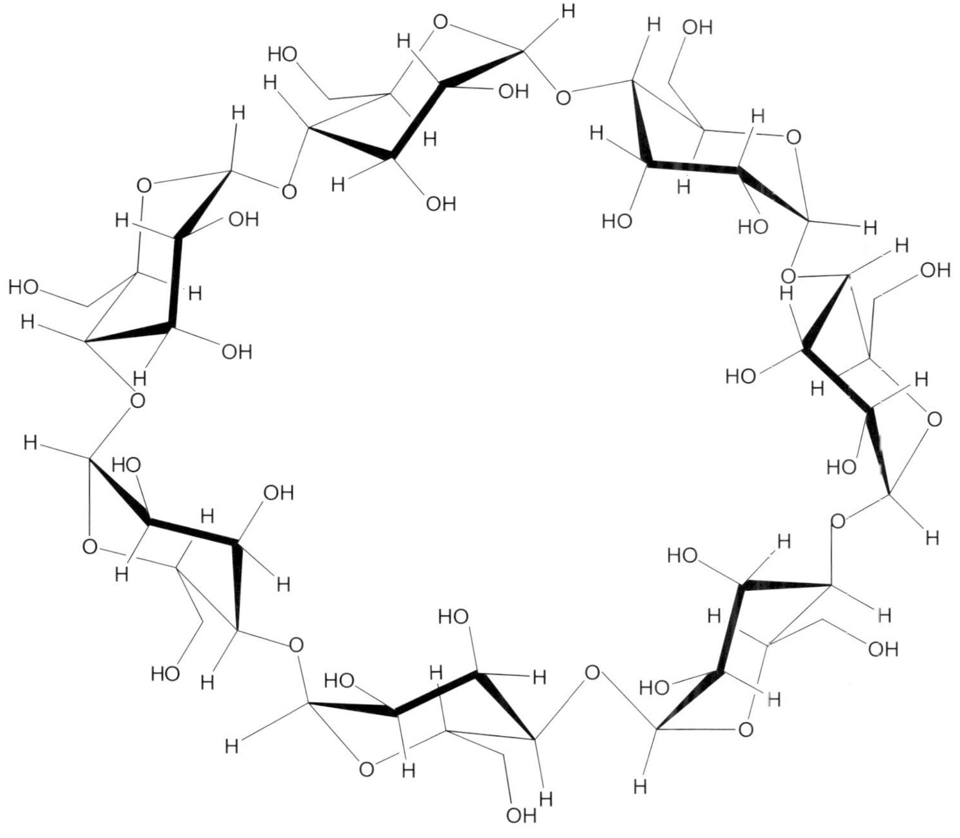

Figure 12.13 β-Cyclodextrin.

small highly volatile flavour chemicals such as dimethyl sulfide, which can be complexed at a level of 2% and protected from evaporation for considerable time periods [37].

Cyclodextrins have variable affinity for flavour chemicals, and when attempts are made to encapsulate whole flavour systems, the dry finished flavour is usually very different from the original flavour. This property can be used to advantage, and bittering substances have been successfully removed from orange and grapefruit juices using a cyclodextrin polymer [38].

The cyclodextrins, however, have found limited use in the flavour industry because they were approved for use in only a few countries, notably Japan and Eastern Europe. This situation is now changing and β-cyclodextrin has a FEMA-GRAS number (4028) and an E number (E 459), and in the EU can be used as a flavour carrier in a range of foods such as flavoured teas, flavoured powdered instant drinks and flavoured snacks [1].

12.4 Formulation issues for the flavourist

12.4.1 Flavour creation

One might believe that with the advances in analytical techniques, particularly GC/MS, all that was required to produce acceptable flavours was to do the analysis of the food and compound together with the individual chemical components in the quantity present. This is very far from the actual situation for several reasons, as follows:

(1) It is impossible to make an analysable sample of any food without both losing some of the components and introducing artefacts. The flavour of many foodstuffs, particularly raw fruits and vegetables are only stable in the cell structure of the foods themselves and change as soon as this is destroyed. This is most obvious in strawberries but occurs in most raw foods.
(2) It is well nigh impossible to undertake a GC/MS analysis without introducing changes in the product being analysed due to decomposition or polymerisation on the column. Thus, the finished analysis does not faithfully represent the sample being analysed.
(3) Modern GC/MS analysis produces a large number of identified compounds, usually in excess of 500. To produce a commercially viable flavour this must be reduced to a manageable number, say 50–70. Therefore, it is vital that the flavourist makes the decision as to which compounds are important at the analytical stage. This probably means that an odour port has to be manned by a flavourist – a very wasteful use of an expensive resource.
(4) GC/MS analysis produces data on individual pure chemical compounds whereas all commercially available compounds are impure, often containing significantly odorous impurities.

For the above reasons it is impossible to use directly the analytical data in its entirety. This does not imply that it is unimportant – quite the opposite; it provides a starting point from which the flavourist can use his experience and creativity to make up for its inadequacy. The procedure for creating a truly novel flavouring is to obtain the analytical data and mix together the identified constituents in the estimated quantity using commercially available materials. This rudimentary flavouring is then compared with the target foodstuff and modified by the addition of non-identified constituents and the removal or reduction of some of the identified ones. A good example of the latter is in the area of compounded beef flavourings. All the analytical data on beef show significant levels of phenylacetaldehyde, presumably originating from the degradation of phenylalanine. If this compound is used in the formulation of beef flavourings, even at a tenth of the level detected, it makes the flavour unacceptably perfumed.

The art of the flavourist is his ability to compensate for the absence of many of the detected materials, make up for the absence of pure ingredients and use the impurities to his advantage. This may often mean using a natural extract or essential oil in place of some of the detected compounds to get the required flavour nuance.

12.4.2 Influence of foodstuff to be flavoured

Having created the flavouring, as outlined above, it is then necessary to modify it to suit the medium in which it is to be used.

- Is the foodstuff liquid? Does the flavouring have to be a stable emulsion or must it produce a clear drink on dilution? Each of these conditions requires different formulations and restrictions on what ingredients may be used.
- Is the foodstuff a solid? If so, are there any restrictions such as stability and shelflife? If a long shelf life is required it may be able to be achieved by encapsulation and/or the addition of antioxidants.
- Does it contain fat or alcohol? The performance of flavourings in fat- or alcohol-containing media is totally different from their performance in aqueous media, and requires radical reformulation to compensate for the phase distribution of each ingredient.
- Is the substrate likely to react with the flavouring? For most foods this is unimportant, but if the medium is a pharmaceutical product it might be rendered useless by one or more of the flavour ingredients reacting with the active ingredient.
- Is the foodstuff going to absorb flavour ingredients and not release them on consumption? Examples are chewing gum, where only about 40% of the added flavour is released, and textured soy protein, where more than 80% of certain flavour ingredients are unavailable to the consumer.

12.4.3 Influence of legislation

At the time of writing (2003) there is little legislative restriction on natural and nature-identical flavour ingredients in Europe. However, this is likely to change in 2005 with the introduction of a positive list of defined flavouring substances that may be used. This list includes natural, nature-identical and artificial flavouring substances, and does not distinguish between them in acceptability. In spite of attempts to harmonise EU flavouring legislation several member states still retain their own peculiar restrictions, particularly on artificial flavouring substances.

In the EU there will soon be a positive list of additives to flavourings, including solvents, carriers, flavour enhancers, emulsifiers, stabilisers, preservatives and antioxidants. Recently, there have been proposals for legislation on smoke extracts and process flavourings.

In USA there has been a positive list system for all flavouring ingredients since 1960, often referred to as the FEMA 'GRAS' list. This is the result of the food additive amendment which stated that for a food additive to be acceptable it had to be 'generally regarded as safe' by a suitable group of independent experts. This was achieved by the US flavour industry setting up such a panel of internationally recognised toxicologists ('FEXPAN') to examine all proposed new flavouring ingredients.

Because of the differences between national legislations it has become increasingly difficult to create adequate flavourings for world-wide markets. This is particularly problematical with solvents. The only solvents for flavourings that are allowed universally are water, glycerol (providing that it is from a non-animal source) and propylene glycol. None of these are suitable solvents for citrus oils, and poor solvents for many other ingredients. The best solvent potentially for flavourings is ethanol but it is not permitted in many Moslem countries. The best alternative is 2-propanol, but this is not permitted in EU or USA as a flavour solvent.

12.4.4 Influence of customer requirements

Although it might be supposed that individual customers would be satisfied with the national legislation for the territories in which they sell, several of the major international food manufacturers have more restrictive requirements, such as having a list of flavouring substances which they will not allow in their products. In addition, many national and international groups will not accept flavourings that contain artificial flavouring substances, even though they are not required to declare their presence.

Many major groups wish to label their products as containing only 'natural flavourings', even though their performance is usually inferior and they are much more expensive in use than normal flavourings. This has particular difficulties since the definition of 'natural flavouring' is fundamentally different between the EU and the USA, the former being much more restrictive.

Several major groups will not accept flavourings that contain antioxidants, even though they do not have to be declared in the final food.

Recently, the problem of genetically modified ingredients has placed constraints on the use of maize- and soya-derived carriers. In EU it is necessary to be able to provide complete traceability for these ingredients to ensure that they are non-GM. Even so, several food groups will not accept them in their formulations.

At present, EU legislation on allergenic ingredients in food is being considered. This will require foods containing milk, wheat, rye, barley, oats, fish, crustacea, eggs, peanuts, tree nuts, soya beans, celery, mustard, sesame seeds and products thereof to be labelled on finished foods. This is in addition to sulfur dioxide and sulfites at more than 10 ppm. Unless some changes are made, virtually, all manufactured food will have to be labelled as containing allergens. Already, several food groups are refusing to allow flavourings containing any of the above ingredients or their derivatives. In addition to the above, there may be a requirement that the flavour must be suitable for a defined group of consumers, such as 'Kosher', 'Halal', organic, vegetarian or vegan.

As can be seen from the foregoing, the creation of commercial flavourings requires an extensive knowledge of the constraints that may be applied. It is sometimes surprising that successful new flavourings continue to be created but it is to be feared that the EU positive list system, beloved of regulators, will finally check new flavour creation to the detriment of the consumer.

References

1. Directive of the European Parliament and of the Council. Amending Directive 95/2/EC on Food Additives other than Colours and Sweeteners. Brussels, 29.11.2002, COM(2002) 662 final, 2002/0274(COD)
2. Arctander, S. (1969) *Perfume and Flavour Chemicals*, Allured Publishing Corp., Box 318, Wheaton, IL, USA.
3. Ashworth, D.R. (1985) Mixing dry ingredients. *Food*, July, 35–37.
4. Patterson, J.M. (1977) High speed mixing for standardised spice production. *Food Processing Industry*, July, 24–26.
5. King, G.T. (1975) Selection of mixing equipment for the food industry. Food RA Symposium Proceedings, No. 26.
6. Reineccius, G.A., Anandaraman, S. and Bangs, W.E. (1982) Spray drying of food flavours. I. Theory of flavour retention. *Perfumer* & *Flavorist*, 7, 1–6.
7. Reineccius, G.A. and Bangs, W.E. (1985) Spray drying of food flavours. III. Optimum infeed concentrations for the retention of artificial flavours. *Perfumer & Flavorist*, **9**, 27–29.
8. Leahy, M.M., Anandaramanan, S., Bangs, W.E. and Reineccius, G.A. (1983) Spray drying of food flavours. II. A comparison of encapsulating agents for the drying of artificial flavours. *Perfumer & Flavorist*, **8**, 49–58.
9. Reineccius, G.A. (2001) The spray drying of food ingredients. In P. Vilstrup (ed.), *Microencapsulation of Food Ingredients*. Leatherhead Publishing, pp. 151–185.
10. King, C.J. (1995) Spray drying: the retention of volatile compounds revisited. *Drying Technol.*, **13**, 1221–1240.
11. Bomben, J.L., Bruin, S., Thijssen, H.A.C. and Merson, R.L. (1973) Aroma recovery and retention in concentration and drying of foods. *Adv. Food Res.*, **20**, 1.
12. Thijssen, H.A.C. (1971) Flavour retention in drying preconcentrated food liquids. *J. Flavor Appl. Chem. Biotechnol.*, **21**, 372–376.
13. Verhey, J.G.T. (1972) Vacoule formation in spray powder particles I. Air incorporation and bubble expansion. *Netherlands Milk Dairy J.*, **27**, 3–18.
14. Rosenberg, M., Vopelman, I.J. and Talmon, Y. (1990) Factors affecting retention in spray drying microencapsulation of volatile materials. *J. Agric. Food Chem.*, **38**, 1288–1294.
15. Kim Ha, J. and Reineccius, G.A. (1998) Spray drying of food flavours. IV. The influence of flavor solvent on retention of volatile flavors. *Perfumer & Flavorist*, **13**, 1–4.
16. Masters, K. (1991) Spray drying handbook, 5th edn, Harlow, UK, Longman.
17. Vilstrup, P. (2001) Multicore-core encapsulation, modified spray congealing/spray drying of aqueous dispersions. In P. Vilstrup (ed.), *Microencapsulation of Food Ingredients*. Leatherhead Publishing, pp. 151–185.

18. Taylor, A.H. (1973) Encapsulation systems and their application in the flavour industry. *Food Flav. Ingred. Packg. Proc.*, 5, 49–52.
19. AD2 Ltd, European Pat. Application 0 242 135, 1987, assigned to Bestfoods.
20. Bishop, J.R.P., Nelson, G. and Lamb, J. (1998) *J. Microencapsulation*, **15**, 761–773.
21. British Textile Technology Group, UK Pat. Application G.B. 2 234 901, 1990, assigned to Bestfoods.
22. British Textile Technology Group, World Pat. WO 94/22572, 1994, assigned to Bestfoods.
23. Todd, R.D. (1970) Microencapsulation and the Flavour Industry, *Flavor Ind.*, **1**(11), 768.
24. Balassa, L.L. (1970) Flavor encapsulation. US Pat. 3,495,988.
25. Givaudan Roure (International) SA (Switz.), US Pat. Application 1997-947694, World Pat. Application 1998-IB1551.
26. Lou, W.C. and Popplewell, L.M. (2003) Hydroxypropyl cellulose encapsulation material. US Pat. Application 20,030,077,378.
27. Barnes, J.M. and Steinke, J.A. (1987) Encapsulation matrix composition and encapsulate containing same. US Pat. 4,689,235.
28. Miller, D.H. and Mutka, J.R. (1986) Preparation of solid essential oil flavor composition. US Pat. 4,610,890.
29. Miller, D.H. and Mutka, J.R. (1987) Solid essential oil flavour composition. US Pat. 4,707,637.
30. Saleeb, F.Z. and Pickup, J.G. (1989) Fixation of volatiles in extruded glass substrates. US Pat. 4,820,534.
31. Fulger, C.V. and Popplewell, L. (1994) Flavor encapsulation. World Pat. 94/06308.
32. Popplewell, L.M., Henson, L. and Toth, A. (2003) Keeping a full flavour. *Food Ingredients & Analysis*, August, 4–7.
33. Reineccius, G.A. (1989) Flavor encapsulation. *Food Rev. Int.*, **5**, 147–176.
34. Pagington, J.S. (1986) β-Cyclodextrin and its uses in the flavor industry. In G.G. Birch and M.L. Lindley (eds), *Developments in Food Flavors*. Elsevier, London, 131–149.
35. Szejtli, J. Szente, L. and Banky-Elod, E. (1979) Molecular encapsulation of volatile easily labile flavor substances by cyclodextrins. *Acta Chem. Acad. Sci. Hung.*, **101**, 27.
36. Szente, L. and Szejtli, J. (1988) Stabilisation of flavors by cyclodextrins. In S.J. Risch and G.A. Reineccius (eds), *Flavour Encapsulation*, ACS Symposium Series No 370, American Chemical Society, Washington, 148.
37. Pagington, J.S. (1985) Molecular encapsulation with β-cyclodextrin. *Food Flav. Ingred. Packg. Proc.*, 7, 51.
38. Shaw, P.E., Tatum, J.H. and Wilson, C.W. (1984) Improved flavour of navel orange and grapefruit juices by removal of bitter components with β-cyclodextrin polymer. *J. Agric. Food Chem.*, **32**, 832.

Chapter 13
Applications II: Fragrance

Stephen J. Herman

13.1 Introduction

A fragrance is not a single material of clearly defined properties, but rather a mixture of individual chemicals, each behaving according to its own unique attributes. Characterizing these chemicals separately, and then combining their effects, allows the behavior of the complete fragrance composition in diverse media to be understood. Important properties of fragrance chemicals include volatility, polarity, surface activity and stability. Each fragrance component interacts with the chemical and structural nature of the environment to determine the aesthetic and the physical characters of the final system. The combined talents of the perfumer and the technical staff, working closely together, are needed to create a successful commercial product.

13.2 The basic structure of fragrances

Modern perfumers use an extensive palette of natural and synthetic chemicals. Until the end of the nineteenth century, perfumers worked with a relatively small number of materials, which were naturally derived, and they created a correspondingly small range of fragrance types, primarily in the citrus and floral families. Within this limited domain, perfumers could use an inefficient trial-and-error approach to create fragrances. The explosive growth of organic chemistry at the end of that century made innumerable new materials available to the perfumer, necessitating a rational method of fragrance creation. One of the pioneers of this period, Jean Carles, later wrote an invaluable series of papers [1], entitled "A Method of Creation in Perfumery", which describes a widely accepted approach to the process. Carles considered volatility to be a key quality for judging how to best use an aromatic ingredient. Each material is placed on a blotter and evaluated for intensity and character over a period of hours. Very volatile materials, which disappear first, comprise the "top notes" of a finished perfume, those of intermediate volatility and tenacity are the "modifiers" or "middle notes", and those with the lowest volatility, tenacious products constitute the "base notes".

The characteristic odor of a single material is called a note. Mixtures of two or more materials, having a unified olfactory theme, are called accords. If individual chemicals are like words, accords are short phrases, the top, middle and bottom notes are sentences, and the finished fragrance is a paragraph. The ratio of 25% top notes, 20% modifiers and 55% base notes is typical of a well-balanced blend.

13.3 The simplest case: hydroalcoholics

The typical use of the fragrance construction described by Carles is hydroalcoholic applications such as perfume, cologne or eau de toilette. These systems, based on aqueous ethanol, are the least technically hostile of the environments where fragrance is used. Although the hydroalcoholics are chemically the simplest exemplars of fragrance intended for consumer use, there are still technical considerations that must be attended to. Even single perfume ingredients can change in storage, for example, the aerial oxidation of benzaldehyde to benzoic acid. Aldehydes routinely polymerize at room temperature, and citrus products easily oxidize. When the individual materials are combined into perfume oil, further changes can occur. Very few chemical reactions occur between perfume materials to a significant degree, but Schiff bases, which produce a new molecule by combining an aldehyde or ketone with a primary amine (Scheme 13.1), are an important exception.

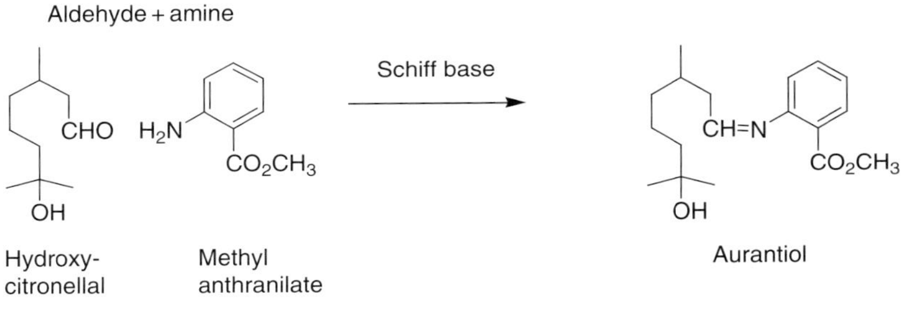

Scheme 13.1

Additives are commonly used to enhance fragrance stability. For hydroalcoholics as well as more challenging systems, UV absorbers such as benzophenone-2 can improve stability toward light. Chelating agents, such as EDTA and its salts, are useful in removing iron and other contaminants before discoloring reactions occur with sensitive fragrance chemicals. Antioxidants such as BHT (butylated hydroxytoluene), citric, tartaric, or oxalic acids help prevent rancidity. A combination of these ingredients often works synergistically. These materials can be added to the perfume if they are oil-soluble. Water-soluble materials must be added to the finished solution. It is best to incorporate the stabilizers as early in the process as possible, since unwanted reactions can rarely be reversed after the damage is done.

To make a perfume, the fragrance oil is added into a solution of alcohol and water. The commercial alcohol is not a single ingredient, but contains ethanol, water and a denaturant. Commercial alcohol is typically 95–96% ethanol and 4–5% water. This mixture is termed an azeotrope, because it boils at a lower temperature than either component separately. Thus, it is impossible to distill it further to increase the proportion of alcohol. To obtain anhydrous alcohol, it is necessary to add a third component to alter the evaporation properties of the mixture. To discourage drinking the ethanol, which is the essential ingredient of alcoholic beverages, it is poisoned with a denaturant. The denaturant is selected from materials that cannot readily be distilled out, which would return the alcohol to its original state. Denaturing exempts the alcohol from the taxes that are normally imposed on the potable version.

The best solubility is obtained if the perfume is added first to the alcohol, mixed, and the water is then added. When the perfume oil is placed in the hydroalcoholic solution, an aging process begins. The container is important; stainless-steel and glass-lined vats are best. Iron should never be in contact with the solution. All hoses, gaskets and stirrers should be carefully chosen to ensure inert response to the product. Aging was much more common in the past than now, and more important when a large amount of natural products are present in the oil. Contemporary mass-market fragrances are frequently not aged at all, with mixing, chilling, filtering and filling done in a continuous process.

A typical formula for a cologne and perfume would be:

	Cologne	Perfume
Alcohol SD39C or 40B (190 proof)	79.60	79.60
Fragrance	8.00	20.00
Distilled water	12.00	–
Benzophenone-2	0.40	0.40
	100.00	100.00

Smaller batches are chilled and run through filter presses. Clarity may be improved by the addition of a filter aid such as diatomaceous earth. Many modern fragrances, containing a high proportion of synthetic aroma chemicals, have excellent solubility in alcohol. Fragrances with a slight haze in solution can frequently benefit from the addition of a solubilizer such as polysorbate 20. The use of solubilizers should be minimal, since they can have an adverse effect on skin feel.

Natural oils are complex mixtures, and often contain waxes that are insoluble in alcohol. If the fragrance contains crystals and resins, some precipitation can occur. The level of crystals in a fragrance should never exceed 30%, although the exact amount that can be tolerated depends on the other materials in the formula and the percentage in the final product. The amount of water that can be incorporated into the system is affected by the nature of the perfume composition and the percentage of oil. Water is desirable because it reduces the sharp odor of the alcohol and lowers the price of the finished product.

Aftershaves are the lowest, least prestigious of hydroalcoholic solutions. The alcohol content is limited to 60–65% to avoid stinging, since minor cuts and abrasions often occur during shaving. Water-soluble emollient oils such as glycerin and propylene glycol are frequently added to modify the feel. Healing agents such as allantoin are also common. The addition of menthol or its ester derivatives is sometimes done for cooling effect.

The fragrance materials can also react with the alcohol to form new compounds. Reaction between aldehydes and alcohol to form a hemiacetal commonly occurs as solutions age. Schiff bases, if they were not intentionally created prior to the fragrance compounding, can slowly form in solution, if the necessary materials are available, although only to a partial degree. The value of the aging process, known by tradition to be necessary to achieve the final effect of the hydroalcoholic product, has a basis of sound science.

Fragrance selection at the point of sale is heavily dependent on its top notes. During the development of a fragrance, the perfumer evaluates the creation as it evaporates off a paper blotter. As a fragrance evaporates it necessarily changes character, but the evolving fragrance must retain a consistency through these changes. A successful fragrance must also exhibit a pleasing effect on the skin. Users of personal fragrances sometimes spray their clothes, but the most important test of a fragrance is its character as it evaporates off the skin over a period of several hours. A challenge for fragrance marketing is that every individual has a different skin chemistry, which changes the quality of the fragrance on each person.

Mookherjee *et al.* [2] (1998) analysed the "aura" above the skin, the molecules in the air above the fragrance. It was found that the standard model of top note, middle note and bottom note did not apply to the aura. Indeed, the aura was not dependent on molecular weight, boiling point, vapor pressure or odor value. Instead, a property of the compound – coined "diffusivity" – was the key, attributed to the tendency of the molecules to pass into the air. Ultimately, evaluations are best obtained by smelling the fragrance on a variety of individuals over a period of several hours.

13.4 Personal care applications: emulsions

It is customary for a successful fragrance to spawn line extensions, traditionally cosmetics and personal care. With increasing frequency, fine fragrances have inspired environmental, household, and even institutional fragrances. Thus, a designer fragrance can evolve into skin- and hair-care products, air fresheners and candles, dish and laundry detergents, and ultimately in janitorial supplies. Obviously, the same fragrance does not find itself in these products, but rather modifications of the odor type.

The major changes involved in altering a fragrance for non-alcoholic vehicles are price, aesthetics, and stability. The challenge for the perfumer is to keep the character of the fragrance as close as possible to the original, while accommodating all the crucial economic and technical issues. Perfumers often learn how to meet these challenges through experience or trial-and-error, but the solutions arise more consistently if the underlying science is understood, and the technical staff is intimately involved in the process. The majority of personal-care products are either emulsions, such as creams and lotions, or surfactant systems, such as shampoos and bath gels. Examining the interaction of these two systems with fragrance reveals most of the essential issues that arise in fragrance applications. The breaking or discoloration of emulsions, viscosity changes in surfactants, and the effects of light or elevated temperatures are key indicators of product integrity.

The first obvious effect of mixing perfume oil in an emulsion is an aesthetic change in odor character. Fragrance oil smelled directly out of the bottle, or in a hydroalcoholic solution, does not accurately represent the olfactory character it will exhibit in an emulsion. Odor changes taking place over time include the loss of strength and character, and the development of sour or rancid notes, particularly at elevated temperatures. An acidic note may also be detected. Furthermore, the fragrance in the emulsion will not smell the same after application to the skin, when the emulsion breaks, and an individual's natural odor and skin lipids become factors.

The quality of the emulsion raw materials can decisively impact performance and stability. Poor grades of emollient oils are generally more prone to rancidity than purer grades. The odor problems associated with raw materials become increasingly obvious with aging at elevated temperatures. These problems can sometimes be overcome by the addition of antioxidants such as BHT. Another important variable in emulsion bases is pH; a fragrance may be fine in a lotion of pH 7 but become unstable at the low pH of an AHA (alpha hydroxy acid) treatment cream. Hydrolysis of esters is common in acidic environments. Fragrance can also trigger changes in viscosity, which become more pronounced as the fragrance level rises. In extreme cases, the fragrance can cause the emulsion to break. Some aroma chemicals trigger the discoloration of a white emulsion, usually toward a brown or yellow hue.

The polarity, solubility, surface activity and stability of fragrance materials can explain their behavior in diverse environments. For example, hydroxycitronellal contains two polar groups, hydroxyl and aldehydic. Its polarity enables it to compete for space on the surface of a micelle, leading to either a decrease in viscosity or a breaking of an emulsion. α-Terpineol has one hydroxyl group on a relatively rigid molecule, less disruptive than hydroxycitronellal, but still capable of destroying the balance of the emulsion. It is also possible for perfume ingredients to produce closer packing on the micelle surface, in which case, emulsion stability will improve.

A useful tool for understanding the location of fragrance materials in an emulsion is to examine the solubility parameter (SP) and surface activity of all the components in the system, combined with knowledge of all the associative structures present in the emulsion. The SP is the sum of all the cohesive forces in a molecule. The SP can be measured by laboratory experiments, calculated based on the molecular structure, or derived using commercially available computer programs. It is much more refined than the simple division into either water-soluble or oil-soluble. Materials used in cosmetics have an SP range of 5–25, with most fragrance materials in the range of 7–13. There can be regions at both extremes of polarity with very limited fragrance solubility. The SP of selected fragrance materials and some common emulsion ingredients are shown in Table 13.1.

The water/octanol partition coefficient (P) is another valuable measure of polarity. The logarithm of P exhibits more linear behavior than P itself, and computer programs are available that derive calculated values of log P, based on chemical structure. The calculated value of the logarithm of the partition coefficient is abbreviated Clog P.

In 1970, Anonis [3] reviewed some empirical quidelines for fragrancing emulsions and summarized the situation as follows: avoid indole, replace methyl anthranilate with methylnaphthylketone when aldehydes are present, replace hydroxycitronellal with cyclamen aldehyde, use aliphatic aldehydes at low levels or replace with acetals

Table 13.1 Solubility parameters and chemical structures of selected cosmetic and fragrance materials

White mineral oil	7.09	$CH_3(CH_2)_nCH_3$	Benzaldehyde	11.00	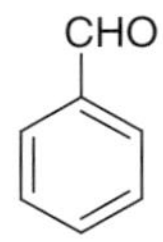
Stearic acid	7.75	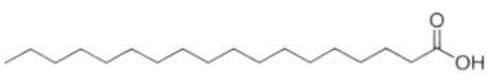	Eugenol	11.12	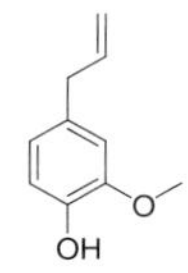
β-Pinene	8.03	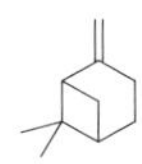	Dipropylene glycol	11.78	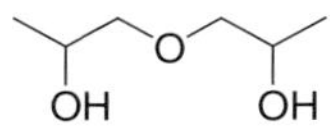
(Iso)amyl acetate	8.43	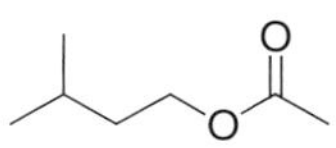	Phenylethyl alcohol	11.79	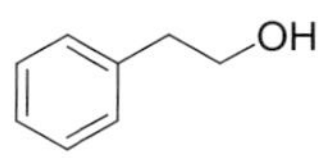
Citronellal	8.83	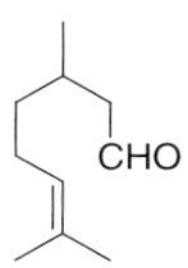	Benzyl alcohol	12.31	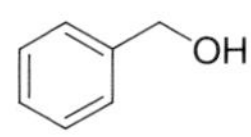
Stearyl alcohol	8.90	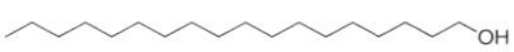	Vanillin	12.34	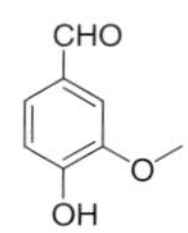
Citral	9.34	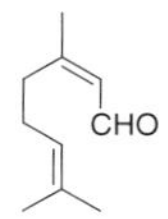	Propylene glycol	14.00	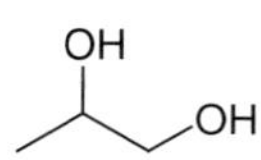
Linalool	9.62	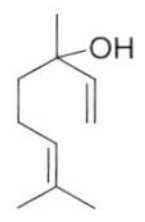	Water	23.40	H_2O
(Iso)amyl alcohol	10.84	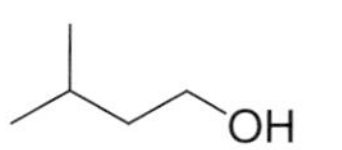			

(always, in the case of phenylacetaldehyde), replace coumarin with dihydrocoumarin, replace vanillin with ethyl vanillin, avoid quinolines, cresols, and animal fixatives.

Anonis and his predecessors established these principles primarily through trial-and-error. By the 1970s, theoretical knowledge of emulsion technology had reached a higher plateau, and it was possible to understand the scientific foundations of fragrancing emulsions. Friberg [4] had established the necessity of liquid crystal structures in the external phase to guarantee stable emulsions, and concepts like the SP and water/octanol partition coefficient made characterizing the interaction of fragrance and emulsion ingredients much more precise.

By 1980, Bonadeo [5] published a paper where K_d, a water/dioxane partition coefficient, was proposed to explain the effect of fragrance on emulsions. This concept relates to the classic hydrophilic/lipophilic balance (HLB) developed by Griffin to systematize the formulation of non-ionic emulsions. The HLB of a surfactant is determined by the weight ratio of the oil-loving and water-loving parts of a molecule. The aromatic material is considered part of the oil phase, and an HLB_r is assigned, corresponding to the required HLB of Griffin. One additional concept was proposed, the critical limit of the water phase (CLWP), which quantifies the water that can be tolerated by the emulsifier system.

Bonadeo concluded that the olfactory properties of a fragrance in a water-in-oil (W/O) or oil-in-water (O/W) emulsion is the same if the system is properly balanced; otherwise, a mixed system is formed which upsets the fragrance distribution. He firmly places the fragrance in the oil phase, thus making fragrance a determinant in calculating the total HLB, and that the CLWP of each perfume component contributes to the limit of water which can be tolerated.

The most common type of emulsion is oil-in-water (O/W), formulated either with anionic or non-ionic emulsifiers. The emulsion has an oil phase of small droplets covered by emulsifier molecules (micelles), an external phase, and liquid crystal structures in the water phase. Four common fragrance materials, pinene, hydroxycitronellal, phenylethyl alcohol and amyl alcohol, can illustrate how a fragrance can partition in an emulsion.

When the fragrance is mixed into the emulsion, the individual components of the fragrance partition in varying degrees into the different structures of the product [6]. Figure 13.1 shows the areas where the fragrance components can preferentially concentrate, although the partitioning is not absolute. Pinene, with a SP under nine and no oxygen groups, will migrate entirely into the internal oil phase. Hydroxycitronellal, with its surface activity, can inhabit the micelle surface, where it can displace some emulsifier, making it possible for it to lower the viscosity or break the emulsion. Phenylethyl alcohol and vanillin, with SPs over 11, can have a substantial presence in the external aqueous phase. The amyl alcohol, with a fatty chain and hydroxyl group, can partition into the liquid crystal structure. The partitioning takes time, accounting for the aging necessary before an accurate evaluation can be made of the effect of the base on the fragrance's olfactory performance.

Studies of the location of fragrance in emulsions by Friberg [7] and his colleagues confirm the results deduced from the partitioning of ingredients. Friberg's phase diagrams are valuable in illuminating basic principles, but only apply to simple, clearly defined systems. It seems unlikely that the phase diagram approach will fully elucidate the interactions of an emulsion, with 20 ingredients, interacting with a fragrance oil, of 200 aroma chemicals, in the near future, although computer programs are making progress in this area.

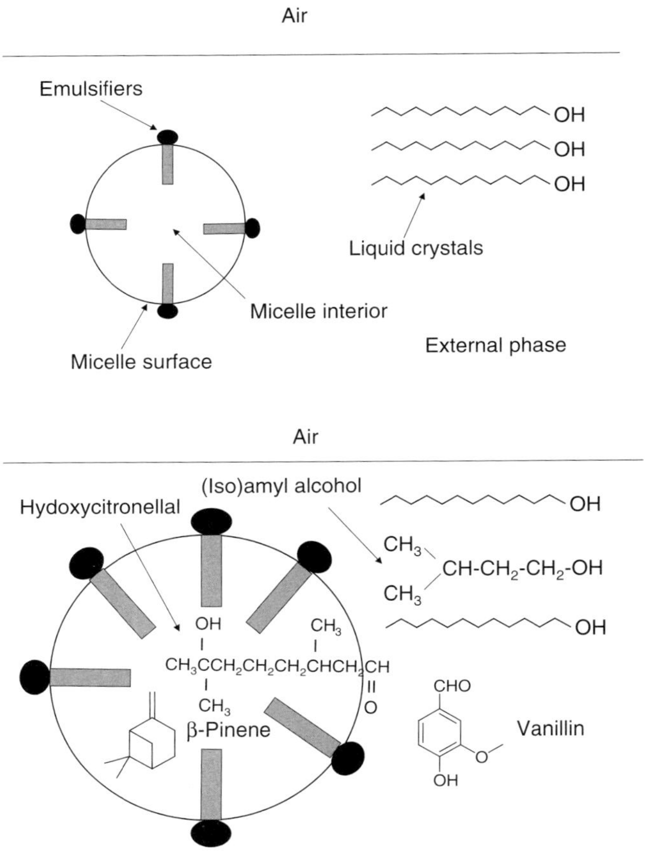

Figure 13.1 Partitioning of selected fragrance materials in an emulsion system.

Traditional fatty-acid soaps exhibit many of the fragrance problems found in emulsions. In addition to its fundamental similarity to emulsions, soap also has a high pH, which immediately restricts the fragrance palette. Vanillin is a notable cause of discoloration, as are eugenol and isoeugenol, heliotropin, indole, citral and Schiff bases. Any qualified perfumer must be aware of the materials that are appropriate for use in a conventional soap.

By understanding the relation of the chemistry of fragrance oil components to the structure of emulsions, it is possible to create a complete picture of the process of developing a successful product. Dellas and Lenoci [8] published an example of how a fragrance is modified from a cologne to a cream version, requiring attention to price, safety, stability and aesthetics. They showed the thought process of an experienced perfumer, rather than that of a chemist. Economics becomes a concern, since normally a fragrance for a lotion must cost significantly less than a hydroalcoholic creation. Some price reduction changes can have almost irrelevant aesthetic impact, since some of the delicate nuances of a fine fragrance are lost in a functional product. Residual fixative notes, which tend to be large, costly ingredients, can also be reduced or eliminated. Some materials must be eliminated because of stability concerns, and replaced with stable ingredients. The balancing of the fragrance to achieve the same olfactory effect in the new base either involves boosting the levels of ingredients muted by the emulsion, or adding new materials with a similar odor profile but greater impact.

13.5 Personal care applications: surfactants

The end of Second World War saw the creation of a mass market for surfactant systems based on new synthetic anionic and non-ionic materials. As the market expanded and matured, the importance of fragrance as a sales and marketing tool increased. Indeed, fragrance became perhaps the most important attribute of brand differentiation. New product formulations and significant fragrance levels made it crucial to understand the effect of fragrance on the surfactant systems, and the influence of the surfactant systems on fragrance performance.

Basic research clarified the concepts of the micelle and critical micelle concentration, and it became clear that different perfume components could concentrate in different areas of a micelle or the continuous phase. Some materials are solubilized by incorporation in the micelle, and any increase in the capacity of the micelle core is helpful. Materials with some true water solubility, such as phenylethyl alcohol or vanillin, become less soluble if the polarity of the external phase is increased by the addition of salt.

The most common product forms of surfactant systems for personal care are shampoos and bath gels, and their composition is similar. A typical formula contains a primary surfactant, secondary surfactant, foam boosters, viscosity builders and specialty additives. The primary surfactants are usually anionics. Amphoterics, non-ionics, betaines, amides and amine oxides can be incorporated to obtain special properties such as foam or viscosity boosting. Common additives for conditioning hair shampoos are silicones and proteins. Medicated shampoos also contain actives such as coal tar or zinc pyrithione.

Viscosity can be built in different ways. The best, but the most expensive, approach is with a properly chosen blend of primary surfactant, secondary surfactant, and a betaine or alkanolamine. The inexpensive way is to add salt or a polymer network such as carbomers or similarly performing materials. Such systems have a minimum of surfactant activity. It may be hard to maintain clarity while adding fragrance to salt-built systems for two reasons: the low level of surfactant, and the excessive polarity of the external phase.

It is now possible to view the problems of fragrancing surfactant systems using the current state of theoretical knowledge of micellar aggregates. The fragrance can be considered an additive, and its effect on the critical micelle concentration (CMC) will determine the consequences for the viscosity. The shape of the micelle can be determined by a relationship between three parameters: the volume occupied by the hydrophobic groups in the micellar core, the length of the hydrophobic group in the core, and the area occupied by the hydrophilic group at the micelle surface. There are two classes of organic compounds that can affect the CMC: Class I materials are incorporated in the micelle and Class II materials modify solvent–micelle or solvent–surfactant interactions. Fragrance materials illustrate this phenomenon, as they have a range of polarities. Whether a material is Class I or Class II will depend on the specific system involved.

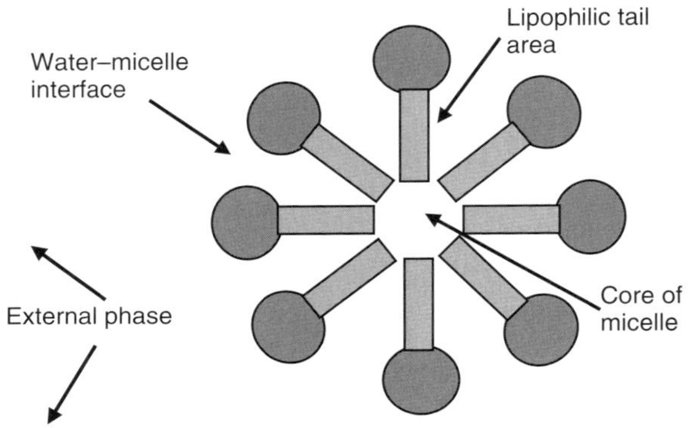

Figure 13.2 Fragrance partitioning in surfactant systems.

The individual aroma chemicals in a fragrance will partition into different areas of the surfactant system, as indicated in Figure 13.2. Some materials will migrate into the core of the micelle, some will align along the hydrophobic tails, some will be near the micelle surface, and a small amount will be in the external aqueous phase. The number, shape, and size of the micelles, and any thickening that has been created in the external phase, determine the viscosity of the system. The fragrance materials can change any of these parameters, and thus make the viscosity increase or decrease. The viscosity changes can be extreme and difficult to correct.

Munden [9] published a study on the effect of fragrances on viscosity. He used a perfume consisting mainly of phenylethyl alcohol, lillial, citronellol, hexycinnamic aldehyde, methyl dihydrojasmonate and Lixetone Coeur. The surfactant system consisted of 10% sodium lauryl ether sulfate, some minor additives, and sodium chloride to adjust viscosity. It was noted that fragrances usually decrease viscosity, but occasionally increase it. Of the materials considered, citronellol has the greatest effect, and thus was selected for detailed study.

Similar to the way in which a salt curve results from adding increasing amounts of sodium chloride to a surfactant blend, a viscosity curve can be established using increasing amounts of citronellol. The peak was found to occur at approximately 1% citronellol. In addition, the presence of citronellol affects the response of the system to salt. Without citronellol, a viscosity of 15 000 cts was achieved with 6% NaCl, while with 0.5% citronellol present, the highest possible viscosity was 7500 cts. Munden concluded that formulating a surfactant system before selecting the fragrance is ill-advised. If the fragrance is selected late in the development cycle, and a viscosity adjuster is used, then the effect of fragrance on the adjuster efficacy should be conducted to establish optimum values.

Blakeway [10] concluded that the odor of a solubilized perfume must be proportional to its concentration in the aqueous phase, not the fraction contained in the micelle. The greater the number of micelles, the less the odor intensity, and at higher surfactant concentrations the micelles must burst, releasing aroma chemicals into the solution.

A study by Behan and Perring [11] examined the vapor-phase concentration over a shampoo system using headspace analysis. Sodium dodecyl sulfate (SDS) was used at a range of 5–20%, but with 10% as the standard level. Aroma chemicals from 0.5 to 3% we used, with 1% as the standard. The aroma chemicals studied were octan-2-one, benzyl acetate, anethole, limonene and heptan-1-ol. A number of solvents/solubilizers were also considered: diethyl phthalate, dioctyl adipate, dipropylene glycol, 2-phenoxyethanol, triacetin, and Triton X100. It was found that the solvent could profoundly influence the headspace composition of the volatile components. Triton X100 was best for depressing vapor pressures. The perfume components will partition between the different structures in the SDS/water system. Increasing the surfactant concentration reduces the quantity of aroma chemicals in the headspace.

Fragrances are not intrinsically water-soluble. Some aroma chemicals, such as vanillin and phenylethyl alcohol, have moderate water solubility, but usually fragrances have to be solubilized by combining with non-ionic surfactants. The phrase "water-soluble fragrance" is misleading, since the fragrances are actually located in microemulsions. Different solubilizers require testing to identify the one with optimum performance for each fragrance and vehicle. Once solubilized, water is a perfectly acceptable vehicle for fragrance delivery in terms of odor quality. It is best to maintain the pH at $5.5 +/- 0.5$ for optimum stability.

With water-soluble fragrances, the ethoxylated groups in the solubilizer are responsible for the hydrogen bonding that bridges the oil and water phases. The hydrogen bonds weaken with increasing temperature, resulting in a reverse cloud point, also referred to as the inverse solubility coefficient. When the temperature rises, the solution clouds, and upon cooling, clarity returns. Clouding normally occurs at 35–80°C, with a lower solubility range at 5–15°C.

Surfactant systems for skin and hair have similar chemistry, but the substrates have very different odor-retaining properties. The skin has fairly low odor retention, particularly when the fragrance is present in a wash-off product. Hair has much greater odor retention, but a marketing question immediately arises – Does the consumer want residual odor? Combining customer preference with knowledge of the retention characteristics of aroma chemicals on hair makes the design of appropriate fragrances more predictable. Hair exhibits more absorptive character based on its non-polar nature. The fragrance molecules penetrate the hair, much like dye intermediates before neutralization. There is competition for space in the keratin, leading to non-additive behavior. As a consequence, two materials do not exhibit their combined strength, unless a unique synergistic effect is present.

Just as fatty soaps are similar to emulsions, transparent and glycerin soaps show kinship to surfactant systems. Clarity and discoloration are crucial issues, although pH is not a contributing problem factor in these bases. Syndet (synthetic detergent) and combo bars (combinations of fatty soaps and synthetic detergents) have formulations significantly different from other soap formats, and their fragrancing must be attacked as unique situations. With all these alternative soap variations, it is essential to work with the exact customer base and with an awareness of the exact production processing employed.

13.6 Air fresheners

Air fresheners provide a significant category of fragrance use, and the products are dispensed in a constantly expanding number of forms. Room temperature systems include aerosols, pumps, gels, beads, pads, powders and plastic. Some dispensing involves

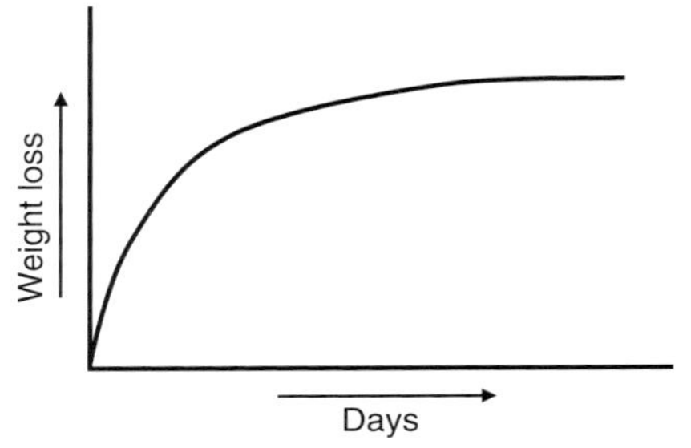

Figure 13.3 Typical fragrance evaporation curve.

heat, such as plug-ins and lamp rings. The simplest system involves the evaporation of a liquid, often through a wick or pad. Desirable features are fragrance impact, odor integrity during the life of the product and long-lasting performance. Marketers typically request 30- or even 60-day performance from these products. Evaporation curves generally show the most pronounced weight loss in the first 2–3 days (Figure 13.3). Since the weight of fragrance material present continually decreases, it is impossible to have significant impact at the end of the product's life cycle, unless some sustained release technology has been employed.

A typical fragrance will invariably change character as it evaporates. By focusing attention on the evaporation properties rather than on the odor character, it is possible to design a fragrance that will maintain an essentially uniform composition as it evaporates. These mixtures are called linear fragrances. The evaporative weight losses of different aroma chemicals are proportional to their vapor pressures.

The weight loss of four fragrance ingredients (1, 2, 3, 4) with vapor pressures $p_1 < p_2 < p_3 < p_4$ is shown in Figure 13.4(a). To create a linear fragrance with these materials, the proportions must be adjusted to conform to Figure 13.4(b). Of course, when other factors such as wick and pads enter the calculations, these linear effects are disturbed, but the basic principles are still valid. The interaction of ingredients also causes non-linear response during evaporation. Linearity is easier to achieve if the various components have similar vapor pressure, and some odor types are more adaptable than others to linearity. It is always true that formulating for linearity involves compromising olfactory quality for technical performance.

An early application of liquid air fresheners was a simple combination of a bottle containing a fragrance solution and a thick wick. A later refinement was to use a capillary to draw the liquid into a pad, usually composed of thick, porous cellulose. The fragrance must not clog the capillary; so testing of aroma chemicals, diluents and surfactants with the customer's components is necessary. The evaporation from pad will differ from the free liquid evaporation curve due to differing interactions of the aroma chemicals with the substrate.

The same laws that govern other combinations of liquids in physical chemistry rule the properties of fragrances. The vapor pressure is calculated by Raoult's law, which states that the vapor pressure of true solutions is dependent on the proportion of each component of the blend. The Phase rule of thermodynamics states that the vapor pressures of two immiscible liquids are not affected by each other. Thus, in a system with water and oil in

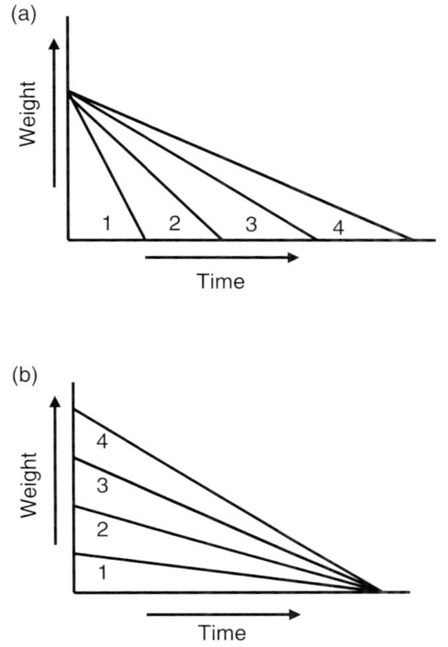

Figure 13.4 (a) Free evaporation of fragrance materials; (b) Evaporation of linear fragrance.

a gel, each behaves as if the other were not present. Altering the solvents in a fragrance is one method for changing the evaporation characteristics.

Water-based solid air fresheners based on carrageenan, a marine-derived polymer used as a thickener, have been manufactured for many years; the first US patent appeared in 1954. A key patent [12] disclosed a gellant mixture of:

Carrageenan	0.75–1.80%
Locust bean gum	0.20–0.75%
Potassium chloride	0.10–0.75%
Sodium carboxymethylcellulose	0.10–0.70%

During times of high carrageenan cost, stearates have sometimes been substituted.

Chlorophyllin was often incorporated in gels as a deodorizer, although it is unlikely that it was efficacious. These products are typically packaged in tightly sealed plastic containers that pull or twist open. The gels form hard surfaces during evaporation, significantly decreasing fragrance release with time. Their base color, a tannish brown, made attractive coloring impractical, and they shriveled into an unattractive, irregular lump. Fragrances require the addition of non-ionic surfactants to become soluble in these bases.

Clear water-based gels became possible with refinements in the quality of gellan gum. Gellan gum is a polymer with a linear tetrasaccharide repeat unit comprised of four sugar units: 1,3-β-D-glucose, 1,4-β-D-glucuronic acid, 1,4-β-D-glucose and 1,4-α-L-rhamnose. High levels of fragrance are employed, necessitating the use of surfactants, glycols and alcohol for achieving complete clarity. Even these materials cannot always make a clear product with certain difficult fragrance types such as citrus. Since they are transparent, the gels can be easily produced in a variety of colors, and objects such as botanicals and fruit slices can be suspended in the base for added marketing appeal. These gels, while visually appealing, unfortunately, have poor fragrance release.

Pennzoil patented a hydrocarbon gel [13] for air fresheners in 1999. The typical hydrocarbon of this invention is mineral oil. Whereas fragrances need water compatibility in other gel systems, the Pennzoil form, being non-aqueous, is clear with non-polar fragrances, or on some occasions solubilized with an amphiphilic molecule such as glyceryl oleate. The fragrance evaporation from this base is superior to that of a water-based gel, and the crystal-clear appearance has excellent consumer appeal.

Air fresheners are always most effective when the fragrance can be efficiently distributed in the environment, and the best means of accomplishing this has always been the aerosol. The best performing sprays are solvent-based, which creates the finest particles and the best suspension. In response to VOC regulations, aerosol air fresheners are usually water-based. Hydrocarbon propellants such as propane, butane and isobutane are not good solvents for polar aroma chemicals. Crystals such as coumarin and vanillin, and resins such as oakmoss, must be kept to a minimum in the formulation.

The basic components of an aerosol are shown in Figure 13.5. The possible interactions of the fragrance with the other components of an aerosol are extensive. The concentrate for a water-based formula resembles an emulsion, and can exhibit the same stability problems, such as the discoloration of vanillin. Hydrocarbons such as A46 are supplemented with other propellants, such as dimethyl ether (DME), for reduced VOC products. DME does not react adversely with aroma chemicals. The propellants and fragrance materials can also react with the valve and gasket material. DME, for example, requires butyl inner gaskets.

Fragranced polymers can be used for a variety of applications, from novelty jewelry to air fresheners. Fragrance levels vary from 0.5 to 1% for masking, 4–8% for a long-lasting but low-impact impression, 20–40% for intense effect in jewelry and toys, and to 50%

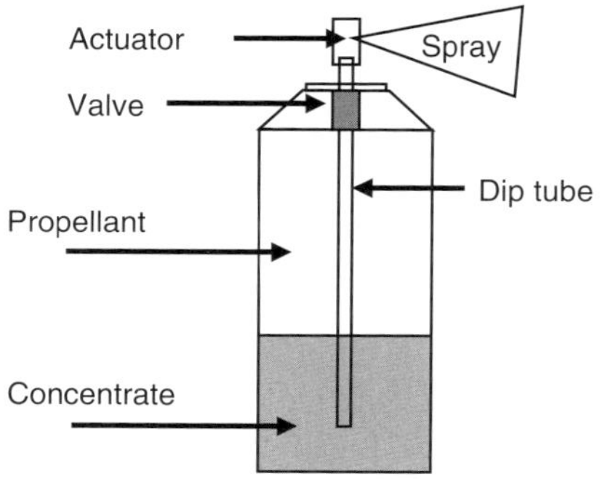

Figure 13.5 Aerosol components.

in air fresheners. Appropriate base resins for fragrance concentrates are LDPE (low-density polyethylene), LLDPE (linear low-density polyethylene), HDPE (high-density polyethylene), PP (polypropylene), EVA (ethyl vinyl acetate), PVC (polyvinyl chloride), styrene copolymers, cellulose acetate, TPE (thermoplastic elastomer) and TPO (thermoplastic olefin). Resins that do not fragrance well are polystyrene, ABS (acrylonitrile butadiene styrene), acrylic, PC (polycarbonate), nylon and polyester.

Meder [14] used the SP to determine the compatibility of a fragrance with a polymer. PVC was an example because it is a major polymer based on tonnage produced, and it is usually compounded with a plasticizer. The fragrance can be added to the plasticizer with no additional processing. PVC has a solubility of 19.4 $MPa^{1/2}$. Since compatibility exists when the plasticizer SP is within 2 $MPa^{1/2}$ of the plastic, the range of acceptable perfume materials is 17.4–21.4.

The characteristics of a good plasticizer are: high boiling point, high molecular weight, low color, low odor and good solubility in the polymer. Odor aside, fragrances fit these parameters, and ethylene brassylate was an example of a plasticizer for PVC. In a reverse example, a polymer compatible with a vanilla was determined. Three common primary vanilla chemicals were considered: vanillin, ethyl vanillin and ethyl vanillin PG acetal. Since these chemicals are polar, water-soluble polymers were candidates; namely, PVOH (polyvinylalcohol), PVP (polyvinylpyrrolidone) and PEOX (polyethyloxazoline). After a series of experiments, the following was determined for a vanilla-fragranced label adhesive: 51 g of vanillin, 56 g of ethyl vanillin, 357 g of PEOX and 536 g of water. This indicates that by matching the polymer to the fragrance, polar crystalline materials can be incorporated successfully.

13.7 Candles

Candles are a key segment of the United States home fragrance market. Fragrance is what drives the sale of candles, followed by color and packaging. It is essential that the fragrance translation into candles is executed with the same care that personal-care products receive. Candles appear to be simple systems, but they can present great technical and aesthetic challenges to the formulator. The correct balance of base, fragrance, color and wick is essential to create a satisfactory product (Figure 13.6). Consumers are increasingly demanding products that are attractive, economical, safe, and perform well. Achieving this requires close cooperation between the candle manufacturer and the fragrance supplier.

The evaluation of a candle fragrance, besides the assumed fidelity of the odor character to the customer's expectations, hinges on "cold throw" and "hot throw". Cold throw is the impact of the fragrance in the candle before burning. A candle is not customarily lit in a retail store; so a purchaser almost invariably selects a candle based on its cold throw. It is crucial that the fragrance balance emphasizes the top and the middle notes. Bottom notes, often the most expensive part of a fragrance, have little place in a candle. Hot throw is the odor impact when the candle is burning, particularly when the pool of molten wax on the top has completely formed. The fragrance should give a strong and characteristic odor, filling an average room with scent.

It is essential that the fragrance performs well in the customer base. Most candles are paraffin-based. Paraffin is a complex mixture, and the detailed chemical composition of a particular material is rarely available. The standard specifications for wax

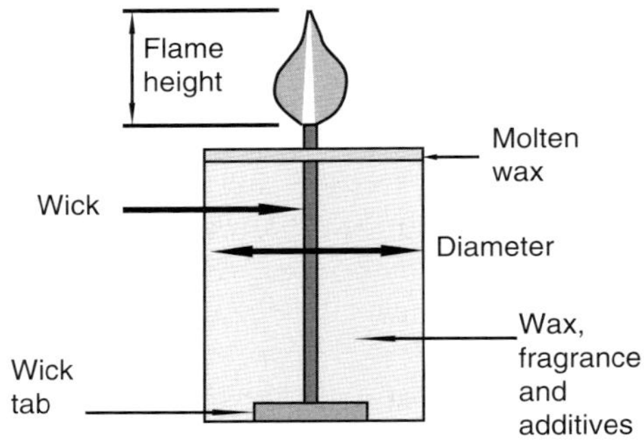

Figure 13.6

are melting point and penetration. Changing suppliers of 145°F paraffin – based on the assumption that all 145°F paraffins are essentially the same – has been the cause of countless difficulties.

The key determinants of paraffin chemistry are the source of the petroleum oil and the refining process. Some oil fields have aromatic or cyclic components, while others do not. Most paraffins are straight chains, but some paraffins contain a significant amount of branched-chain compounds. The melting point rises with increasing chain length, and branched chains are lower-melting than straight chains. The boiling range, melting point, and paraffin composition can all be controlled, in theory, to fairly tight specifications by the refining operations, but rarely is the paraffin purchased by candle companies actually held to these rigid standards.

Candles are made in different forms. The most common types are tapers, votives, pillars and jars. Each should have a different wax blend. For example, tapers must have higher melting points than glass jars. The oil content of the paraffin (excluding the additional fragrance) should be no higher than 0.5% in straight paraffin for a free-standing candle, while a jar candle can contain as much as 1.5% oil. It is important that candles in glass do not pull away from the surface upon cooling. It is equally necessary that a votive has good mold release. Materials added to paraffin to change the properties include stearic acid, petrolatum, mineral oil and microcrystalline wax.

Regardless of the exact composition of the paraffin, the key fact is that paraffin is non-polar. There are a few non-polar aroma chemicals, but the majority of the fragrance ingredients have some polarity: aldehydes, esters and alcohols are a few examples. Bleeding out, or syneresis, occurs when the wax matrix of the candle cannot hold the fragrance, like a sponge holding too much water. An oil that seems perfectly soluble initially may bleed out over time. Observation after a minimum of one day is necessary to ensure that no solubility problem exists.

A second polarity-related problem is sublimation. Materials that are totally incompatible with paraffin, particularly crystalline materials such as vanillin, can sublime. These materials can migrate up the wick and deposit on the glass cover if it is a jar candle, resulting in an observable film.

The perfume percentage in a candle is frequently at a very high level, compared to most household products. In modern candles, where color and fragrance are prime selling features, 5% fragrance is often used. The fragrance must be completely soluble in the molten wax blend. Some manufacturers will accept a slight haze, but there should certainly not be insoluble droplets or particles settling to the bottom.

The solvent for the fragrance must be compatible with the wax. DPG, the solvent of choice for many personal-care applications, is totally inappropriate. DEP, dioctyl adipate, Hercolyn D, benzyl benzoate, mineral oil, isopropyl myristate, IsoPar, NorPar and capric/caprylic triglycerides are all possible candidates. Low HLB solubilizers are occasionally used for some problem fragrances. Which solvent performs best is determined by testing the individual fragrance oil and candle base. Proper choice of solvent can also be used to adjust the flash point of the fragrance, which is a critical factor for shipping.

Besides solubility, the fragrance must not adversely affect the burning characteristics. Burn tests are commonly performed by manufacturers, both on new submissions and on each individual lot of fragrance oil received for production. The flame must burn well both initially and when relit. In a properly burning system, the paraffin will totally combust, yielding carbon dioxide and water. Most aroma chemicals are primarily carbon, hydrogen and oxygen, and do not create significant quantities of byproducts upon combustion, especially since the bulk of fragrance chemicals evaporate intact from the molten pool, rather than burn in the flame.

Once the diameter, wax base, and percent perfume are all fixed, only one variable remains to achieve proper burn – the wick. In general, the thicker the wick, the more rapidly the candle will burn. Wicks come with and without cores, coated and uncoated, and with different weaves. A large selection from wick suppliers should be tested, and the most appropriate one chosen through burn tests. The flame should burn clean. Holding a clean broad metal spatula over a flame will quickly reveal the amount of carbon residue produced. The desired pool of molten wax should form in about an hour. The candle should also perform well when relit.

The addition of color increases the potential stability problems. Candles are not restricted in the US to the certified-type colorants (FD&C, D&C, ext D&C), and other colors are available as powders, liquids and waxes. Generally, these colors contain nitrogen in the form of an azo bond (–N=N–), which can react with aroma chemicals. Stray metal contamination can also come from the wick, wick tab, a poor grade of wax, or iron present in the processing equipment. Dark perfume oils, usually yellow, orange or amber, when incorporated at 5%, can add significant color to a base, making a blue into a green or a white candle yellow. Light, either from the sun, fluorescence or UV, can trigger unwanted color reactions. UV absorbers, chelating agents and antioxidants, alone or in combination, may eliminate or alleviate some unwanted reactions – sometimes. Raw material studies on the fragrance in the colored base may be necessary.

Clear candles based on polyamide resins have been available since the 1960s. The clear candles did not burn well, and were at best novelty items. By the early 1990s, the clear gel, invariably in a jar due to their soft structure, became popular, although never popular enough to displace the wax candle. Unfortunately, a major company marketed unsafe clear gel candles, which caused several fires. This resulted in a well-publicized recall and consequent loss of consumer acceptance. The flare problem involved fragrances that were not perfectly soluble and a solvent base that was not properly engineered. These fragrances also had flash points that we now know to be too low. When areas of fragrance concentration encountered the molten pool, and the wick dipped into the liquid, a fire could result.

In response to this problem, Penreco developed a new, patented base [15]. A special cut of mineral oil, having a high flash point, was gelled. The composition is approximately 92% mineral oil and 8% polymer. Fragrances, which have specific solubility and flash point characteristics, are specified. The fragrances must be completely soluble in mineral oil in blends of 25% fragrance and 75% mineral oil, and 75% fragrance and 25% mineral oil. The fragrance must thus be as non-polar as possible. The flash point must be at least 140°F, and preferably 170°F. A special self-extinguishing base developed by Penreco incorporates silica in the gel. The wick clogs when the flame reaches the bottom; this removes a safety issue, since the users frequently forget to blow out the candle at that point.

13.8 Reactive hair care

Products that chemically interact with hair provide great technical challenges for perfumers. Permanent waves, hair dyes, depilatories and relaxers are diverse in composition and purpose. As a group, they have very high pH. Permanent waves and depilatories also have extraordinary base odor, hair dyes are unusually reactive chemicals, and relaxers have an extreme pH coupled with a base formulation that is not very stable even without fragrance, due to the sensitive nature of the active ingredients. Some of these products require a neutralizer, usually ammonia, which is not strictly an odor but a chemical irritant. The chemical interactions with hair provide an additional source of odor problems.

Permanent waves have a severe base odor, both in product and in use. The characteristic smell comes from thioglycolates, usually ammonium thioglycolate ($C_2H_4O_2S{\cdot}H_3N$) or glyceryl thioglycolate (1).

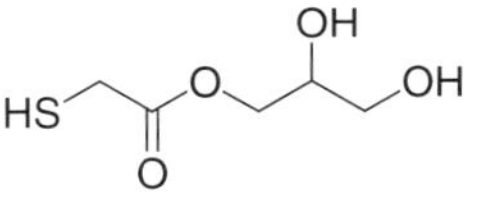

(**1**)

No fragrance can mask the odor completely. Sagarin and Balsam published the basic reference [16] concerning perfume materials in these bases in 1956, offering data on 200 aroma chemicals, and surprisingly little information has since been published.

Sagarin and Balsam found it impossible to draw conclusions based solely on structure. Aldehydes would be expected to react with thioglycolates, but some proved stable, such as methyl nonyl acetaldehyde. Some materials mask but do not retain their odor value; cinnamon oil and citronellal change odor but have good masking ability. Phenylethyl alcohol maintains its character but covers poorly. Some materials can maintain odor and coverage but exhibit other problems, such as trichlorostyralyl acetate, which is useful, except for the deep yellow color it produces.

The odor stability and coverage achieved by blends of aroma chemicals are hard to predict. Some materials perform better in mixtures; others require careful blending to be effective. Certain materials that tested well, such as cade oil and methyl acetophenone, are unpleasant by themselves and must be carefully added in small proportions to be useful.

An approach to reducing the thioglycolate odor was disclosed in US patent 5,554,363 [17]. In 1888, Bongartz observed that thioglycolic acid combines with benzaldehyde over a span of a few hours, and with cinnamic aldehyde in minutes, with the release of significant heat. The patent builds on these observations. It specifies benzaldehyde, methyl hydrocinnamic aldehyde, and mixtures of the two, for being more reactive with the cysteamine reducing agent, than with the aldehydes that naturally occur in hair. The reducing agent thus attacks the fragrance aldehydes rather than the hair aldehydes, resulting in reduced odor during the permanent waving process.

13.9 Depilatories

Depilatories use calcium thioglycolate as the active of choice, combined with a non-reactive filler (typically calcium carbonate) to reduce irritation, a wetting agent to improve contact with skin, and an emollient such as cetyl alcohol. Raising the pH to 10–12 can reduce the time required for effective depilation. Above pH 12, there is a high risk of skin damage. Below pH 10, the depilatory action falls off rapidly.

Barry [18] provides some guidelines on perfuming depilatories. Alcohols can form mercaptans, esters may be hydrolyzed to odorless alcohols, and aldehydes undergo aldol condensations. Many aroma chemicals are discolored by alkali. Linalyl acetate maintains floral note after hydrolysis, making it a valuable odorant. Generally, the best performing ingredients are ionones and rose alcohols (citronellol, geraniol).

Deplilatories can be important as an alternative to blade shaving for men of African ethnicity. The hair is typically curly and wiry, and conventional shaving leaves the ends of the hair shafts with points, and these can penetrate the skin, a condition named *pseudofolliculitis barbae*. A depilatory system should work quickly, finishing its task in 3–7 minutes. Barium sulfide and calcium thioiglycolate are effective actives, but the former is banned in the EU. Fragrances used in perms and depilatories can be used in these products, but the base odor during use is so severe that no coverage is truly effective, and the consumers seem resigned to its offensive odor.

13.10 Dyes and perms

Hair dyes are possibly the most difficult systems to fragrance. Besides the high pH, the active ingredients of the dye are extremely aggressive to aroma chemicals. Some oxidative dyes contain perfume, while others do not. Ammonia adds a strong impression that masks the fragrance, but when the product is applied, the ammonia dissipates, and it is possible to smell the fragrance. Any aroma chemical should be able to resist alkali and hydrogen peroxide during the brief time when it is combined with the dye. No aldehydes are appropriate, because they react with amino compounds.

A typical relaxer base is shown in Table 13.2. The pH is very high, and the base has only moderate stability even without fragrance. The base odor out of the container and

Table 13.2 Cream hair relaxer

Ingredient	% by wt
Water	48.30
Propylene glycol	3.00
Petrolatum	20.00
Mineral oil	15.00
Cetearyl alcohol	8.00
Ceteareth-20	2.50
Steareth-10	1.00
Sodium hydroxide	2.20
	100.00

during use is not offensive, and hence many commercial products use no fragrance. The most successful fragrance is a simple rose, essentially phenylethyl alcohol, which is an exceptionally rugged molecule. Comber [19] examined some of the difficult chemistry involved in the bases: sodium, lithium or calcium hydroxide in relaxers, guanidine carbonate in the activators for relaxers, and guanidine in the activated relaxer. Most relaxers, about 85% of the market, use sodium hydroxide; they are termed "lye relaxers". Lye relaxers have a pH of 13.5, putting a severe restriction on the aroma chemicals that can be used. The so-called no-lye, no mix lithium hydroxide relaxers have a slightly more moderate pH range, 12.5–12.8. The no-lye, guanidine/hydroxide relaxers are back to pH 13.5.

The consumer use of the guanidine/hydroxide relaxer involves mixing a calcium hydroxide containing cream emulsion base (pH 12.4–12.6) with a liquid guanidinium carbonate activator (pH 11.0–11.2) to produce the activated relaxer base (pH 13.5). The cream emulsion base is sometimes fragranced. It is necessary that the fragrance be stable not only in the emulsion base, but also in the final activated base.

13.11 Bleach

There are three common types of bleach products available, with significantly different levels of activity. A 5.25% sodium hypochorite solution sold principally for whitening laundry can be termed simply "bleach". Concentrations up to 15% available chlorine with reasonable stability are available for commercial use. Cleaners containing sodium hypochorite as a functional ingredient but also possessing other surfactants and/or thickeners comprise a second category. The final group consists of non-hypochorite bleaching products, such as peroxides.

Sodium hypochlorite poses a unique combination of challenges for the perfumer. Not only must the fragrance itself be stable; it also must not adversely affect the chlorine stability. Few solubilizers have satisfactory stability, and those available are not highly effective, compared to the solubilizers used in other applications. The solubilizers can also have a negative effect on stability and create unwanted foam. The high pH required for hypochlorite would alone limit the palette of allowed raw materials. The packaging, including the cap and cap lining, can create problems. The cumulative constraints result in a narrow range of creative options.

The key factors effecting bleach stability are hypochlorite concentration, temperature, pH, chloride ions and metallic impurities. When the pH drops below 10.5, the decomposition becomes precipitous. A sealed, full container stored in a cool environment will have the best shelf life. As the product is used, the increased presence of air becomes problematic. Currently, the most frequently used solubilizer for fragrance in bleach is Dowfax, a series of alkylated diphenyl oxide disulfonates (ADPODS). While Dow was the innovator with these products and their trade name is familiar to the industry, other companies now make similar chemicals. The standard member or the series in bleach solubilization is Dowfax 2A1, which consists of a C_{12} branched hydrophobe, sodium salt form, with 45% activity. Phosphate esters have also found use as bleach compatible surfactants.

Liquid bleach is usually, but not always, fragranced with solubilized perfume oil. Some manufactures have successfully marketed products where the insoluble fragrance oil lays on the top as a film. When the product is used, the top layer does not pour off with the first use. Rather, the fragrance percentage stays relatively uniform during the life of the product. There is considerable psychological resistant to this approach, despite any advantage it may have in terms of economics or foam reduction during processing.

Testing bleach stability requires great care. Chlorine has the potential of adding to unsaturated bonds in aroma chemicals. The resulting adduct may still have an acceptable odor; thus, a chemical may smell fine, yet adversely effect the chlorine content. Conversely, an aroma chemical may have no effect on the chlorine content, but may have its odor value destroyed. Analytical and olfactory testings must both be satisfactory before a product is appropriate for use.

Available chlorine is the basic measure of the oxidizing power, present as hypochlorite. To measure available chlorine, the bleach solution with potassium iodide is acidified, and the iodide titrated with sodium thiosulfate.

Test samples are prepared with either complete fragrances or individual raw materials. Fragrance levels are very low; a starting point could be 0.02% fragrance solubilized with 0.06% Dowfax 2A1. Unfragranced samples are used as a control, as hypochlorite levels in bleach fall with time, even in the absence of fragrance. Samples require testing by titration, each week, for four weeks. Samples at room temperature, elevated temperature (35 or 40°C), refrigeration (4°C) will give a complete stability profile. There is a non-linear degradation at higher temperatures, so test conditions should not be higher than 40°C.

Odor evaluations of the samples are an essential part of the stability testing. Assuming the odor test is satisfactory, the chlorine content of the fragranced product should be 85–90%, compared to that of the unperfumed sample. At what point a material is unfit to use is a matter of judgment. If an aroma chemical is used at a low level in the fragrance, and has an important odor contribution, perhaps, its stability can have a lesser standard.

Many cleaning products have lower levels of bleach. Bleach deteriorates more slowly at lower levels, and the bases may contain surfactants that help solubilize the fragrance. In products where the bleach gains protection from an encapsulation process, the perfuming problems are much easier. Powder products protect the fragrance more than liquids.

13.12 Malodor counteractants

Fragrance companies are frequently required to reduce or eliminate unpleasant odors. Intrinsically unpleasant molecules cause some malodors. Others arise from the products of organic decay. A common characteristic of many offensive smells is the presence of sulfur, nitrogen, or halogens. Some osmophoric groups to be focused on are $-NH_2$, –SH, and C=C. Thioglycolates, mercaptans and pyrazines are examples of foul smelling compounds used in personal care and perfumery. Sour milk and underarm odor are some common products of organic decay.

The earliest use of fragrance in malodor control was the use of a large dose of aromatic materials to overwhelm a foul smell. The incense used in churches helped to reduce the cumulative olfactory impact of the assembled masses of unwashed individuals. These materials were not designed to be effective malodor counteractants, and the effect could be achieved only by employing a heavy application of odor-masking materials.

The first modern approach was the design of fragrances composed of aroma chemicals known to be particularly efficacious against malodors. Aldehydes and esters were the key components; the neutroleum fragrance of the 1960s was a characteristic example of this approach. US4840792 [19] suggests a mixture of citral, eugenol, coumarin, and helional for odor control. US5683979 [20] prefers 20–60% musk, 30–70% citrus and 1–20% mint.

An extension of this idea is the patented use of two aldehydes selected from certain defined groups. There must be a minimum of 20% of each aldehyde in the mixture. It is required that one aldehyde must be selected with a double bond in the alpha position, R–C=C–CHO, and one from all the remaining aldehydes used in perfumery. Examples of materials with the double bond in the alpha position are citral, cinnamic aldehyde, benzaldehyde and vanillin. The finished fragrance must contain a minimum of 10% of the aldehyde blend with a total fragrance level of 1% in the finished product, with the proportion adjusted appropriately as the fragrance level goes up or down.

The next technical advance was the use of non-odorous materials that could complex with the malodor molecules and effectively eliminate their odor. Not only are these ingredients not aroma chemicals, they could potentially neutralize selected components of standard fragrances. A number of proprietary materials, generally recognized by trade names, are typical of this category of modern malodor counteractants.

Metazene is one of the first malodor counteractants. It originated around 1950 and is still available. The structure of Metazene shows it to be a mixture of fatty alcohol esters of methacrylic acid, primarily lauryl methacrylate. The molecule can react with osmophores such as –SH, $-NH_2$ and –CHO to produce large molecules, whose vapor pressures are too low for olfactory detection.

Meelium is a blend of polyhydroaromatic sulfonates. Neutrolair D-7 is a mixture of geranyl crotonalte and dihexyl fumarate, Vandor B is principally 3,5,5-trimethylhexanal. Forestall is soyaethyl morpholinium ethosulfate.

Grillocin® contains a clearly defined substance, zinc ricinoleate, synergistically combined with other zinc compounds made up of multiple hydroxylated sebatic acids, oxamines and resinic acids. Other metal salts, such as copper, cobalt or nickel show similar deodorizing properties but have toxicological problems. The properties of the ricinoleic anion are important, and similar fatty acids are not effective. Grillocin® reacts with low molecular weight fatty acids, amines, mercaptans and isovaleric acid, a prime component of body odor.

Some structures form cavities that can effectively encapsulate an odor molecule. One aspect claimed for Odorone® is the entrapment of the odorant in a cavity. Cyclodextrin is a notable example of this phenomenon, as expounded in various patents. US5783544 [21], assigned to Proctor & Gamble, is the basis of Febreze™, a popular fabric deodorizing spray. Cyclodextrin is available in different forms, α, β and γ, which have different cavity sizes. The α and β forms are available from several suppliers, but the γ version is only available from Wacker [22]. The inside of the cavity is hydrophobic, while the outside is hydrophilic. The solubility characteristics allow cyclodextrin to disperse in an aqueous system, while trapping oily materials such as fragrance or malodors in its cavity.

The fragrance materials preferred in patent US5474917 [23] have a Clog P of 3 or less. This value indicates some hydrophilic tendency, which the patent claims enhances the performance in the base and during product use. Some examples cited in the patent are benzaldehyde, benzyl acetate, coumarin, dihydromyrcenal, ethyl vanillin, eucalyptol, eugenol, geraniol, hydroxycitronellal, linalool, methyl anthanilate, nerol, phenyl ethyl alcohol, terpineol and vanillin. The patent also prefers fragrances with at least 25% of these materials, more preferably 50%, ideally 75%.

13.13 Stability testing

A product must be stable if it is to succeed. The more rigorous the stability testing during the development process, the more likely that success will be. The base, fragrance, color, packaging, and even labeling, must all be carefully scrutinized. Time is needed to carry out these tests, and a last-minute or haphazard approach to stability and compatibility studies is an invitation to commercial failure.

Products being tested should be as representative of the final marketed version as possible. In addition to the exact base and fragrance, it is necessary to have the packaging, including details such as the label and closure. A control run in glass will show if any troublesome situations are due to the product itself or the package. Different plastics vary in qualities such as thermal resistance, color fastness and oxygen permeability. Some fragranced products may be packaged in paper, like sachet stones, or wrapped in cellophane, like cellulose car air fresheners, and unwanted permeation is always a possibility.

Among the parameters requiring close monitoring during the test period are pH, physical separation, color, odor, particle size, viscosity and weight loss. Of course, not all test parameters are appropriate for all product types; particle size is important for emulsions, not bath gels. Stability requirements are generally more difficult for emulsions, and more tests are applicable for them. Certain changes of appearance are more critical in clear packaging, such as a slight haze in the base or slight color shifts. Often perfection is unobtainable; comparison with a successful product that has been on the market for years reveals a great deal about the acceptable range of key variables.

Elevated temperature is the most universal stability test. As a rough rule, reaction rates double for every 10°C. Assuming room temperature to be 25°C, then one year at room temperature would correspond to three months at 45°C or one month at 50°C.

These conditions, while useful, are nevertheless misleading. Reactions can occur at 50°C, which will never happen to the product in its normal lifecycle. Confidence can reasonably be placed in a product that passes a 50°C test; however, a product that fails at 50°C may still be suitable for the marketplace. Other commonly employed temperature conditions include low temperatures (4°C), subfreezing temperatures (−20°C) and freeze–thaw cycles. These tests attempt to simulate all the shipping and storage conditions that can occur in all seasons through the entire region to be served. The centrifuge test is also useful for emulsions. Separation of emulsions can occur due to differences in specific gravity between the discrete and continuous phases, and the centrifuge increases this gravitational effect. The test sample is first warmed to 50°C, and then centrifuged at 3000 rpm for 30 minutes.

The effect of fragrances on the stability of emulsion and surfactant systems and in a number of other systems has been dealt with in other sections. Bath gels and shampoos often combine color and high fragrance level to enhance consumer appeal. Color molecules have color because of structural groups called chromophores. Chromophores are configurations that can alter the energy levels of delocalized systems. Benzene absorbs around 200 nm, but our eyes see radiation only between 400 and 700 nm. Thus, benzene is colorless, despite its delocalized electrons. The addition of a chain with conjugated double bonds or an azo group can shift the radiation into the visible region.

Fragrance materials can change properties when exposed to light. Benzaldehyde is difficult to stabilize with red dyes. Benzaldehyde is a relatively unreactive molecule in its ground state, but UV readily converts it into a free radical (Scheme 13.2), which attacks the chromophore structures of the red dye molecule. One way to combat this phenomenon is to introduce UV absorbers into the system.

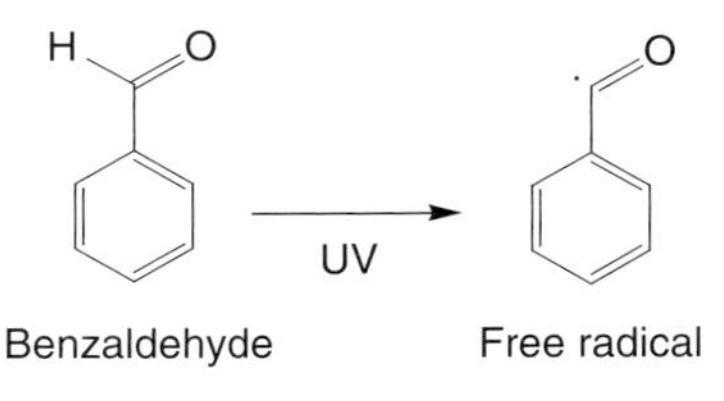

Scheme 13.2

13.14 Conclusion

The fundamental key to dealing with the technical aspects of fragrances is to treat them as mixtures of chemicals – and, of course, to know what those chemicals are and their properties. The environment contributes additional chemistry plus larger structures susceptible to interactions, such as micelles and liquid crystals. It is essential to have clearly defined technical parameters at the beginning of product development, such as specific stability requirements or the presence of sensitive base materials.

Open cooperation between perfumer, chemist and customer is necessary for the most effective resolution of problems, since each can hold a piece of the solution. Since aesthetic considerations enter into most fragrance applications, marketing concerns impose yet another level of issues. Taking every possible variable into account, encouraging input from everyone contributing to the product development chain, and allowing adequate time for testing and the resolution of problems, will maximize the opportunities for success.

References

General reference: Herman, S.J. (2002) *Fragrance Applications: A Survival Guide.* Allured, Carol Stream, Il.

1. Carles, J., A. Method of creation in perfumery, *Researches*, 11 Dec. 1961, pp. 8–25; 12 Dec. 1962, pp. 18–29; 13 Dec. 1963, pp. 92–103.
2. Mookherjee, B.D. *et al.* (1998) Fragrance emission from the skin. *C&T*, **113** (7), 53–60.
3. Anonis, D. (1970) Perfumes and creams. *DCI*, **107**(3), 44–48, 164–166.
4. Friberg, S., Mandell, L. and Larsson, M. (1969) Mesomorphous phases, a factor of importance for the properties of emulsions. *J. Colloid Int. Sci.*, **29**(1), 155–156.
5. Bonadeo, I. *et al.* (1980) Hydrophilic properties of aromatics. Int. *J. Cos. Sci.*, **2**, 215–229.
6. Herman, S.J. (1994) Fragrancing emulsions. *C&T*, **109**(8), 71–75.
7. Friberg, S. *et al.* (1999) Stability factors and vapor pressure in a model fragrance emulsion system, *JSCC*, **50**(4), 203–219.
8. Dellas, J.G. and Lenoci, Vito, G. (1980) The EST principle: maintaining a fragrance accord in diverse media. *Cos. Tech.*, **2**(2), 33–35.
9. Munden, D.R. (1988) Effect of perfumes on the viscosity of surfactant systems. *C&T*, **103**(5), 65–67.
10. Blakeway, J.M. *et al.* (1979) Studies in perfume solubilization. *Int. J. Cos. Sci.*, **1**, 1–15.
11. Behan, J.M. and Perring, K.D. (1987) Perfume interactions with sodium dodecyl sulphate solutions. *Intl. J. Cos. Sci.*, **9**, 261–268.
12. Lanzet, M. (1960) Air treating gel and method of preparing the same. US Pat. 2 927 055.
13. Johnson, R.L. and Morrison, D.S. (1999) Non-aqueous controlled release pest and air care gel composition. US Pat. 5 871 765.
14. Meder, M. A primer on perfuming polymers, *Perf. & Flav.*, **11**(1) Feb/Mar 1986: Part II, **12**(6) Dec. 1987/Jan. 1988.
15. Morrison, D.S. and Heilman, W.J. (1999) Transparent gel candles. US Pat. 5 879 594.
16. Sagarin, E. and Balsam, M. (1956) The behavior of perfume materials in thioglycolate hairwaving preparations. *JSCC*, **6**, 480–486.
17. Nandagiri, A., Solka, B.H. and Kocis, J.A. (1996) Method of reducing malodors in permanent waving. US Pat. 5 554 363.
18. Barry, R.H. (1972) Depilatories. In M.S. Balsam and E. Sagarin (eds), *Cosmetics: Science and Technology*, 2nd edn, Wiley-Interscience, New York.
19. Comber, R. (1995) The Fragrancing of African-American Cosmetic Products, SCC Carolina and Southeast Chapters Joint Symposium.
20. Joulain, D., Racine, P. and Maire, F. (1989) Agent neutralizing bad smells from excretions and excrements of animals. US Pat. 4 840 792.
21. Trinh, T. *et al.* (1998) Composition for reducing malodor impression on inanimate surfaces. US Pat. 5 783 544.
22. Schulz, G.E. and Candussio, A. (1995) Modified cyclodextrin glycosyltransferases for producing gamma-cyclodextrins. US Pat. 5 474 917.
23. Schreck, L. *et al.* (1997) Malodor counteractant composition and process for using same. US Pat. 5 683 979.

Appendix
Common names

David J. Rowe

In the Introduction one of the themes in the historical development of the flavour and fragrance industry was the end of alchemy, with its mystical mumbo-jumbo, and the creation of chemistry – a rational 'hard science'. Of course, life is never quite that simple, and most areas of chemistry have kept a certain character reminiscent of the mediaeval guilds, with certain traditions and arcane knowledge only familiar to those 'in the trade' and meaningless to those outside of it. The flavour and fragrance industry is no exception to this. A key ritual for neophytes in the industry is The Learning Of The Names. Whilst individual areas of chemistry, and indeed individual companies, have their own 'shorthand' names for common chemicals, the area of aroma chemistry is particularly rich in names that are unsystematic, obscure, misleading, and, in some cases, outright lies! A simple example might be maple lactone pyrazine, which is a pyrazine, but not a lactone; nor is it found in maple; it is called this because it is made from maple lactone, found in maple, but which is not a lactone either . . . and this is as clear as crystal, compared with aldehyde C19, which is not an aldehyde, nor does it have nineteen carbons – it is actually benzyl butyrate.

Table A1.1 includes some of the odd names used in the industry. I have not included registered trademark names, as to list everyone's trade name would be an impossible task, and I should have to constantly apologise if I included one company's trade name and missed off another's for the same material! Instead, I have put together my personal favourites, especially those which evoke the aroma of the material, to give, as it were, a flavour of the odd nomenclature used in the industry.

Table A1.1

Industry name	Understandable name
Aldehyde C14 so-called	γ-Undecalactone
Aldehyde C16 so-called	Ethyl 3-methyl-3-phenylglycidate
Aldehyde C18 so-called	γ-Nonalactone
Aldehyde C19 so-called	Benzyl butyrate
Aldehyde C20 so-called	Ethyl 3-phenylglycidate
Oenanthic ether	Ethyl heptanoate
Maple lactone	2-Methylcyclopent-3-onolone
Maple lactone pyrazine	Methyldihydrocyclopentapyrazine
Bell pepper pyrazine	2-Methoxy-3-isobutylpyrazine
Strawberry furanone	2,5-Dimethyl-4-hydroxy-3[2*H*]-furanone
Pineapple ketone	2,5-Dimethyl-4-hydroxy-3[2*H*]-furanone
Fenugreek lactone	4,5-Dimethyl-3-hydroxy-2[5*H*]-furanone
Caramel furanone	4,5-Dimethyl-3-hydroxy-2[5*H*]-furanone
Peach aldehyde	γ-Undecalactone
Coconut aldehyde	γ-Nonalactone
Pineapple mercaptan	Ethyl (3-methylthio)propionate
Raspberry ketone	4-(4-Hydroxyphenyl)-2-butanone
Bramble ketone	4-(4-Methoxyphenyl)-2-butanone
Zingerone	4-(4-Hydroxy-3-methoxyphenyl)-2-butanone
Cat ketone	4-Methyl-4-mercapto-2-pentanone
Leaf alcohol	*cis*-3-Hexenol
Leaf aldehyde	*trans*-2-Hexenal
Mushroom alcohol	1-Octen-3-ol
Coffee furanone	2-Methyltetrahydrofuran-3-one
Grapefruit mercaptan	*para*-Menthene-8-thiol
Valeric acid	Pentanoic acid
Caproic acid	Hexanoic acid
Oenanthic acid	Heptanoic acid
Capryllic acid	Octanoic acid
Pelargonic acid	Nonanoic acid
Capric acid	Decanoic acid
Lauric acid	Dodecanoic acid

Index